# Understanding and Crafting the Mix

*Understanding and Crafting the Mix, 3rd edition* provides the framework to identify, evaluate, and shape your recordings with clear and systematic methods. Featuring numerous exercises, this third edition allows you to develop critical listening and analytical skills to gain greater control over the quality of your recordings. Sample production sequences and descriptions of the recording engineer's role as composer, conductor, and performer provide you with a clear view of the entire recording process.

Dr. William Moylan takes an inside look at a range of iconic popular music, thus offering insights into making meaningful sound judgments during recording. His unique focus on the aesthetic of recording and mixing will allow you to immediately and artfully apply his expertise while at the mixing desk. A companion website features recorded tracks to use in exercises, reference materials, additional examples of mixes and sound qualities, and mixed tracks.

New to this edition:

- Coverage of basic listening skills and techniques, with integrated, practical exercises
- A chapter on surround sound productions
- Fully updated content reflecting current production concepts and workflows

**Dr. William Moylan** has worked with leading artists across the full spectrum of jazz, popular, and classical genres. His recordings have been released by major and independent record labels, resulting in wide recognition, including several Grammy award nominations. A leading educator and an active recording engineer and producer for over 30 years, he is a Professor and Coordinator of Sound Recording Technology at the University of Massachusetts Lowell.

# Understanding and Crafting the Mix

## *The Art of Recording*

Third Edition

William Moylan

LONDON AND NEW YORK

First published 2002 by Focal Press

This edition published 2015
by Focal Press

Published 2019 by Routledge
2 Park Square, Milton Park, Abingdon, Oxon OX14 4RN
52 Vanderbilt Avenue, New York, NY 10017

*Routledge is an imprint of the Taylor & Francis Group, an informa business*



*Notices*
Knowledge and best practice in this field are constantly changing. As new research and experience broaden our understanding, changes in research methods, professional practices, or medical treatment may become necessary.

Practitioners and researchers must always rely on their own experience and knowledge in evaluating and using any information, methods, compounds, or experiments described herein. In using such information or methods they should be mindful of their own safety and the safety of others, including parties for whom they have a professional responsibility.

Product or corporate names may be trademarks or registered trademarks, and are used only for identification and explanation without intent to infringe.

*Library of Congress Cataloging in Publication Data*
Moylan, William.
Understanding and crafting the mix: the art of recording/
William Moylan. — Third edition.
pages cm
Includes bibliographical references and index.
Discography: p.
1. Sound—Recording and reproducing. 2. Acoustical engineering.
3. Music theory. 4. Music—Editing. I. Title.
TK7881.4.M693 2015
621.389'3—dc23
2014021953

ISBN 13: 978-0-415-84281-5 (pbk)
ISBN 13: 978-0-415-84280-8 (hbk)

Typeset in Univers
by Florence Production Ltd, Stoodleigh, Devon, UK

*To the Mysterious Energies that carry creativity*

Bound to Create
You are a creator.
Whatever your form of expression — photography, filmmaking, animation, games, audio, media communication, web design, or theatre — you simply want to create without limitation. Bound by nothing except your own creativity and determination.
Focal Press can help.
For over 75 years Focal has published books that support your creative goals. Our founder, Andor Kraszna-Krausz, established Focal in 1938 so you could have access to leading-edge expert knowledge, techniques, and tools that allow you to create without constraint. We strive to create exceptional, engaging, and practical content that helps you master your passion.
Focal Press and you.
Bound to create.
We'd love to hear how we've helped you create. Share your experience:

# Contents

# Figures

# Tables

# Exercises to Develop Listening, Evaluation, and Production Skills

# Foreword

Nowhere within the mystery of creation is the concept of infinity more closely demonstrated than in the human response to sound.

**Sounds** barely audible in the quiet solitude of a forest glade contain information about direction, height, distance, and character, that unconsciously provide awareness of our surroundings.

**Sounds** as you step into a great cathedral, hearing nearby soft footfalls on ancient flagstones and the singing of a distant choir, that provide clues of distance and perspective, and even invite sharing of mood.

**Sounds** in a concert hall, an office, a bathroom, or a recording studio, each with its own message prompting our response.

Blindfold, the smallest sample provides amazing awareness of our environment.

The way sound behaves within a space paints a picture. We don't have to analyze, measure, or evaluate. Created in the image of a communicating God, we communicate with speech to express what we think and with music to express what we feel.

Music is as old as man. References to music, song, poetry, and musical instruments go back thousands of years. We see Egyptian *bas-reliefs* and know that music played an enormous part in their culture and religion. Near the end of King David's reign, the Hebrews had professional temple choirs and a 4,000-strong orchestra.[1] We read of strolling minstrels and court musicians of later ages. But what did they sound like? Every age of man is recorded in writing, painting, and sculpture, but there are no sound recordings.

Today we can choose from vast libraries of music from any part of the world, which have become the benchmarks of artistic and technical quality for many thousands of people who may never have even been to a concert. Technical quality has improved enormously over the years but, surprisingly, we can still enjoy the very earliest recordings with all their imperfections of noise, distortion, limited bandwidth, and poor dynamic range. Why, then, is *The Art of Recording* important?

As technology advances, realism has become more accurate, but it seems that the artistic qualities of a performance have become more elusive. Neither accuracy nor artistry can be tabulated in some book of rules. Unlike hardware design—for example, computers, where precision and speed predominate and the figures accurately define the performance—in the field of sound recording, specifications and measurements cannot describe the sounds that we hear and neither can they predict the effect of those sounds on us. The recordist must provide the vital link as our interpreter.

In the 1953 fourth edition of his *Radio Designer's Handbook*, Langford-Smith states: "It is common practice to regard the ear as the final judge of fidelity, but this can only give a true judgment when the listener has acute hearing, a keen ear for distortion and is not in the habit of listening to distorted music. A listener with a keen ear for distortion can only cultivate this faculty by making frequent direct comparisons with the original music in the concert hall." [2]

We must cultivate "A Point of Reference."

Truly successful recordists and producers have developed very high degrees of refinement and can perceive qualities (or the lack of them!) in the sound recording and reproducing chain that seem to defy reason and sometimes to contradict our current state of knowledge.

In 1977, Geoff Emmerick, who with George Martin recorded The Beatles at Abbey Road and later at Air Studios in London, showed me that he could hear a difference between two identical channels on a recently delivered new console. After some hours of listening with him, I agreed that I could hear a subtle difference. When we measured, I found that out of 48 channels, 3 had been incorrectly terminated and displayed a rise of 3 dB at 54 kHz. The limit of hearing for most humans does not extend beyond 20 kHz, and this small resonance, whilst obviously an oversight in the factory, would not normally have been regarded as important.

One of the significant features of this episode was that Geoff was deeply "unhappy," even "distressed" at what he was hearing or perceiving. Since then I have seen much more evidence that the range beyond 20 kHz is part of human awareness. Newly introduced designs that transmit frequencies to beyond 100 kHz (with low distortion and noise) surprisingly sound warmer, sweeter, and fuller.

In 1987, when addressing the Institute of Broadcast Sound in London, I carried out a simple experiment for the first time, with the object of discovering what effect frequencies above 20 kHz might have on a professionally aware audience.

A generator capable of switching between a sine and a square wave was fed through an "ordinary" amplifier to an "ordinary" monitor loudspeaker. The frequency was set at 3 kHz. The audience confirmed that they could hear the third harmonic as a superimposed 9 kHz tone or "whistle" when the generator was switched to square wave. (A square wave contains predominantly odd harmonics. When the first of these exceeds the limit of hearing, the sine and square wave should sound the same.)

The frequency was then progressively raised and older members soon admitted that they could no longer hear the third harmonic "whistle." But all could still hear a *difference* in quality as the generator was switched between sine and square. As the frequency continued to be slowly raised, some could still identify a difference when the fundamental had reached 15 kHz.

This experiment has been repeated many times in different parts of the world without any real attempt at scientific control. "Ordinary" equipment provided by my host was used on every occasion. Results have been surprisingly consistent: some 35 to 45 percent of those present were able to identify a quality difference when the fundamental was as high as 15 kHz.

There was one exception: At the University of Massachusetts Lowell, some 60 percent of the audience were still identifying a difference when the fundamental frequency had exceeded 17 kHz.

At the 83rd Convention of the Audio Engineering Society, Dr. William Moylan proposed a "Systematic Method for the Aural Analysis of Sound Sources."[3] Dr. Moylan uses his method at the University of Massachusetts Lowell Department of Music. I think this points to the success of Dr. Moylan's training in aural analysis! This same method has led to this book.

No measurements or formulae can ever replace the recordist, but he must develop a reliable point of reference and learn "the Art of Recording." The learning process is endless. We never arrive at an ultimate state of knowledge. We are, even now, only scratching the surface, and this is especially so as we explore new formats to convey not only technical excellence but a whole listening experience of music and its environment.

Can there be a perfect recording?

Only if we can arrive at perfect knowledge. How, then, should such knowledge be used? Could it change, for example, the world in which we live?

Stephen Hawking, in *A Brief History of Time*,[4] seeks a unified theory—drawing together the general theory of relativity and of quantum mechanics—which would lead to a complete understanding of the events around us and of our own existence. He says: "If we find the answer to that, it would be the ultimate triumph of human reason—for then we would know the mind of God." We need open minds to *envision* direction, responsible minds to *choose* direction, and a point of reference beyond ourselves to whom we are ultimately answerable.

Well over one hundred years ago, Lord Rayleigh told us that the ears are the final arbiter of sound: "Directly or indirectly, all questions connected with this subject must come for decision to the ear, as the organ of hearing; and from it there can be no appeal. But we are not, therefore, to infer that all acoustical investigations are conducted with the unassisted ear. When once we have discovered the physical phenomena which constitute the foundation of sound, our explorations are, in great measure, transferred to another field lying within the dominion of the principles of Mechanics. Important laws are in this way arrived at, to which the sensations of the ear cannot but conform."[5]

To follow William Moylan's "important laws" will prepare your ears and your mind for the true "Art of Recording." His approach is proven and will lead the reader's ears to the refined levels of "keen-ness" and "acuteness" required today. Dr. Moylan's book also gives us insight into what turns "recording" into art, and provides ways to bring artistic sensibility into our work.

Rupert Neve
Wimberley, TX
September 2001

## Endnotes

1 *Bible*. 1 Chronicles 23:5. " . . . and four thousand are to praise the LORD with the musical instruments I have provided for that purpose."

2 F. Langford-Smith, editor, *Radio Designer's Handbook,* 4th edition, Sydney, New South Wales: Wireless Press for Amalgamated Wireless Valve Company Pty. Ltd, 1953, Ch.14, Section 12, (iii).

3 William Moylan, "A Systematic Method for the Aural Analysis of Sound Sources in Audio Reproduction/Reinforcement, Communications and Musical Contexts," paper read at the 83rd Convention of the Audio Engineering Society, October, 1987.

4 Stephen W. Hawking, *A Brief History of Time* (Toronto: Bantam Press, 1992), pp. 173–175.

5 Lord Rayleigh, The Theory of Sound, 1st edition, 1877, New York: Dover Publications, 1945.

# Preface

This third edition is expanded and largely reworked from the second edition. Perhaps only 20% of the second edition remains untouched here.

This edition improves the functioning of the previous editions with simplified language and more detailed coverage of some subjects and layout of materials. It adds new and important features, and seeks to be more useful and pertinent to the intended audience: you.

These are the greatest differences. First, the book has been expanded to four parts, with a new Part Two on background fundamental listening skills. A new chapter dedicated to surround sound brings extensive coverage of the sound stage dimensions of surround, and examines the musical impacts of various approaches to surround. The number of exercises has expanded, more thoroughly covering both developing listening skills and becoming aware of the creative potentials of production. These latter exercises support the greatly expanded coverage of the artistic elements in crafting the mix.

A companion website has been established for this edition, and replaces the CD of the second edition. It can be found at www.focalpress.com/cw/moylan. This website will allow additional and more detailed evaluations of recordings and graphics for the reader. It will supplement the text in supplying templates for certain graphs and exercises. There will be more tracks available on the website, as well as the original tracks from the CD. It is expected that other sound files will be added over time to refresh the content without revising the text in the short term.

While the purpose and goals of *Understanding and Crafting the Mix: The Art of Recording* have not changed, I believe this writing is now significantly more thorough and effective, clearer and more readable.

## Background

*Understanding and Crafting the Mix: The Art of Recording* is the product of my experiences as an educator in sound recording technology (music and technology), and of my thought processes and observations as a composer of acoustic music and of music for recordings (recording productions and electronic music). It includes what I have learned through my creative work as a record producer and through my attempts to be a facile and transparent recording engineer. It is informed by my perspective as a music theorist and a teacher of music analysis. This book is an outgrowth of my on-going and in-depth research into how we hear music and sound as reproduced through loudspeakers (aural perception, music cognition) and aural skills pedagogy, and into the analysis of recording related to the dimensions of recorded music and their impact on the music, its materials, expression, and meaning.

This book has evolved substantially since my initial research and writings in the early 1980s. It has been greatly shaped by years of devising instructional methods and materials in my courses at the University of Massachusetts Lowell, and by my interaction with many other audio educators and observations of other recording programs worldwide. This evolution has been enhanced by many other people and experiences as well; I have been fortunate to have interacted with numerous talented production professionals in the audio industry, representing many different types and sizes of facilities (in both major and idyllically isolated locations) and many production styles—with very diverse and deeply impressive credits and credentials.

The concepts and methods of this book have gone through many stages of development over the past three decades. They will continue to be refined as we learn more about listening and what we hear, how we perceive sound and understand art, and how we learn.

## Purpose and Intended Audience

*Understanding and Crafting the Mix: The Art of Recording* seeks to bring the reader to understand how recorded sound is different from live sound, and how those differences can enhance music. It will bring the reader to explore how those sound characteristics appear in significant recordings. The book also presents a system for developing critical and analytical listening skills necessary to recognize and understand these sound characteristics.

This leads to the production process itself. *Understanding and Crafting the Mix* seeks to move the reader to engage audio recording as a creative process. Techniques and technologies are purposely not covered. Instead, the book explores the recording process as an act of creating art, and helps the reader envision recording devices as musical instruments. It seeks to develop an artistic sensitivity that will lead the reader to find and create their own unique artistic voice in shaping and creating music recordings—neutral to the continual shift of technologies and techniques, devices and software, fashion and convention.

Unchanged from the first edition, *Understanding and Crafting the Mix: The Art of Recording* is intended to be used: (1) as a resource book for all people involved in audio; (2) as a textbook for courses in recording analysis, critical listening (listening skills-related courses), audio-production-related courses of all types in sound recording technology, music engineering, music technology, media/communications, or related programs; or (3) as a self-learning text for the motivated student, beginning professional, or interested amateur. It can also provide a unique look at some of the most important recordings by The Beatles, for all who may be interested. It has been written to be accessible to people with limited backgrounds in acoustics, engineering, physics, math, and music.

The intended reader might be an active professional in any one of the many areas of the audio recording industry, or a student studying for a career in the industry. The reader might also be learning recording through self-directed study, or be anyone interested in learning about recording and recorded sound, perhaps an audiophile.

The portions of the book addressing sound-quality evaluation will be directly applicable to all individuals who work with sound. Audio engineers in technical areas will benefit from this knowledge, and the related listening skills, as much as individuals in creative, production positions. Even those involved in consumer audio and pro audio sales can benefit. All those people who talk about sound can make use of the approach to evaluating sound that is presented.

As a resource book, *Understanding and Crafting the Mix* is designed to contribute to the professional development of recordists. The book seeks to clearly define the sonic dimensions of audio recording, and brings the reader to approach recording production creatively. It offers to expand the current professionals' creative thinking, their skills in critical and analytical listening, and their skill in and sensitivity to accurate and meaningful communication about sound quality.

Many excellent books exist on recording techniques and audio technologies. Articles in many excellent magazines, journals, and serials exist that cover recording devices and techniques, audio technologies, and acoustic concerns. These areas are not addressed here.

Little has been written on the aesthetic and creative aspects of recording music or audio production. Few books exist that discuss and develop the listening skills required to evaluate recordings for technical quality, and none to evaluate the artistry of recordings or to analyze the artistic content of recordings. This book seeks to fill this void. It may be used as a sourcebook or a textbook in all of these areas, from beginning through to the most advanced levels.

The book is well suited to developing the student's music-production skills and artistry. It is designed to stimulate thought about the recording process as being a collection of creative resources, and to invite a creative spirit. These skills and creative ideas can then be applied to the act of crafting the music recording artistically.

Graphing the activity of the various artistic elements is important for developing listening and evaluation skills, especially during beginning studies. It is also valuable for performing in-depth evaluations of recordings that allow us to study how accomplished record producers have used the artistic elements to achieve their "sounds." This process of graphing the activity of the various artistic elements is also a useful documentation tool—functional even with the less exacting detail that would make it more practical for daily use. Working professionals through to beginning students will find something useful in the approach.

Finally, it is hoped this book will clarify communication in some small way and in some small segment of our industry.

All audio professionals communicate "about" sound in some way, for some reason. Still, we often do not communicate well, clearly, or accurately. Perhaps this book might serve as a meaningful point of departure for an "audio recording syntax" to be devised.

## Selection of Recorded Examples: The Beatles

Over the years I have received many well-intentioned comments that my writings would benefit from more current musical and recorded examples than my almost exclusive use of Beatles records. Most comments have been very articulate, and made great sense. I have always seen much to support this idea, on its surface. Bringing this into practice is something else. I have struggled with this almost daily in my teaching, as I wish my examples to be of relevance to my younger audience.

Current music has many additional social and musical dimensions that make this difficult to navigate. I find when I bring current recordings to listening sessions, people have difficulty separating the artist (and what the artist represents to those individuals) from the sounds they hear (they turn off their ears, or distort their perception). Further, preference for certain types of music can run deep with young people; deeper, still, are their dislikes. Then the question becomes: What artists, or types of music, or recording styles should I use? Should it be Taylor Swift, Eminem, Will-i-am, U2, Dixie Chicks, James Taylor, Linkin Park, Red Hot Chili Peppers, Neil Diamond, or [place *your* favorite artist here]? Which of these would you like to listen to, perhaps 30 times, to fully appreciate the supplied graphs?

I venture to estimate that any recording selection may alienate 75% of this readership for one or more personal reasons.

The first book I published that discussed many of this book's topics was released in 1991. Musical examples included tracks from the then current and now classic albums *The Joshua Tree, So, Face Value, Every Breath You Take*, as well as important songs by Elton John, Tracy Chapman, and Kate Bush. It was interesting (and disconcerting) to recognize: Only three years after its publication the book *looked* out of date because of the musical examples, though its content was not. The recordings that brought it to continue to appear most relevant were those that were not fixed in the time of publication: those of The Beatles. This gave me much to consider.

Now I keep returning to The Beatles, for a great many reasons that are deeper than surface. First, as teaching tools and sonic examples, they continue to work perfectly well in demonstrating fundamental concepts and sound dimensions that are permanent in the field. As early stereo and multitrack recordings, many sound qualities and dimensions of sound are clearer, and more easily perceived by novice listeners, than those found in many later recordings.

The matter of preferences to listen to certain types of music can be mostly circumvented with The Beatles. The Beatles are simply part of "our" common experience and common language—across generations, across time periods. While rooted in the 1960s, they reach far beyond—and well before. Just as the music of Coltrane or Bach, they are timeless in their expression, though a product of their time. They simply "are," in much the same way as the music of Beethoven just is—whether you like it or do not like it is irrelevant, these recordings are important and should be known. They are touchstones of music, and of the recording industry. Everyone knows them, and can talk intelligently about them—if you don't, you are at a disadvantage in this field.

Of course, new production techniques are not found in The Beatles' records. Exceptions are the surround mixes of *LOVE* reflecting more current production practice and the reissue of The Beatles' catalog remastered in 2009 with its more contemporary timbral balances. This writing, however, is not "about" presenting the reader with examples of modern recordings; it is about educating the ears and providing knowledge to understand "any" recording—most importantly, to prepare the reader to engage those recordings and styles that have yet to be created, especially their own.

And, the songs, recordings, and performances really are well constructed on a great many levels. Stylistically diverse, they offer a great many vantage points into production sound and techniques, as well as music. The public's curiosity has brought many reissues, each unique in some way. All of this allows us to study the topics of this book with some consistency of material—material that can simply be accepted as a given, that is well crafted, and that was masterfully recorded. I do not believe these statements to be subjective or personal opinion.

On a personal level, I do not consider myself an unbridled "fan" of The Beatles, blinded by their mystique and fanatically obsessed with the Fab Four. This is often assumed of me.

Rather, I am immensely indebted to their work, and what it has taught me and brought me. And not only the four composer/performers: Sir George Martin, Geoff Emerick, and all the other production staff of all the releases educated my ears and my awareness of the impossible world of records. I learned a significant portion of what I know as a recordist by studying the production techniques and sounds of their records. Beginning with experiencing their appearance on *The Ed Sullivan Show* live at age eight and listening to their records in my formative years, through struggling to figure out their unusual sounds and to play their newly released "Strawberry Fields Forever" at age eleven, to poring over their recording processes as a fledgling assistant engineer in my early twenties, their music has helped shape my musicianship, my career, my artistic sensibilities, and my sense of self—and more. The intersections of this music, of these records, and my career continue to this day. I hold a deep sense of gratitude for all they have provided me.

# Acknowledgments

It has been nearly 14 years since I joined the list of Focal Press authors, and I have been blessed to have worked with many fine editors. My profound thanks to all of my editors during these years, most of whom have moved on: Terri Jadick, Marie Lee, Tricia Tyler, Catharine Steers, Emma Baxter, Beth Howard, Diane Wurzel, and of course Meagan White and Megan Ball. My deepest appreciation to the other wonderful editorial staff and production people at Focal Press—who might be too numerous to mention but who have all made my work easier and more enjoyable by their professionalism and constant willingness to help. I am honored to be affiliated with this extraordinary organization and group of people.

For over 30 years now, my students in the Sound Recording Technology program at the University of Massachusetts Lowell have worked through the materials and concepts of this book in any number of forms, and have taught me a great deal. I extend a special thanks to these serious and gifted young people for their (largely unknowing) contributions to my work.

Many other people have provided support for this book, or assisted in shaping my thoughts about these materials. While they are far too many to mention, they are all very important to me. Only a very few follow.

My gratitude to Paul Angelli, Patrick Drumm, Erh-Chaun Lai, Gavin Paddock, Phil Reese, Brad Swanson, Michael Testa, and Mark Whittaker, who gave much time and extended considerable effort working through specific graphs and exercises.

Finally, I am deeply indebted to Mr. Rupert Neve for generously providing his esteemed insights to this book by writing its foreword.

# Introduction

The recording process shapes music. Recording techniques and technologies change the acoustic sounds they capture, and impart new sound characteristics. These sound qualities are under the control of an individual who shapes the music recording—the recordist.

The sound qualities created by recording do not occur in nature, though some qualities simulate natural dimensions or relationships. These qualities are sonic dimensions that are unique to audio recordings, and give recorded music (or music reproduced over loudspeakers) a set of characteristic sound characteristics. While these dimensions create an experience that is different from a live, unamplified performance, these sound qualities have become accepted as part of the experience of listening to recorded and reproduced sound, and of that music itself. Considering that vastly more recorded or amplified music is heard today than live, entirely acoustic music, it is readily understood that these sound qualities of recorded sound are now integral parts of today's music—and of the listening experience of today and of recent decades.

These sound qualities have become part of the fabric of recorded music. The way a piece of music might be shaped has thus been extended to include the unique sound qualities of music recordings.

The person controlling, or creating, these sound qualities (the recordist) is therefore functioning as a creative artist. This person is a musician of the technology and process—"conducting" by encouraging and ensuring quality performances, "performing" recording, mixing, and processing devices, and "composing" the mix's sound qualities and relationships.

As this book unfolds, the answers to the following questions will gradually become evident, as we begin *understanding the mix*, consider *crafting the mix*, and navigate the dimensions of *the art of recording*:

- What makes recording music an art?
- What makes the music recording a unique medium for artistic expression?
- What is different between a music recording and a live, acoustic-music performance?
- What has brought the recordist recognition as an artist?
- How does the recording process contribute to the sound of a piece of music?

## Overview of Organization and Materials

This book will define the unique aspects of sound that are captured, created, and shaped in the recording process, and will examine their appearance in well-known recordings by The Beatles. A system for developing listening skills and understanding the artistry of music recordings will then be presented, unfolding throughout the middle chapters of the book. Through this process and its many related exercises the reader will be brought to gain understanding of artistic aspects of audio recordings, and control over shaping them in the recording process. Ultimately, the reader will learn to better understand and utilize these qualities to establish their own artistic voice as a recordist.

Accordingly, the book is divided into four parts:

- Part One defines the artistic dimensions and sound characteristics of music recordings.
- Part Two builds fundamental listening skills towards hearing these dimensions of recorded sound.
- Part Three leads to an understanding of how those characteristics are commonly used in production practice, and carries the reader through a system for listening-skill development to recognize nuance in those dimensions.
- Part Four explores the artistry of recording processes, and leads the reader conceptually through the journey and the artistry of crafting a music recording.

### Part One. Defining the Art of Recording: The Sound Characteristics and the Aesthetic Qualities of Audio Recordings

Part One is divided into three chapters. To begin defining the artistic dimensions of music recordings, sound itself will be explored. The states of sound in air, in human perception, and as applied to music are followed in the sequence of understanding the meaning of sounds. The processes of moving from one state of sound to another are covered. The anomalies that occur in these transfer processes are recognized and evaluated.

Sound as a resource for artistic expression is the basis for Chapter 2. This chapter explains the unique sound qualities of recordings, and the potential of those qualities to be used in artistic expression.

This is followed by an examination of the musical message itself, leading to how musical materials are perceived by the listener, and how that perception leads to the understanding of musical messages and to communication. In many ways this foreshadows the last chapter of the book.

### Part Two. Learning to Listen, Beginning to Hear: Acquiring Fundamental Listening Skills and Establishing an Effective Approach to Listening

Part Two is also divided into three chapters. The focus of this part is to bring the reader to learn the skill of listening, so they may begin to consistently hear the dimensions presented in Part One. Chapter 4 discusses the listening process, its importance and role, personal development, and "talking about sound."

Chapter 5 leads the reader through a good number of fundamental listening skills. This is a core set of skills and experiences that will bring the reader/listener the background needed to engage the more complex tasks that follow. As a whole, acquiring these skills will require some significant time and attention. The chapter's eleven exercises are carefully crafted to effectively guide the reader through acquiring these skills. The skills are all important, as they prepare the reader for Part Three—and their career.

The book's system for evaluating sound is outlined in Chapter 6. A typical sequence for exploring materials, found to be most functional, is offered. Graphing dimensions of sound against time is central to this analytical approach and system. Creating a timeline and recognizing sound sources against time and structure presented here will be used throughout Part Three.

The exercises presented in this chapter can be undertaken in any order. Some readers will find some exercises much easier than others; mastery of one is not required before moving on. The reader is strongly encouraged to continue returning to exercises that pose a challenge, until the material is learned and the reader is comfortable that the skill is controlled.

### Part Three. Understanding the Mix: Developing Listening and Sound-Evaluation Skills

The six chapters of Part Three present the complete system for evaluating the dimensions of sound in music and audio recordings that was outlined in Part Two.

Each element of sound is evaluated separately, with stereo and surround separated into two chapters. A method of evaluation has been specifically devised for each individual element, and reflects how the elements appear in music recordings. Commercially available and widely celebrated recordings by The Beatles provide the examples of how the unique sound qualities of recordings have enhanced the music and have at times contributed in fundamental ways to shaping musical ideas.

The methods of evaluation for each individual sound quality are accomplished in relation to a complete, interrelated system of evaluating sound in music recordings. A series of listening exercises is presented throughout the course of these chapters to guide the reader in developing sound-evaluation and listening skills. This culminates by examining the evaluations of all elements, at their various levels of detail, for their interrelationships.

Starting in Part Two and continuing throughout Part Three, the reader progresses from simple concepts and listening processes to the most complex. The sequencing builds on experiences that are likely the most easily learned and evolves systematically to those typically most challenging. Listening experiences that many audio professionals or intermediate-level musicians may have already acquired (at least intuitively) are incorporated.

As the system develops the reader's listening skills, it will establish a basis for meaningful and accurate communication on sound content and quality. This is a significant outcome from learning this system.

An objective vocabulary and a way to evaluate sound have been devised; this allows precise information about sound to be recognized and communicated. This information will reflect the actual states or values of the sound material. Subjective impressions and subjective terms about the sound's quality (a very common way musicians and recordists attempt to communicate about sound) are always avoided; they do not allow communication to be accurate and severely limit its value. The method avoids any personal impressions about the qualities of sound by addressing only the physical dimensions of sound as they are perceived and as they appear in the music. The method for evaluating sound will allow individuals to talk "about" sound in meaningful ways. It will allow people to exchange precise and conclusive information about the sound qualities, once they acquire the skills required to recognize those qualities.

The diligent reader will gradually gain the experience and the knowledge that will allow them to perform quite complex listening and evaluation tasks, and to describe sounds to others.

Time, practice, understanding, repetition, and concentrated effort are all required to develop listening skills. Auditory memory will increase as the listener becomes more accustomed to the sound material, more aware of patterns of levels and changes within all of the artistic elements, and more skilled at focusing attention on the aspects of sound we are conditioned from birth to ignore. Developing refined listening skills is a long-term project; the individual's listening skills will likely continue to develop throughout their career.

## Part Four. Crafting the Mix: Shaping Music and Sound, and Controlling the Recording Process

Part Four brings the reader to apply these learned sound qualities to crafting the mix. How these sound qualities contribute to mixes is explored, including how they interrelate in practice. These five chapters bring the reader to make use of the production process to shape their own recordings with artistic sensitivity.

Part Four begins by examining the roles of the recordist, both functional and artistic. The first chapter continues with the production process shaping the recording aesthetic, its relationship to the live listening experience, and the altered realities of performance created by the production process.

Next we examine constructing and crafting the mix by bringing attention to individual artistic elements, and groupings of elements, as musical materials. The contributions of elements are examined for how they shape and impact the music and its expression. These are the concepts and materials engaged in composing the mix.

The intersections of practical, technical considerations with artistic decisions are encountered by examining microphones, equipment selection and technologies, monitoring and selecting timbres within Chapter 15. How these play out in practice is mirrored in the next chapter's coverage of production sequencing: Every project is unique, and requires unique choices of sound and methodology; some stray further from typical choices than others. Tracking and recording; editing, compilations, and signal processing; and preparing for the mix take us through preparation and planning the mix.

Part Four concludes the text with "Crafting the Mix, and Finalizing the Production." Here mix considerations meet musical materials and expression, and the story of the song. The mix is conceptualized to support the form and structure of the music and the text. Mastering and the listener's experience finish the chapter and the book. Many readers might have purchased this book specifically for this chapter; should the reader wish, this last chapter could be read first—though reading in the presented sequence is recommended, and the reader would do well to return to the first chapter after reading the last first.

It is a goal of *Understanding and Crafting the Mix: The Art of Recording* to bring the reader to explore the creative potentials of the medium's tools (equipment) as musical instruments. Exercises are contained throughout Part Four to engage the reader in learning the subtleties of performing the mix, and crafting its sound qualities. The concepts are encountered singly and build, thus allowing for their exploration for artistry as well as function.

## Companion Website

A companion website has been constructed to support this book at www.focalpress.com/cw/moylan. On it is a core set of tracks to illustrate many key concepts of the book. The reader will be referred to the website for these tracks throughout the text. Text boxes, such as the one below, are inserted throughout the book to identify specific tracks to hear and the purpose for listening.

The uncompressed audio files on the website are the intended versions for use. It is imperative the reader download these uncompressed files for use in listening exercises. Compressed files are present only to allow the reader the option of auditioning a track before downloading the larger, uncompressed file. The compressed files will alter or eliminate much of the material that is being discussed, and are unusable for study and understanding the material. The compressed files could make for advanced listening exercises, though, in learning to recognize the differences created by compression by comparing them to the uncompressed files.

The many recorded examples on the website will assist in developing important listening skills, and some tracks will present dimensions of recorded sound in ways that are more easily recognized than in many commercial recordings. The reader will often want to listen to tracks while reading specific sections in the text; they are meant to clarify materials and make some of the more abstract concepts much more tangible.

The website also contains additional materials, including graphs and descriptions of recordings other than those explained in the text. Materials will continue to be added over time to refresh the website, as well as to enhance the book's content and the reader's experience.

Three tracks are included on the website to aid in establishing an accurate playback system. Readers are strongly encouraged to carefully evaluate their sound systems. Accurate and quality playback of the website tracks, and of the commercial recordings cited throughout the book, is necessary if the reader is to understand the materials being presented.

## Establishing an Accurate Playback of Recordings

A reasonable-quality, accurate playback system located in an acceptable listening environment is needed to experience and learn the material discussed in this book. It is important that the recorded examples be heard as intended—with a minimum of change from the playback system and the room—in order to learn the materials being discussed. The reader should have access to such a system in order to benefit from the readings. Headphones are not a suitable substitute for accurate listening to nearly all of the examples.

High-resolution monitors in an acoustically neutral environment and an impeccable signal chain are required for many activities performed by professional recordists. It is not at all realistic to expect a beginner in the industry, a student, or an interested amateur to have such a system. It is, however, necessary for the reader to hear the website tracks and the commercial recordings cited throughout the book with some semblance of accuracy.

Putting together a quality sound system and establishing an accurate playback environment are covered in Chapter 15. Topics presented are:

- Components and specifications for a high-quality playback system to provide unaltered, detailed sound.
- Loudspeaker placement and listening-room interaction.
- Listener distance and orientation to loudspeakers.
- Monitoring levels.

If readers are uncertain about having a suitable playback system, they are encouraged to look over that material now, or certainly before playing any of the listening examples.

The sound-system evaluation tracks (tracks 54–56) on the website can help the reader evaluate their current system. It is strongly suggested that readers do so in order to know the performance of their system, and to ensure it is properly set up. It is also recommended they acquire a sound-level meter. The meter will not only assist in evaluating their system, but will also help the reader to work through a number of exercises in this book and will lead the reader to establish accurate and healthy monitoring practices.

**Listen**

to tracks 54–56

for playback system set-up and calibration material; read the track descriptions for assistance.

If a reader cannot establish a reasonable quality of playback on their own, they are strongly encouraged to find a suitable location where they can listen to the recordings that are cited. This is an important part of learning the materials presented.

## Summary

Part One provides the necessary background to recognize the qualities of sound. It defines the sound characteristics and aesthetic qualities of audio and music recordings. It brings the reader to understand that the recording process enhances music recordings, and has the potential to shape their artistic qualities substantially.

Parts Two and Three provide a framework and a method to develop high-level listening skills. It will lead the listener to develop listening and sound-evaluation skills and to learn how to communicate objective and meaningful information about sound. Important commercial recordings are examined to illustrate the sound characteristics being studied.

Part Four brings focus to the art and aesthetics of recording, provides some context to understand the recordist's place within any project, and guides the creative processes of conceiving, composing, performing, and crafting the mix. While this section will not present prescriptions for production techniques to achieve certain results, or provide information on selecting and using specific equipment or technologies, it will lead the reader to directed discovery and channeled creativity. It will bring the reader to gain control of the craft of recording music and shaping sound, and to use the recording process in an artistic way—no matter the technology, techniques, music genre: those of yesterday, today, or tomorrow.

*Understanding and Crafting the Mix: The Art of Recording* encourages the reader to think broadly and deeply about the creative, artistic, and technical processes, and it provides the guidance to enable them to do so. The reader will be brought to conceive the music recording as a self-contained piece of music, rather than a product of technological processes.

The goal of *Understanding and Crafting the Mix: The Art of Recording* is to bring the reader to an awareness of the dimensions of recorded sound, and to an understanding of the dimensions as they appear in music recordings, through the development of listening and evaluation skills. With these acquired, it seeks to lead readers to find, establish, and refine their own unique artistic voice in shaping and crafting music recordings.

# Part One

## Defining the Art of Recording:

### The Sound Characteristics and the Aesthetic Qualities of Audio Recordings

# 1 The Elements of Sound and Audio Recording

Audio recording is the recording of sound. It is the act of capturing the physical dimensions of sound and then reproducing those dimensions. The dimensions are reproduced either immediately or from a storage medium (magnetic, vinyl, electronic, digital), and thereby returning those dimensions to their physical, acoustic state.

Audio recording is a process that transforms sound from its physical acoustic state, through the recording/reproduction chain, and back to physical sound.

The "art" of recording emerges with the artistically sensitive application of the recording process. The recording process is being used to shape or create sound as an artistic statement (piece of music), or supporting artistic material. To be in control of crafting the artistic product, one must be in control of the recording process, be fluent in the ways the recording process modifies sound, and be skilled in executing well-defined creative ideas.

These all closely involve interaction with sound. Inconsistencies between the various states of sound are present throughout the audio-recording process. Many of these inconsistencies are the result of the human factor: the ways in which humans perceive sound and interpret or formulate its meanings. In order for the artist (audio engineer/recordist) to be in control of the artistic processes they must understand the substance of their material: sound, in all its inconsistencies.

## The States of Sound

In audio recording, sound is encountered in three different states. Each of these three states directly influences the recording process. These three states are:

1. Sound as it exists physically (having physical dimensions);
2. Sound as it exists in human perception (psychoacoustic conception): sound perceived after being transformed by the ear and interpreted by the mind (the perceived parameters of sound being human perceptions of the physical dimensions); and
3. Sound as idea: sound as an abstract or a tangible concept, as an emotion or feeling, or representing a physical object or activity (this is how the mind finds meaning from its attention to the perceived parameters of sound); sounds as meaningful events, capable of communication, provide a medium for artistic expression; sounds hereby communicate, have meaning.

The audio-recording process ends with sound reproduced over loudspeakers, as sound existing in its physical state, in air. Often the audio-recording process will begin with sound in this physical state, to be captured by a microphone.

The *recordist*—the person making the recording—and all others involved in the industry listen to the sound in all facets of the audio-recording process. They evaluate the audio signal at all stages while the recording is being made. This, of course, results in the recording that is heard by the end listener, and is the reason for making the recording. In the listening process we translate the physical dimensions of sound into the perceived parameters of sound through the listening process (aural perception).

This translation process involves the hearing mechanism functioning on the physical dimensions of sound, and the transmission of neural signals to the brain. The process is nonlinear and alters the information; the hearing mechanism does not produce nerve impulses that are exact replicas of the applied acoustic energy.

Certain aspects of the distortion caused by the translation process are, in general, consistent between listeners and between hearings; they are related to the physical workings of the inner ear or the transfer of the perceived sounds to the mind/brain. Other aspects are not consistent between listeners and between hearings; they relate to the listener's unique hearing characteristics and their experience and intelligence.

The final function occurs at the brain. At a certain area of the cortex, the neural information is processed, identified, consciously perceived, and stored in short-term memory; the neural signals are transferred to

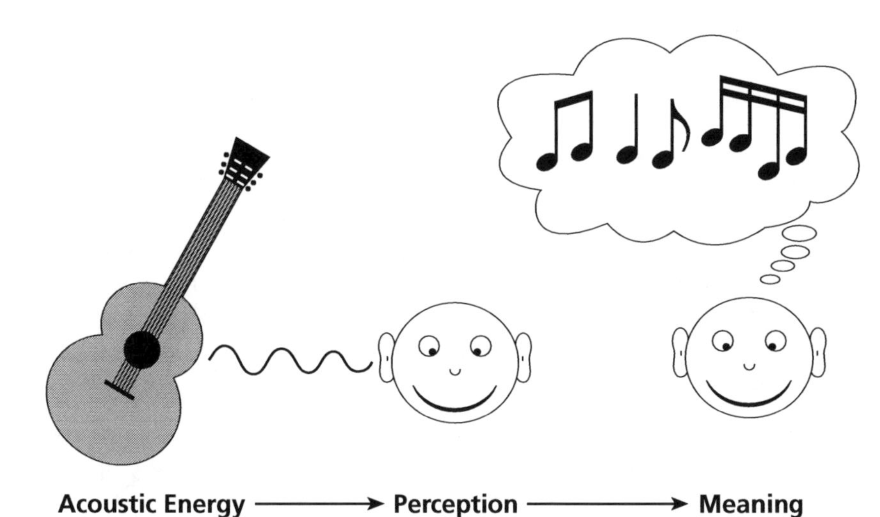


**Figure 1.1**
Three states of sound: in air, in perception, as message.

other centers of the brain for long-term memory. At this point, the knowledge, experience, attentiveness, and intelligence of the listener become factors in the understanding and perception of sound's artistic elements (or the meanings or message of the sound). The individual is not always sensitive or attentive to the material or to the listening activity, and the individual is not always able to match the sound to their previous experiences or known circumstances.

Sequentially, the physical dimensions (1) are interpreted as perceived parameters of the sound (2). The perceived parameters of sound (2) provide a resource of elements that allow for the communication and understanding of the meaning of sound (and artistic expression) (3).

The audio-recording process communicates ideas, and can express feelings and emotions. Audio might take the form of music, dialog, motion-picture action sounds, whale songs, or others. Whatever its form, audio is sound that has some type of meaning to the listener. The perceived sound provides a medium of variables that are recognizable and have meaning when presented in certain orders or patterns. Sound, as perceived and understood, becomes the resource for creative and artistic expression. The artist uses the perceived parameters of sound as the artistic elements of sound, to create and ensure the communication of meaningful (musical) messages.

In the next section the individual states of sound as physical dimensions and as perceived parameters will be discussed individually. The interaction of the perceived parameters of sound will follow the discussion of the individual parameters. These discussions provide critical information for understanding the breadth of the "artistic elements of sound" in audio recording, presented in the next chapter.

## Physical Dimensions of Sound

Five physical dimensions of sound are central to the audio-recording process. These physical dimensions are: the characteristics of the sound waveform as (1) *frequency* and (2) *amplitude* displacements, occurring within the continuum of (3) *time*; the fusion of the many frequency and amplitude anomalies of the single sound to create a global, complex waveform as (4) *timbre*; and the interaction of the sound source (timbre) and the environment in which it exists creates alterations to the waveform according to variables of (5) *space*.

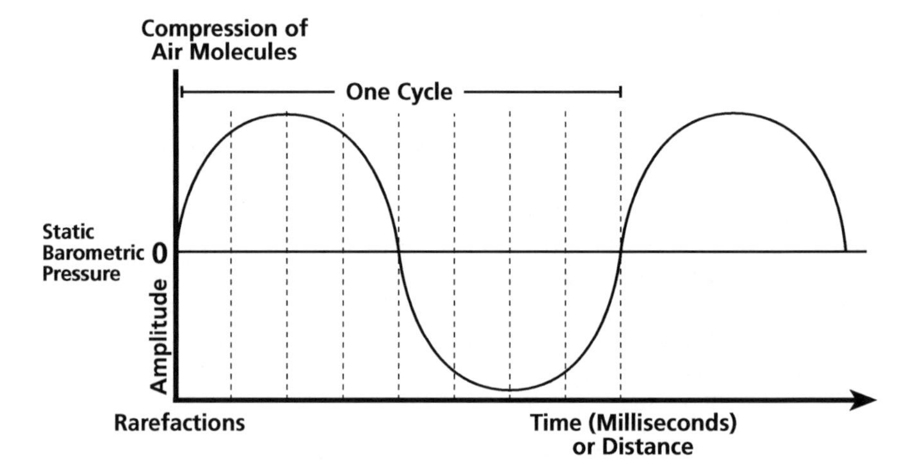


**Figure 1.2**
Dimensions of the waveform.

Frequency is the number of similar, cyclical displacements in the medium, air, per time unit (measured in cycles of the waveform per second, or Hz). Each similar compression/rarefaction combination creates a single cycle of the waveform. Amplitude is the amount of displacement of the medium at any moment, within each cycle of the waveform (measured as the magnitude of displacement in relation to a reference level, or decibels).

### Timbre

Timbre is a composite of a multitude of functions of frequency and amplitude displacements; it is the global result of all the amplitude and frequency components that create the individual sound. Timbre is the overall quality of a sound. Its primary component parts are the dynamic envelope, spectrum, and spectral envelope.

The *dynamic envelope* of a sound is the contour of the changes in the overall dynamic level of the sound throughout its existence. Dynamic envelopes of individual acoustic instruments and voices vary greatly in

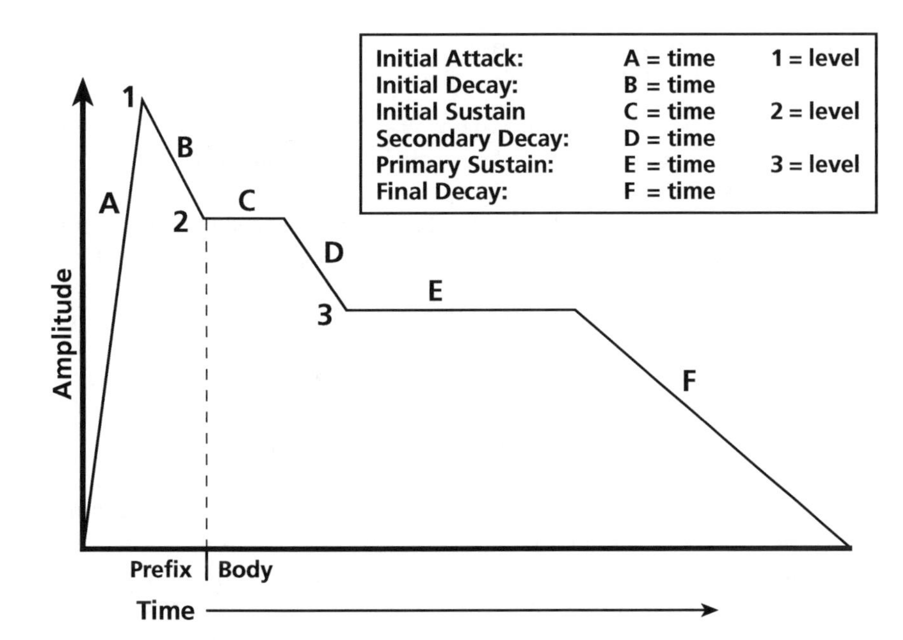


**Figure 1.3**
Dynamic envelope.

content and contours. The dynamic envelope is often thought of as being divided into a number of component parts. These component parts may or may not be present in any individual sound. The widely accepted components of the dynamic envelope are: attack (time), initial decay (time), initial sustain level, secondary decay (time), primary sustain level, and final decay (release time).

Dynamic envelope shapes other than those created by the above outline are common. Many musical instruments have more parts to their characteristic dynamic envelope, and many instruments have fewer. Further, vocalists and the performers of many instruments have great control over the sustaining portions of the envelope, providing internal dynamic changes to sounds. Musical sounds that do not have some variation of level during the sustain portion of the envelope are rare; the organ is one such exception.

The *spectrum* of a sound is the composite of all of the frequency components of the sound. It is comprised of the fundamental frequency, harmonics, and overtones, sometimes including subharmonics and subtones.

The periodic vibration of the waveform produces the sensation of a dominant frequency. The number of periodic vibrations, or cycles of the waveform, that repeat its characteristic shape is the *fundamental frequency*. The fundamental frequency is also that frequency at which the sounding body resonates along its entire length. The fundamental frequency is often the most prominent frequency in the spectrum, and will often have the greatest amplitude of any component of the spectrum.

In all sounds except the pure sine wave, frequencies other than the fundamental are present in the spectrum. These frequencies are usually higher than the fundamental frequency. They may or may not be in a whole-number relationship to the fundamental. Frequency components of the spectrum that are whole-number multiples of the fundamental are *harmonics*; these frequencies reinforce the prominence of the fundamental frequency (and the pitched quality of the sound). Those components of the spectrum that are not proportionally related to the fundamental we will refer to as *overtones*. Traditional musical acoustics studies define overtones as being proportional to the fundamental, but with a different sequence than harmonics (first overtone = second harmonic, etc.); this traditional definition is herein replaced by a differentiation between *partials* that are proportional to the fundamental (harmonics) and those that are not (overtones). This distinction will prove important in the evaluation of timbre and sound quality in later chapters. All of the individual components of the spectrum are partials. Partials (overtones and harmonics) can exist below the fundamental frequency as well as above; they are accordingly referred to as subharmonics and subtones.

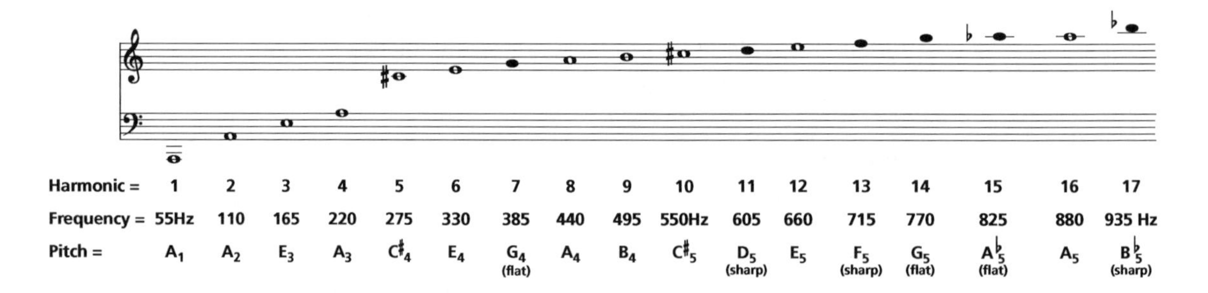


**Figure 1.4**
Harmonic series.

For each individual instrument or voice, certain ranges of frequencies within the spectrum will be emphasized consistently, no matter the fundamental frequency. Instruments and voices will have resonances that will strengthen those spectral components that fall within these definable frequency ranges. These areas are called *formants, formant regions*, or *resonance peaks*. Formants remain largely constant, and modify the same frequency areas no matter the fundamental frequency. Spectral modifications will be present in all occurrences of the sound source with harmonics or overtones in the formant regions. Formants can appear as increases in the amplitudes of partials that appear in certain frequency bands, or as spectral components in themselves (such as noise transients caused by a hammer striking a string). They can also be associated with resonances of the particular mechanism that produced the source sound. Formants are largely responsible for shaping the characteristic sounds of specific instruments; they allow us to differentiate between the instruments of different manufacturers, or even to tell the difference between two instruments of the same model and maker.

**Listen**

to tracks 1 and 2

for the harmonic series played in individual frequencies and pitches, and as a chord.

A sound's spectrum is composed primarily of partials that create a characteristic pattern. This pattern is recognizable as being characteristic of a particular instrument or voice. This pattern of spectrum will transpose (change level but maintain the same distances between frequencies/partials) with every new fundamental frequency of the same sound source and remain mostly unchanged. In this way, this consistent pattern will form a similar timbre at different pitch levels. Formants establish frequency areas that will be emphasized for a particular instrument or voice. These areas will not change with varied fundamental frequency, as they are fixed characteristics (such as resonant frequencies) of the device that created the sound. Formants may also take the form of spectral information that is present in all sounds produced by the instrument or voice.

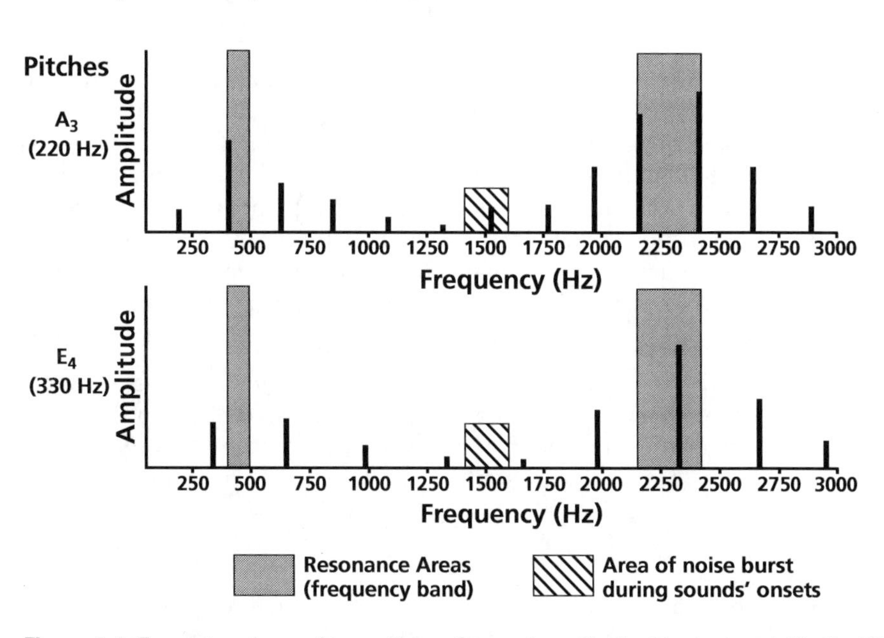


**Figure 1.5** Format regions of two pitches from a hypothetical instrument. Vertical lines represent the partials of the two pitches, placed at specific frequencies and at specific amplitudes.

The frequencies that comprise the spectrum (fundamental frequency, harmonics, overtones, subharmonics, and subtones) all have different amplitudes that change independently over the sound's duration. Thus, each partial has a different dynamic envelope. Altogether these dynamic envelopes of all the partials make up the *spectral envelope*. The spectral envelope is the composite of each individual dynamic level and dynamic envelope of all of the components (partials) of the spectrum.

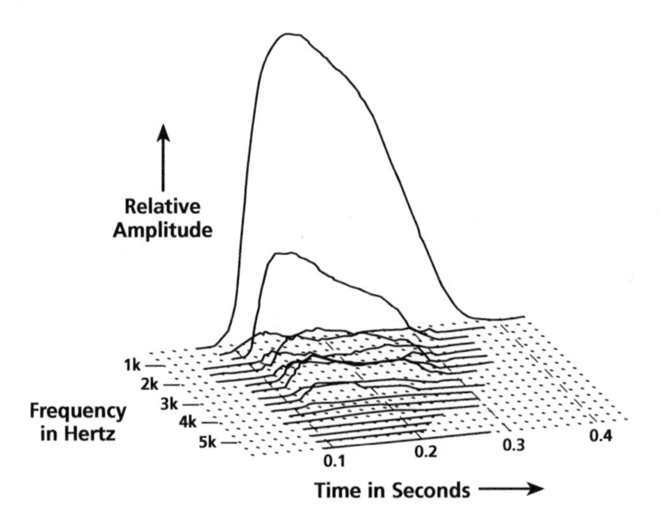


**Figure 1.6**
Spectral envelope.

The component parts of timbre (dynamic envelope, spectrum, and spectral envelope) display strikingly different characteristics during different parts of the duration of the sound. The duration of a sound is commonly divided into two time units: the *prefix* or *onset*, and the *body*. The initial portion of the sound is the prefix or onset; it is markedly different from the remainder of the sound, the body. The time length of the prefix is usually determined by the way a sound is initiated, and is often the same time unit as the initial attack. The actual time increment of the prefix may be anywhere from a few microseconds to 20–30 ms.

The prefix is defined as the initial portion of the sound that has markedly different characteristics of dynamic envelope, spectrum, and spectral envelope than the remainder of the sound. The body of the sound is usually much longer in duration than the prefix. (See Figure 1.3.)

## Space

Sounds are changed by the environments in which they are produced. The interaction of the sound source (timbre) and the environment will create alterations to sound. These changes to the sound source's sound quality are created by the acoustic space. The nature of these alterations is directly related to (1) the characteristics of the acoustic space in which the sound is produced and (2) the location of the sound source within the environment.

Space-related sound measurements must be performed at a specific physical location. The measurements are calculated from the point in space where a receptor (perhaps a microphone or a listener) will capture the composite sound (the sound source within the acoustic space). The location of the listener (or other receptor) becomes a reference in the measurement of the acoustic properties of space.

The aspects of space that influence sound in audio recording are: (1) the *distance* of the sound source to the listener, (2) the *angle* of the sound source to the listener, (3) the *geometry of the environment* in which the sound source is sounding, and (4) the *location* of the sound source *within the host environment*.

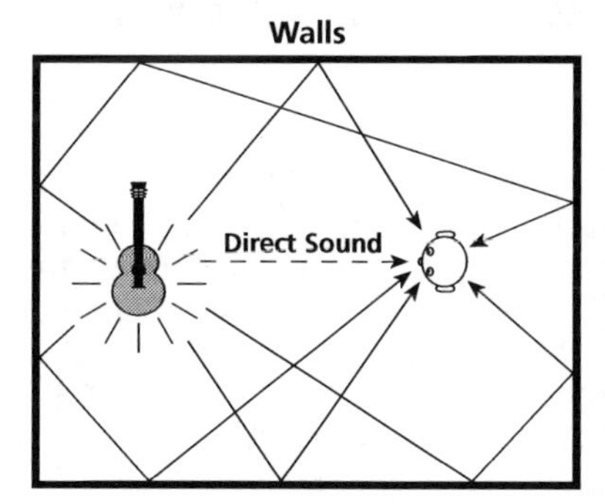


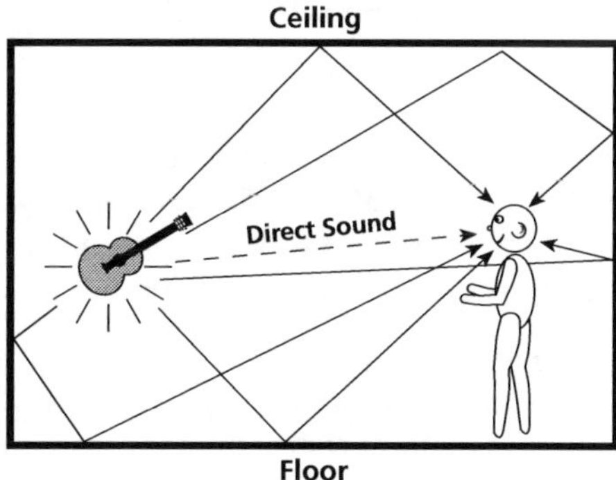


**Figure 1.7**
Paths of reflected sound within an enclosed space.

The environment in which the sound source is sounding is often referred to as the *host environment*. Within the host environment, sound will travel on a direct path to the listener (as *direct sound*) and sound will bounce off reflective surfaces before arriving at the listener (as *reflected sound*).

*Reverberant sound* is a composite of many reflections of the sound arriving at the listener (or microphone) in close succession. The many reflections that comprise the reverberant sound are spaced so closely that the individual reflections fuse into a dense set of reflections and create a single sonic impression. These many reflections are therefore considered as a single entity. As time progresses, these closely spaced reflections become even more closely spaced and of diminishing amplitude, until they are no longer of consequence. *Reverberation time* (often referred to as RT60) is the length of time required for the reflections to reach an amplitude level of 60 dB lower than that of the original sound source.

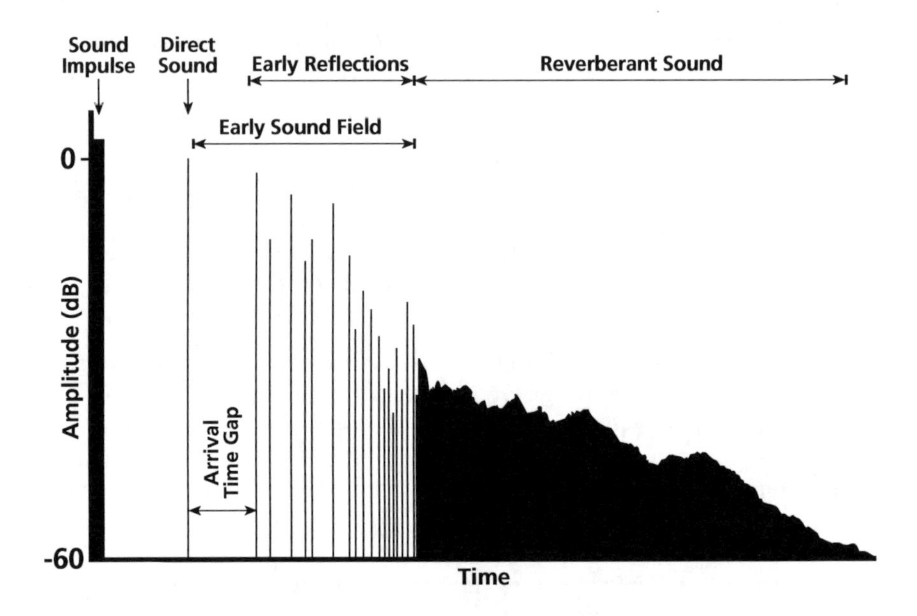


**Figure 1.8**
Reflected sound.

*Early reflections* are those reflections that arrive at the ear or microphone within around 50 ms of the direct sound. As a collection, the reflections that arrive at the receptor within the first 50 ms after the arrival of the direct sound comprise the *early sound field*.

Varying the *distance* of the sound source from the receptor (ear or microphone) alters the sound at the receptor. The sound at the receptor will be a composite of the direct sound and the reflected sounds (reverberation and early sound field). That composite sound is affected by the distance of the sound source from the receptor in two ways: (1) low-amplitude portions of the sound's spectrum (usually high frequencies) are lost with increasing distance of the sound source to the receptor and (2) reflected sound increases in prominence to the direct sound as distance increases. Figure 1.9 illustrates the loss of *timbral detail* (the subtle partials and/or changes in the content of a sound's timbre, also called *definition of timbre*) with increasing distance as well as the change of the proportion of direct to reflected sound.

The characteristic changes to the composite sound, caused by the geometry of the host environment and by the location of the sound source within the host environment, are also influenced by the changes caused by distance.

These two dimensions of the relationship of the sound source to its acoustic space may alter the composite sound in four additional ways: (1) timbre differences between the direct and reflected sounds; (2) time differences between the arrivals of the direct sound, the initial reflections,

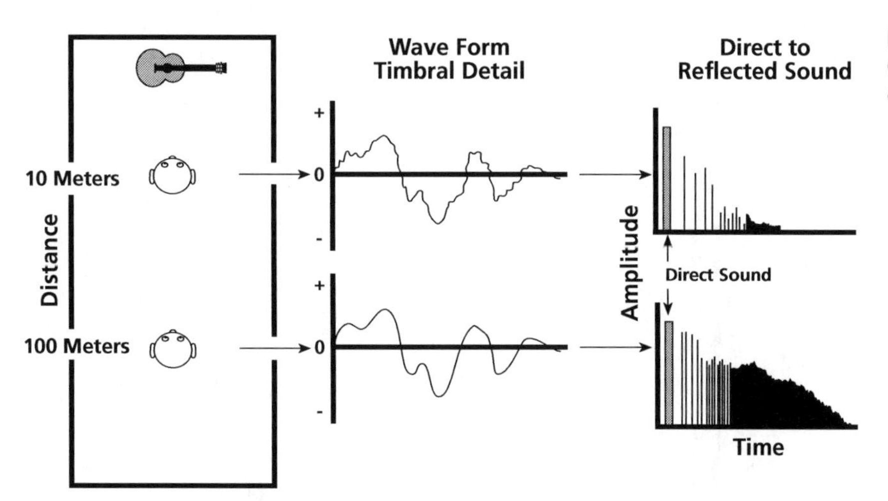


**Figure 1.9**
Changes in sound with distance.

and the reverberant sound; (3) spacing in time of the early reflections; and (4) amplitude differences between direct and reflected sounds.

The geometry of the host environment greatly influences the content of the composite sound. The dimensions and volume of the space, the angles of boundaries (walls, floors, ceilings), materials of construction, and the presence of openings (such as windows) and large objects within the space will all alter the composite sound. Host environments cover the gamut of all the physical spaces and open areas that create our reality (from a small room to a large concert hall, from the corridor of a city street to an open field, etc.).

Unique sequences of reflected sound are created when a sound is produced within an environment, and sequences are shaped by the location of the sound source within the host environment. These unique sequences contain patterns of reflections that are defined by the spacing of reflections over time and the amplitudes of the reflections. A "rhythm of reflections" exists and will form the basis of important observations in later chapters. By altering the early time field and reverberant sound, the location of the sound source within the host environment may cause significant alterations to the composite sound at the receptor (ear or microphone).

The location of the sound source within the host environment may strongly influence the composite sound. The amount of influence will be directly related to the proximity of the sound source to the walls, ceiling, floor, openings (such as windows and doors), and large objects reflecting sound within the host environment.

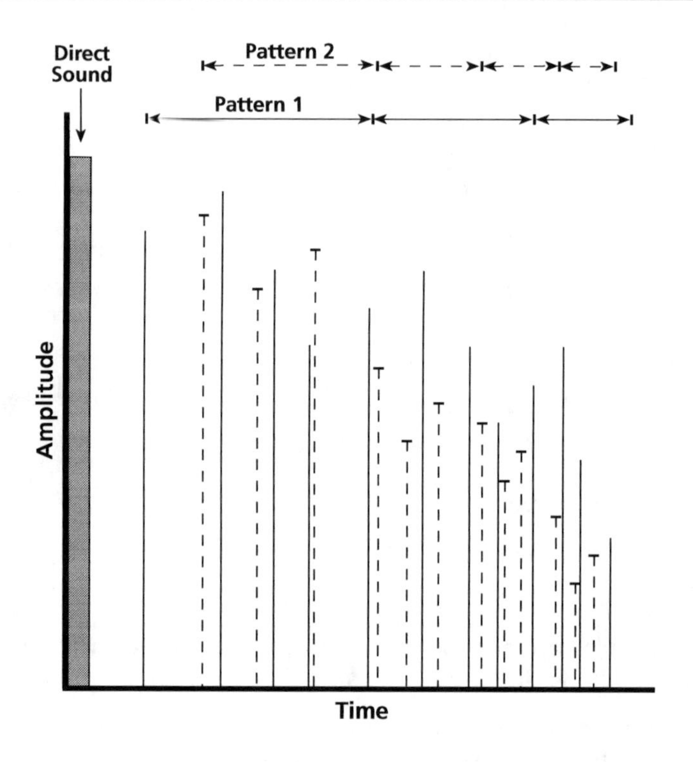


**Figure 1.10**
Patterns of reflections.

**Listen**

to tracks 34–36

for a realization of Figure 1.10, where the rhythms of these patterns of reflections are sounded separately and together.

In audio production, the spatial properties of host environments and the location of a sound source within the environment can be generated artificially. It is common to use reverberation units and delays (or related plug-ins) to create environmental cues. These cues may be very realistic representations of natural spaces or they may be environmental characteristics that cannot occur in our physical reality.

The angle of the sound source to the receptor is an important influence in audio recording. The sound source may be at any angle from the receptor (listener or omnidirectional microphone) and be detected. The sound source may be present at any location in the sphere surrounding the receptor. The location is calculated with reference to the 360° vertical and horizontal planes that encompass the receptor.

The angle of the sound source to the receptor may be calculated against the horizontal plane (parallel to the floor), the vertical plane (height), or by combining the two (in a way very similar to positioning locations on a globe). Defining elevation (vertical plane) and direction (left, right,

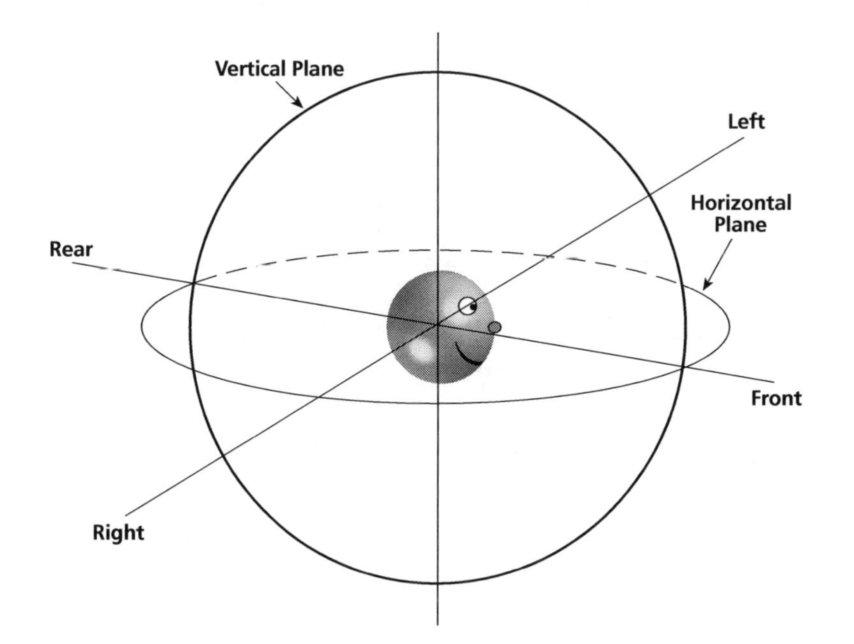


**Figure 1.11**
Horizontal and vertical planes.

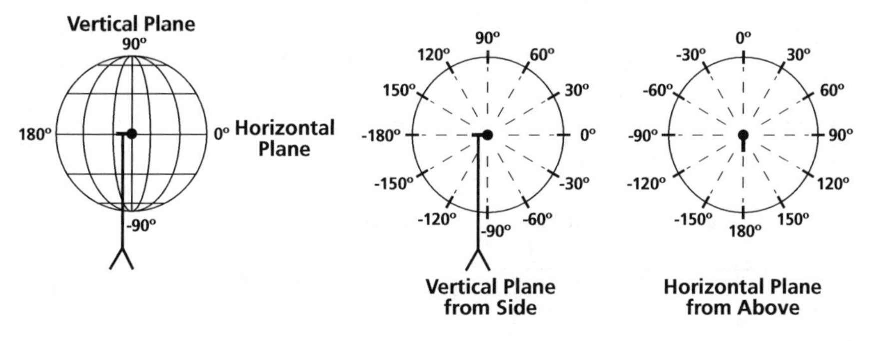


**Figure 1.12**
Defining sound source angle from a microphone.

front, rear) can determine the precise location of the sound source within our three-dimensional space by precise increments of degrees.

Angles of source locations on the horizontal plane are captured or generated in audio recording to provide stereo and surround sound. To date, the vertical plane has received limited attention because of playback format difficulties. Recent surround-sound advances have produced formats that provide these cues in ways that can strikingly enhance programs.

## Perceived Parameters of Sound

The five physical dimensions of sound translate into respective perceived parameters of sound. Sound as it exists in human perception is quite different from sound in its physical state, in air. Our perception of sound is a result of the physical dimensions being transformed by the ear and interpreted by the mind. The perceived parameters of sound are our perceptions of the physical dimensions of sound.

This translation process from the physical dimensions to the perceived parameters is nonlinear, and differs between individuals. The hearing mechanism does not directly transfer acoustic energy into equivalent nerve impulses. The human ear is not equally sensitive in all frequency ranges, nor is it equally sensitive to sound at all amplitude levels. This nonlinearity in transferring acoustic energy to neural impulses causes sound to be in a different state in our perception than what exists in air. Thus, the physical states of sound captured by recording equipment will be heard by the recordist in ways that may be unexpected, without knowledge of these differences.

Complicating this further, there is no reason to believe any two people actually hear the characteristics of sound in precisely the same way. If it were possible for all conditions for two sounds to be identically sent to two listeners, the two people likely would hear slightly (or strikingly) different characteristics. We only need notice the different ear shapes around us to recognize no two people will pick up acoustic energy in precisely the same way.

**Table 1.1** The physical dimensions and the perceived parameters of sound

| ***Physical Dimensions*** | ***Perceived Parameters*** |
|---|---|
| Frequency | Pitch |
| Amplitude | Loudness |
| Time | Duration |
| Timbre (physical components) | Timbre (perceived overall quality) |
| Space (physical components) | Space (perceived characteristics) |

### Pitch

*Pitch* is the perception of the frequency of the waveform. It has been defined as the perceived position of a sound on a scale from low to high, and as an attribute of hearing sensation by which sounds may be ordered on a musical scale. Pitch is a subjective attribute and cannot be measured. We assign values to pitches, to allow us to create and understand their relationships to other pitches. We organize these values into tuning and harmonic systems, which give rise to melody and harmony. The creation of these systems may have been different (and are therefore subjective). Indeed these organizational systems differ markedly between cultures and have evolved substantially over the centuries.

A clear pitch sensation is perceived when a sound wave regularly repeats a wave shape of very similar characteristics. The number of periodic repetitions of the shape per second allows a fundamental frequency to be perceived, and an unambiguous pitch sensation results. This is further solidified by the presence of frequency components that are integer multiples of the fundamental (harmonics). Thus, the physical dimensions of sound are transformed into our perception of pitch.

The frequency area most widely accepted as encompassing the hearing *range* of the normal human spans the boundaries of 20–20,000 Hz (20 kHz), though humans are sensitive to (if not actually able to hear) frequencies well below and above this range.

Most humans cannot identify specific pitch levels. Some people have been blessed with, or have developed, the ability to recognize specific pitch levels (in relation to specific tuning systems). These people are said to have "absolute" or "perfect pitch." The ability to accurately recognize pitch levels is not common even among well-trained musicians.

It is commonly within human ability, however, to determine the relative placement of a pitch within the hearing range. A *register* is a specific portion of the *range*. It is entirely possible to determine, within certain consistent limits of accuracy, the relative register of a perceived pitch level. This skill can be developed and accuracy improved significantly.

We are able to consistently perform the estimation of the approximate level of a pitch, associating pitch level with register. With practice, this consistency can be accurate to within a minor third (within three semitones). This skill in the "estimation of pitch level" will be an important part of the method for evaluating sound presented later.

Humans perceive pitch most accurately as relationships. We perceive pitch as the relationship between two or more soundings of the same or related sound sources. We do not perceive pitch as identifiable, discrete increments; we do not listen to pitch material to define the letter names (increments) of pitches. Instead, we calculate the distance (or interval) between pitches by gauging the distance between the perceived levels of the two (or more) pitches.

The interval between pitches becomes the basis for all judgments that define and relate the sounds. Thus, melody is the perception of successively sounded pitches (creating linear intervals), and chords are the perceptions of simultaneously sounded pitches (creating harmonic intervals). We often perceive pitch in relation to a reference level (one predominating pitch that acts as the key or pitch-center of a piece of music), or to a system of organization to which pitches can be related (a tonal system, such as major or minor).

Our ability to recognize the interval between two pitches is not consistent throughout the hearing range. Most listeners have the ability to accurately judge the size of the semitone (or minor second, the smallest musical interval of the equal-tempered system) within the range of 60 Hz and 4 kHz. As pitch material moves below 60 Hz, a typical listener will have increased difficulty in accurately judging interval size. As pitch material moves above 4 kHz, the typical listener will also experience increased difficulty in accurately judging interval size.

The smallest interval humans can accurately perceive is not consistent throughout our hearing range. It changes with the register of the two pitches creating the interval. The size of the minimum audible interval varies from about 1/12 of a semitone between 1–4 kHz, to about half of a semitone (a quarter-tone) at approximately 65 Hz. These figures are dependent upon optimum duration and loudness levels of the pitches; sudden changes of pitch level are up to 30 times easier to detect than gradual changes. It is possible for humans to distinguish up to 1,500 individual pitch levels by spacing out the appropriate minimum-audible intervals throughout the hearing range.

With all factors being equal, the perception of harmonic intervals (simultaneously sounding pitches) is more accurate than the perception of melodic intervals (successively sounded pitches). Up to approximately 500 Hz, melodic and harmonic intervals are perceived equally well. Above 1 kHz, humans begin to be able to judge harmonic intervals with greater accuracy than melodic intervals; above 3,500 Hz, this difference becomes pronounced.

### Loudness

*Loudness* is the perception of the overall excursion of the waveform (amplitude). Amplitude can be physically measured as a sound pressure level. In perception, loudness level cannot be accurately perceived in discrete levels.

Loudness is referred to in relative values, not as having separate and distinct levels of value. Traditionally, loudness levels have been described by analogy ("louder than," "softer than," etc.) or by relative values ("soft," "medium loud," "very soft," "extremely loud," etc.). Humans compare loudness levels and conceive loudness levels as being "louder than" or "softer than" the previous, succeeding, or remembered loudness level(s).

A great difference exists between loudness as perceived by humans and the physical amplitude of the sound wave. This difference can be quite large at certain frequencies. In order for a sound of 20 Hz to be audible, a sound pressure level of 75 dB must be present. At 1 kHz, the human ear will perceive the sound with a minute amount of sound pressure level, and at 10 kHz a sound pressure level of approximately 18 dB is required for audibility. The unit *phon* is the measure of perceived loudness established at 1 kHz, based on subjective listening tests.

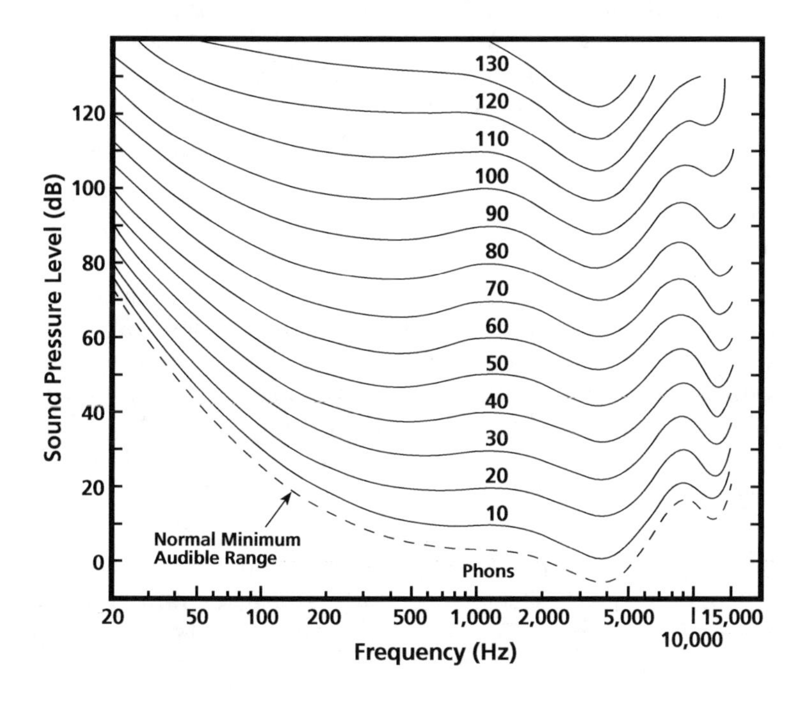


**Figure 1.13** Equal loudness contour.

The nonlinear frequency response of the ear and the fatigue of the hearing mechanism over time contribute to further inaccuracies of the human perception of loudness in three ways:

1. With sounds of long durations and steady loudness level, loudness will be perceived as increasing with the progression of the sound until approximately 0.2 seconds of duration. At that time, the gradual fatigue of the ear (and possibly shifts of attention by the listener) will cause perceived loudness to diminish.
2. As loudness level of the sound is increased, the ear requires increasingly more time between soundings before it can accurately judge the loudness level of a succeeding sound. We are unable to accurately judge the individual loudness levels of a sequence of high-intensity sounds as accurately as we can judge the individual

loudness levels of mid- to low-intensity sounds; the inner ear itself requires time to reestablish a state of normalcy, from which it can accurately track the next sound level.

3. As a sound of long duration is being sustained, its perceived loudness level will gradually diminish. This is especially true for sounds with high sound pressure levels. The ear gradually becomes desensitized to the loudness level. The physical masking (covering) of softer sounds and an inability to accurately judge changes in loudness levels will result from the fatigue. When the listener is hearing under listening fatigue, slight changes of loudness may be judged as being large. Listening fatigue may desensitize the ear's ability to detect new sounds at frequencies within the frequency band (frequency area) where the high sound-pressure level was formerly present.

### Duration

Humans perceive time as *duration*. Sound durations are not perceived individually. We cannot accurately judge time increments without a reference time unit. Regular reference time units are found in musical contexts and rarely in other types of human experiences. Even the human heartbeat is rarely consistent enough to act as a reliable reference. The underlying metric pulse of a piece of music does, however, allow for accurate duration perception. This accuracy cannot be achieved in any other context of the human experience.

In music, the listener remembers the relative duration values of successive sounds, in a similar process to that of perceiving melodic pitch intervals. These successive durations create musical rhythm. The listener calculates the length of time between when a sound starts and when it ends, in relation to what precedes it, what follows it, what occurs simultaneously with it, and what is known (what has been remembered). Instead of calculating an interval of pitch, the listener proceeds to calculate a span of time, as a durational value.

#### *Metric Grid*

A *metric grid*, or an underlying pulse, is quickly established in the perception of the listener, as a piece of music unfolds. This creates a reference pulse against which all durations can be defined. The listener is thereby able to make rhythmic judgments in a precise and consistent manner. The equal divisions of the grid allow the listener to compare all durations and to calculate the pulse-related values of the perceived sounds. Durations are calculated as being in proportion to the underlying pulse: at the pulse, half pulses, quarter of the pulse, double the pulse, etc.

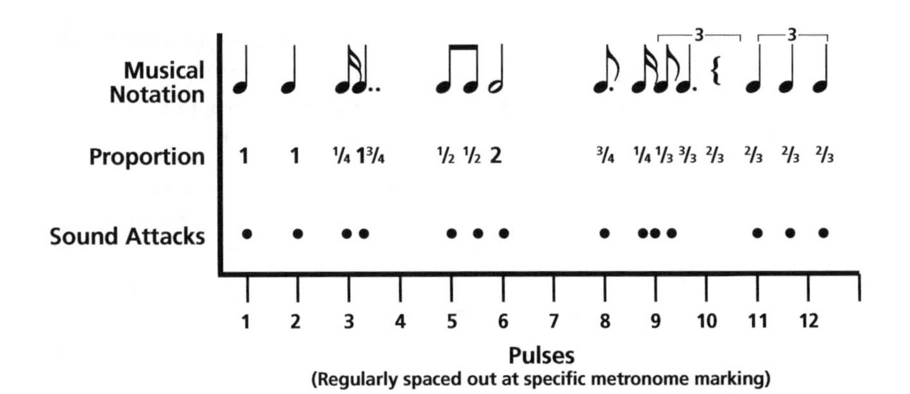


**Figure 1.14**
Metric grid.

In the absence of the metric grid, durational values cannot be accurately perceived as proportional ratios. Humans will not be able to perceive slight differences in duration when a metric grid is not established.

The listener is only able to establish a metric grid within certain limits. Humans will be able to accurately utilize the metric grid between 30 to 260 pulses per minute. Beyond these boundaries, the pulse is not perceived as the primary underlying division of the grid. We will instead replace the pulse with a duration of either one half or twice the value, or the listener might become confused and unable to make sense of the rhythmic activity.

The metric grid is the dominant factor in our perception of tempo, as well as musical rhythm. In most instances, the metric grid itself represents the steady pulsation of the tempo of a piece of music.

### *Time*

The listener's *time perception* plays a peripheral role in the perception of rhythm. Time perception is significant, however, to the perception of the global qualities of a piece of music, and to the estimation of durations when a metric grid is not present in the music. The global qualities of aesthetic, communicative, and extra-musical ideas within a piece of music are largely dependent on the living experience of music—on the passage of musical materials across the listener's time perception of their existence.

Time perception is distinctly different from duration perception. The human mind makes judgments of elapsed time based on the perceived length of the present. The length of time humans perceive to be "the present" is normally two to three seconds, but might be extended to as much as five seconds and beyond.

The "present" is our window of consciousness, through which we perceive the world and listen to sound. We are at once experiencing the moment of our existence, evaluating the immediate past of what has just happened and anticipating the future (projecting what will follow the present moment, given our experiences of the recently passed moments, and our knowledge of previous, similar events).

Human time judgments are imprecise. The speed at which events take place and the amount of information that takes place within the "present" greatly influences time judgments. The amount of time perceived to have passed will change to conform to the number of events experienced within the present; the listener will estimate the amount of time passed in relation to the number of experiences during the present, and make time judgments accordingly.

Time judgments are greatly influenced by the individual listener's attentiveness and interest in what is being heard. If the material stimulates thought within the listener, the event will seem shorter; if the listener finds the listening activity desirable in some way, the experience will seem to occupy less time than would an undesirable experience of the same (or even shorter) length. Expectations, boredom, interest, contemplation, and even pleasure caused by music can alter the listener's sense of elapsed time.

The time length of a piece of music (or any time-based art form, such as a motion picture) is separate and distinct from clock time. A lifetime can pass in a moment, through the experience of a work of art. A brief moment of sound might elevate the listener to extend the experience to an infinite span of existence.

### Timbre in Perception

The overall quality of a sound, its *timbre*, is the perception of the mixture of all of the physical aspects that comprise a sound. Timbre is the global form, or the overall character of a sound, which we can recognize as being unique.

This overall form (timbre) is perceived as the states and interactions of its component parts. The physical dimensions of sound, discussed previously, are perceived as dynamic envelope, spectral content, and spectral envelope (perceived values, not physical values). The perceived dimensions are interpreted and shape an overall quality, or conception, of the sound.

We remember timbres as entities, as single objects having an overall quality (that is comprised of many unique characteristics), and sometimes as having meaning in themselves (as a timbre can bring with it associations in the mind of the listener). We recognize the sounds of hundreds of human voices because we remember their timbres.

We remember the timbres of a multitude of sounds from our living experiences. We remember the timbres of many musical instruments and their different timbres as they are performed in many different ways.

The global quality that is timbre allows us to remember and recognize specific timbres as unique and identifiable objects.

Humans have the ability to recognize and remember a large number of timbres. Further, listeners have the ability to scan timbres and relate unknown sounds to sounds stored in the listener's long-term memory. The listener is then able to make meaningful comparisons of the states and values of the component parts of those timbres. These skills will serve as meaningful points of departure for the method for evaluating timbre in Parts Two and Three.

Sufficient time is required for us to process the many characteristics of a sound in order to recognize and understand its overall image. The time required to perceive the component parts of timbre varies significantly with the complexity of the sound and the listener's previous knowledge of the sound. For rather simple sounds, the time required for accurate perception is approximately 60 ms. As the complexity of the sound is increased, the time needed to perceive the sound's component parts will also increase. All sounds lasting less than 50 ms are perceived as noise-like, since a specific timbre cannot be identified at that short a duration; exceptions occur when the listener is well acquainted with the sound, and the timbre can be recognized from this small bit of information.

The partials of the timbre's spectrum fuse to create the impression of a single sound. Although many frequencies are present, the tendency of our perception is to combine them into one overall texture. We fuse partials that are harmonically related to the fundamental frequency, as well as overtones that are distantly related to the fundamental, into a single impression.

It is especially important for the recording professional to note the following related to the fusing of timbres:

- Fusion can also occur between two separate timbres (two individual sound sources) if the proper conditions are present.
- Timbres that are attacked simultaneously, or are of a close harmonic relationship to each other, are most likely to fuse into the perception of a single sound.
- The more complex the individual sound, the more likely that fusion will not occur.
- When the listener recognizes one of the timbres, fusion will be far less likely to occur.

- Related to recognizing timbre, synthesized sounds are more likely to fuse with other sounds than are known sounds of an acoustic origin.

## Spatial Characteristics

The perception of the *spatial characteristics* of sound is the impression of the physical location of a sound source in an environment, together with the modifications the environment itself places on the sound source's timbre.

The perception of *space* in audio recording (reproduction) is not the same as the perception of space of an acoustic source in a physical environment. In an acoustic space, listeners perceive the location of sound in relation to the three-dimensional space around them: distance, vertical plane, and horizontal plane. Sound is perceived at any possible angle from the listener, and sound is perceived at a distance from the listener; both of these perceptions involve an evaluation of the characteristics of the sound source's host environment.

Illusions of space are created in audio recordings. Sound sources are given spatial characteristics through the recording process and/or through signal processing. This spatial information is intended to complement the timbre of the music and/or sound source. The spatial characteristics may simulate particular known, physical environments or activities, or be intended to provide spatial cues that have no relation to our reality. In theory, all of the interactions of the sound with its host environment are captured with, or can be simulated and applied to, the sound source; upon playback through two or more loudspeakers, these spatial cues are reproduced.

### *Playback Environment*

Recordings and their sound sources (combined with their spatial characteristics) are heard through two (or more) loudspeakers. The loudspeakers themselves are placed in, and interact with, a playback environment—such as a living room or automobile. The playback environment is nearly always quite unrelated to the spaces on the recording. Thus, the listener ultimately perceives spatial characteristics applied to the sound source after they have been altered by:

- The characteristics of loudspeakers
- The interaction of the loudspeakers and the environment (caused by placement of loudspeakers within the playback environment), and
- The playback environment itself.

Listeners perceive the reproduced spatial characteristics of the sound source within the three-dimensional space of their listening environment (headphone monitoring is not a workable solution, as will be discussed later).

To accurately perceive the spatial information of an audio recording, the listening environment must be acoustically neutral, and the listener must be carefully positioned within the environment and in relation to the loudspeakers. The listening environment (including the loudspeakers) should not place additional spatial cues onto the reproduced sound. These conditions are goals that are never fully met in actual monitoring conditions. Therefore one must be aware of how the playback system and the listening room are altering the recording and its sound sources.

## Perceived Spatial Relationships and Current Sound Reproduction

We perceive spatial relationships in three primary ways:

1. As the location of the sound source being at an angle to the listener (above, below, behind, to the left, to the right, in front, etc.)
2. As the location of the sound source being at a distance from the listener
3. As an impression of the type, size, and acoustic properties of the host environment.

These perceptions are transferred into the recording medium, to provide a realistic illusion of space, with one major exception. The angular

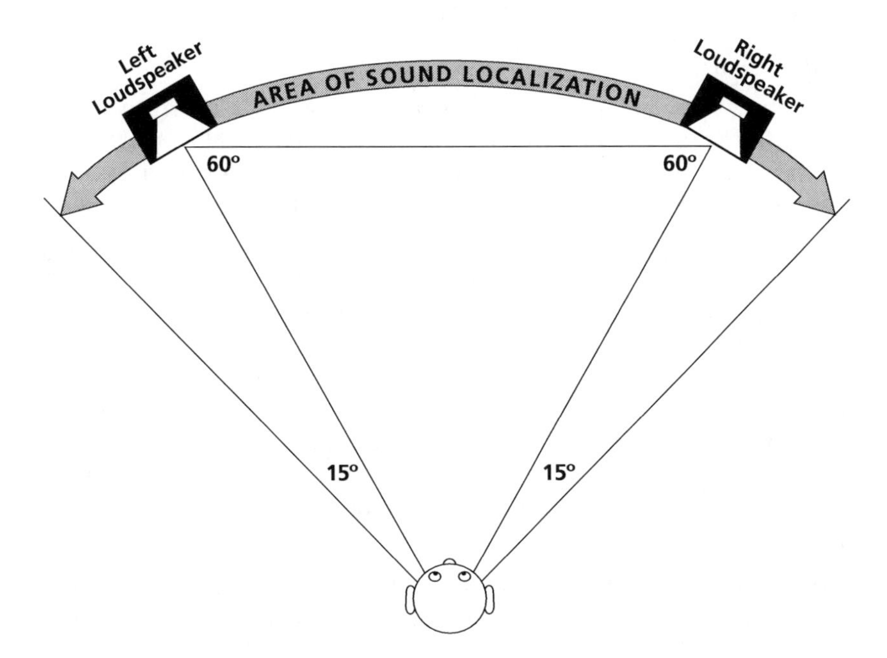


**Figure 1.15** Area of sound localization in two-channel audio playback.

location is severely restricted in audio reproduction, as compared to human perceptual abilities. Currently used audio playback formats can only accurately and consistently reproduce localization cues on the horizontal plane, and then only slightly beyond the loudspeaker array in stereo recordings. The three-dimensional space of our reality is simulated in the two dimensions of audio recording.

Much research and development continues toward reliably extending the localization of sound sources to behind the listener and to the vertical plane, as well as to provide a more realistic reproduction of distance and environmental cues. Significant advances are being made. Surround sound is now widely accepted and provides reproduction around the listener that can be mostly stable and accurate. Control of sound localization on the vertical plane and a more complete simulation of environmental characteristics are not, however, presently feasible, though the technology of the future will likely address these areas as well.

The following discussions of the perception of the spatial dimensions of reproduced sound refer to common two-channel stereo and to the widely accepted 5.1 surround-sound format. With adjustments, the outlined concepts will transfer to other systems such as binaural recordings, head-related transfer function and impulse response algorithm-based approaches, the full-sphere surround of ambisonics, and others, but with different boundaries of the limits of perception inherent with each particular format. Expanding coverage for all of these formats is out of the scope of this writing, until there is more widespread use and consumer acceptance of any of these formats.

***Localization of Direction***

Our ability to localize sounds is one of the survival mechanisms we have retained from our ancient past. This ability has been developed throughout our evolution; we have learned to perceive the direction of even those sounds that our hearing mechanism has difficulty processing.

Humans use differences in the same sound wave appearing at the two ears for the accurate *localization of direction.* Interaural time differences (ITD) are the result of the sound arriving at each ear at a different time. A sound that is not precisely in front or in back of the listener will arrive at the ear closest to the source before it reaches the furthest ear. These time differences are sometimes referred to as phase differences. The sounds arriving at each ear are almost identical during the initial moments of the sound, except the sound at each ear is at a different point in the waveform's cycle (and might contain minute spectral differences).

Interaural amplitude differences (IAD) work in conjunction with ITD in the localization of the direction of the sound source. IAD are also referred to as interaural spectral differences. IAD are the result of sound pressure level differences at high frequencies present at the two ears. The head of the listener, which blocks certain frequencies from the furthest ear (when the sound is not centered), causes the interaural spectral differences. This occurrence has been termed the "shadow effect." Interaural amplitude differences (IAD) will at times consist solely of amplitude differences between the two ears, with the spectral content of the waveform being the same at both ears.

The sound wave is almost always different at each ear. The differences between the sound waves may be time/phase related, amplitude/spectrum related, and/or spectrum differences. These differences in the waveforms are essential to the perception of the direction of the sound source. In addition, these same cues play a major role in perceiving the characteristics of host environments.

Up to approximately 800 Hz, humans rely on ITD for localization cues. Phase differences are utilized for localization perception up to about 800 Hz, as amplitude appears to be the same at both ears.

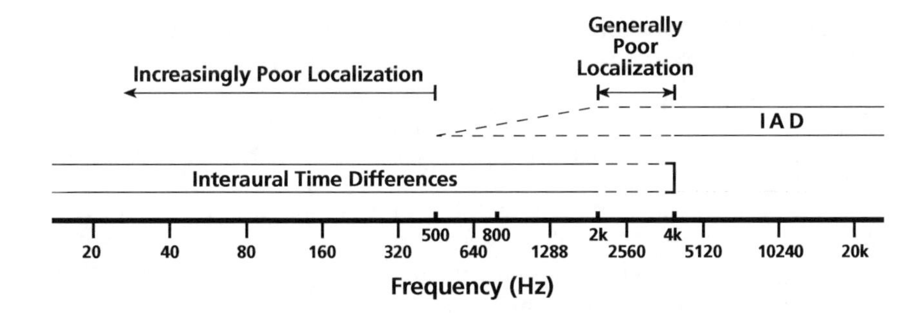


Figure 1.16
Frequency ranges of localization cues.

Between 800 Hz and about 2 kHz both phase and amplitude differences are present between the two ears. IAD and ITD are both used for the perception of direction in this frequency range, with amplitude differences becoming significant around 1,250 Hz.

In general, time/phase differences seem to dominate the perception of direction up to about 4 kHz. Although IAD are present, ITD dominates the perception of direction between 2 kHz and 4 kHz. Humans have poor localization ability for sounds in this frequency band.

Above 4 kHz, interaural amplitude differences (IAD) are the cues that determine the perception of location. Localization ability improves at 4 kHz and is quite accurate throughout the upper registers of our hearing range.

Recent studies have revealed the human body also generates physical cues for localization. The chest, head, shoulders, and outer ears all affect sounds of various wavelengths in different ways. Our body parts' different sizes and their different angles to the hearing canal create a very complex source of reflected and diffracted sound waves. These waves all lead to important interaural spectrum differences between the two ears. These differences are created by a comb-filter effect, comprised of minute cancellations and reinforcements of frequencies. The brain processes these subtle differences to aid in identifying a sound source's direction.

As we have seen, humans do not perceive direction accurately at all frequencies. Below approximately 500 Hz, our perception of the angle of the sound source becomes increasingly inaccurate, to the point where sounds seem to have no apparent focused location. An area exists around 3 kHz where localization is also poor; wavelength similarities between the distance between the two ears and those of the frequencies around 3 kHz cause interaural time/phase differences to be unstable and unreliable.

Humans have a well-refined ability to localize sounds in two approximate frequency areas: 500 Hz–2 kHz, and 4 kHz to upper threshold of hearing (whatever that might be). Within these areas, the minimum discernible angle is approximately one to two degrees, with less accuracy at the sides and back than in the front. Sounds that have fundamental frequencies outside of these frequency areas, but that have considerable spectral content within these bands, will also be localized quite accurately.

Interaural spectral differences occur throughout the frequency spectrum. While they may be subtle, it appears they are important for the localization of objects in frequency ranges where IAD and ITD are ineffective. The makeup of these interaural spectral differences will inherently be unique to individuals, as (obviously) no two people are the same size and shape, and no two outer ears are alike.

The outer ear is called the pinna. This part of our anatomy (an elaborately shaped piece of cartilage) plays several important roles in our perception of direction. The pinna gathers sound and funnels it into the ear canal. As the ridges of the outer ear reflect sound into the ear, the ridges introduce small time delays between their reflections and the direct sound that travels directly to the ear canal. These small time delays vary according to sound-source location, and are important components of the interaural spectral differences described above. The pinna and the delays it generates aid us in differentiating between sounds arriving from the front and those arriving from the rear.

Resonances also appear to be excited in the outer ear. These also alter the frequency response of the sound source in predictable ways that vary between individuals. The brain learns these patterns of spectral changes to assist in localization. Pinna cues play significant roles in direction perception even though each individual has a unique ear-shape, and the resultant spectrum changes are equally unique. Location cues based on spectrum are thus not universal, but unique to the individual and are learned.

Pinnae serve a critical function in front to back localization. When sound arrives at the head from the rear, ridge reflections are not generated. The pinna actually blocks the direct sound from reaching the hearing canal and its ridges when sounds are generated beyond 130° from the front center. The pinna allows us to perceive the sound source as being generated to our rear because of the absence of spectral differences.

It is interesting to note that our distance and location judgments are not as accurate to the sides and the rear. The absence of this spectral information generated and collected by the outer ear may well play a role.

We instinctively move our heads involuntarily to assist in locating sounds—especially those sounds that are not in front. In moving our heads to remove location confusion, we bring the source into our front listening field and thus reintroduce the IAD, ITD, and spectral differences of the pinnae. We also instinctively seek to bring the source into visual view, which eliminates all ambiguity for acoustic sources—but not for phantom images.

Front–back hearing is only partially understood. Little relevant research in spatial perception of sounds arriving from the rear and the sides is available. Certainly more will need to be accomplished before we are able to more thoroughly understand this area. It is increasingly important, however, that we understand how we perceive sound arrival from the rear and the sides, and the different qualities of those sounds, if we are to fully understand and control the differences between surround sound and stereo.

#### *Distance Perception*

*Distance* perception has not been studied as thoroughly as other dimensions. The following information is well documented, and it is likely that numerous subtleties will be discovered in the future.

Two impressions lead to the perception of the distance of a sound source from the listener: (1) the ratio of the amount of direct sound to reverberant sound, and (2) the primary determinant, the loss of low-amplitude (usually high-frequency) partials from the sound's spectrum with increasing distance (*definition of timbre* or *timbral detail*). Both of these functions rely on the listener's knowledge of the sound's timbre for accurate perception of distance location. While sound pressure decreases with distance, loudness itself does not factor into distance location perception.

Low-energy spectral information is lost with the compressions and rarefactions of the waveform over distance. Some information is simply absorbed by the atmosphere due to air friction. This leads to the listener's determination of the level of timbral detail (definition of timbre) that is the major factor in distance perception.

Some timbre-related distance information results from waveform travel and the speed of sound. As high frequencies travel slightly faster than low frequencies, the spectrum of the sound is altered with increasing distance. The partials of complex sounds will become increasingly out of phase with the fundamental frequency, and between themselves, the longer the propagation of the sound.

As the source moves from the listener, the percentage of direct sound decreases while the percentage of reflected sound increases. This pertains to enclosed spaces only. This ratio of direct to reflected sound will play a significant role in distance perception when the reverberant energy begins to mask timbral detail. It will also play a more significant role as timbral detail becomes more and more diminished, or when the sound source is unknown.

The listener must know the timbre of a sound in order to recognize missing timbral detail. If the sound is unknown or not recognized, the listener cannot recognize a loss of low-energy components from its spectrum. With knowledge of the sound source, the listener will be able to calculate how much low-energy information is missing and thus be able to determine the general amount of distance between them and the object.

Knowledge of the timbre of the sound source will assist the listener in recognizing the absence of spectral information and/or perceiving the reiterations of the direct sound and the reverberant sound. These perceptions will provide the listener with the needed information to judge distance. The previous experiences and listening skills of the listener will play a major role in the accuracy of judgments made.

Without prior knowledge of the timbre of a sound, perception of distance location is considerably less accurate, causing judgments to be rough estimates made without a point of reference.

Related to the ratio of direct to reflected sound, the time difference between the ceasing of the direct sound and the ceasing of the reverberant energy will increase with distance. Through *temporal fusion* we perceive the reverberant sound as being a part of the direct sound. This creates a single impression of the sound in its environment (referred to as the composite sound, above). As distance increases, temporal fusion begins to diminish and the ending of the direct sound and the continuance of the reverberant energy become more prominent.

***Perception of Environmental Characteristics***

The perceptions of the *characteristics of the host environment* and the *placement of the sound source within the host environment* are also dependent upon the ratio of direct to reflected sound and the loss of low-level spectral components with increasing distance. In addition, the characteristics of the host environment are perceived through:

1. The time difference between the arrival of the direct sound and the arrival of the initial reflections
2. The spacing in time of the early reflections
3. Amplitude differences between the direct sound and all reflected sound (the individual initial reflections and the reverberant sound), and
4. Timbre differences between the direct sound, the initial reflections, and the reverberant sound.

The time delay between the direct and the reflected sounds is directly related to (1) the distance between the sound source and the listener, (2) the distance between the sound source and the reflective surfaces (which send the reflected sound to the listener), and (3) the distance of the reflective surfaces from the listener. These three physical distances also create the patterns of time relationships (the rhythms) of the early reflections.

Early reflections arrive at the listener within 50 ms of the direct sound. These early reflections comprise the *early sound field*. The early sound field is composed of the first few reflections that reach the listener before the beginning of the diffused, reverberant sound (see Figure 1.8). Many of the characteristics of a host environment are disclosed during this initial portion of the sound. The early sound field contains information that provides clues as to the size of the environment, the type and angles of the reflective surfaces, even the construction materials and surface coverings of the space.

We have the capability to learn to accurately judge the size and characteristics of the host environments of sound sources. This is accomplished by evaluating the sound qualities of the environment. Humans experience and remember the sound qualities of a great many natural environments in much the same way as we recognize and remember timbres. Further, we have the ability to compare the sound qualities of new environments we encounter to our memories of environments we have previously experienced.

The listening skill needed to evaluate and recognize environmental characteristics can be developed to a highly refined level. Some people who work regularly with acoustical environments develop these listening skills to a point where many can perceive the dimensions and volume of an environment, its surface coverings, or even openings within the space (doors, windows, etc.).

## Interaction of the Perceived Parameters

The perception of any parameter of sound is always dependent upon the current states of the other parameters. Altering any of the perceived parameters of sound will cause a change in the perceived state of at least one other parameter.

The parameters of sound interact, causing the perception of the state of one parameter to be altered by the state of another. Certain occurrences of these interactions were noted under individual perceived parameters. The following are additional examples of note and are separated for clarity.

### *Duration for Pitch Perception*

Sufficient duration is required for the ear to perceive pitch. If the duration is too short, the sound will be perceived as having indefinite pitch, as being noise-like. The time necessary for the mind to determine the pitch of a sound is dependent on the frequency of the sound. Sounds lower

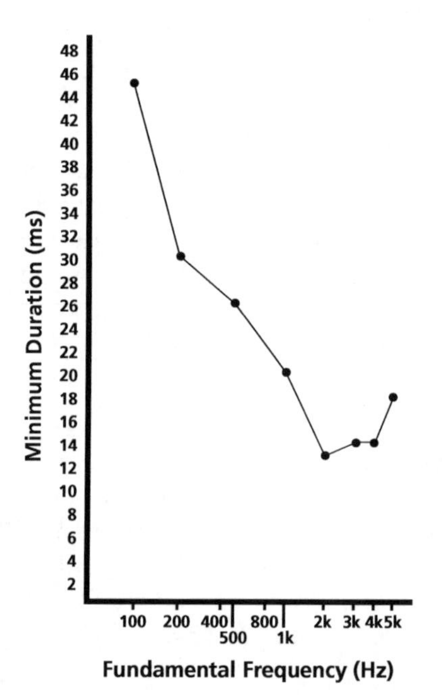


**Figure 1.17**
Minimum durations for pitch perception at select frequencies.

than 500 Hz and sounds higher than 4 kHz require more time to establish pitch quality than sounds pitched between 2 kHz and 4 kHz, where pitch perception is most acute. At the extremes of the hearing range, pitch quality may require as much as 60 ms to become established.

The length of time required to establish a perception of pitch will also depend on the sound's attack characteristics and its spectral content (its timbre). Sounds with complex (but mostly harmonic) spectra and sounds with short attack times will establish a perception of pitch sooner than other sounds.

### *Loudness and Pitch Perception*

Loudness will influence the perception of pitch, as humans will perceive a change of pitch with a change of loudness (dynamic) level. The level 60 dB SPL (or about the loudness level of normal conversation) is considered to be a threshold where increases or decreases in loudness affect pitch perception oppositely. Above 60 dB, for sounds below 2 kHz a substantial increase in loudness level will cause an apparent lowering of the pitch level; the sound will appear to go flat, although no actual change of pitch level has occurred. Similarly, a substantial increase in the dynamic (loudness) level of a pitch above 2 kHz will cause the sound to appear to go sharp; an impression of the raising of the pitch level is created, although the actual pitch level of the sound has remained unaltered. Below 60 dB an increase in loudness will cause sounds below 2 kHz to be perceived as getting sharp and sounds above 2 kHz perceived as going flat.

### *Loudness and Time Perception*

Loudness level can influence perceived time relationships. When two sounds begin simultaneously, they will appear to have staggered entrances if one of the two sounds is significantly louder than the other. The louder sound will be perceived as having been started first.

Perceived loudness level is often distorted by the speed at which information is processed. When a large number of sounds occur in a short period of time, the listener will perceive those sounds as having a higher loudness level than sounds of the same sound pressure level, but that are distributed over a longer period of time. This distortion of loudness level is caused by the amount of information being processed within a specific period of time (the time period is related to the perceived length of the present).

### *Loudness Perception Altered by Duration and Timbre*

Duration can distort the perception of loudness. Humans tend to average loudness levels over a time period of about 2/10 of a second. Sounds of shorter durations will appear to be louder than sounds (of the same intensity) with durations longer than 2/10 of a second.

Timbre can also influence perceived loudness. Sounds with a complex spectrum will be perceived as being louder than sounds that contain fewer partials. Similarly, sounds with more complex spectra with a strong presence of overtones will be perceived as louder than sounds containing mostly proportionally related partials (harmonics) when both are at the same sound pressure level. Following this principle, a change of timbre during the sustained duration of a sound will result in a perceived change in loudness.

#### *Pitch Perception and Spectrum*

As a product of the interaction of the harmonics and closely related overtones of a sound's spectrum, a timbre can create a perception of pitch where a fundamental frequency is not physically present. The harmonics of a sound reinforce its fundamental frequency to enhance the perception of pitch. This phenomenon is so capable of producing the perception of the fundamental frequency that a harmonic spectrum can provide the perception of pitch when the fundamental frequency is not physically present (missing fundamental). A perception of the periodicity of the fundamental frequency is created by the spectrum of the sound, although that frequency itself may not be present.

#### *Amplitude, Time, and Location*

The amplitude of two reiterated sounds (separated in time) can influence location perception. The precedence or Haas effect results when two loudspeakers reproduce the same sound in close succession. The effect works against the principle that when two loudspeakers reproduce a sound simultaneously, and at the same amplitude, the sound appears to be centered between the two loudspeakers.

When two loudspeakers reproduce the same sound source in close succession, normal perception would seem to be to localize a sound source at the earliest sounding loudspeaker, then to shift the image to center when the second loudspeaker is sounded. The Haas effect functions to continue the localization of the sound source at the first speaker location, while adding the second loudspeaker (to reinforce the sound intensity of the first speaker) without losing the localization of the sound source at the location of the first loudspeaker. The time difference between the sounding of the loudspeakers must be at least 3 ms to keep the sound at the leading speaker, with 5 ms being a more effective minimum; a maximum delay of approximately 25–30 ms may be used before the delayed signal is perceived as an echo (echoes will be perceived at all frequencies at a delay of 50 ms). If the leading channel is lowered by 10 dB, or the following channel increased by 10 dB, the sound source will again be centered.

### *Masking*

Masking occurs when a sound (or a portion of a sound) is not perceived because of the qualities of another sound. The simultaneous sounding of two or more sounds can cause a sound of lower loudness level, or a sound of more simple spectral content, to be masked or hidden from the perception of the listener. The masking of sounds is a common problem for people beginning their studies or work in audio recording.

When two simultaneous sounds of relatively simple spectral content have close fundamental frequencies, the sounds will tend to mask each other and blend into a single, perceived sound. As the two sounds become separated in frequency, the masking will become less pronounced until both sounds are clearly distinguishable.

Sounds of relatively simple spectral content tend to mask sounds that are at higher frequencies. This masking becomes more pronounced as the loudness of the lower sound is increased, and is more likely to occur when a large interval separates the two pitch levels of the sounds. This masking is especially prominent if the two pitch levels are in a simple harmonic relationship (especially 2:1, 3:1, and 5:1). A higher-pitched sound can mask a lower-pitched sound if the higher sound is significantly higher in loudness level, and given the same conditions as above; the higher the loudness level, the broader the range of frequencies a sound can mask.

Masking can occur between successive sounds. With sounds separated in time by up to 20–30 ms, the second sound may not be perceived if the initial sound is of sufficient loudness level to draw and retain listener attention, or to fatigue the ear. In a similar way, a sound may not be perceived if it is followed by another sound of great intensity within 10 ms.

Audio equipment can produce "white" and other broadband noise that can mask sounds at all or many frequencies. An entire program might be masked by the noise of the sound system itself, should the loudness level of the noise be sufficiently higher than that of the program. This type of masking problem will first be noticed in the high frequencies, where low loudness levels exist in the upper components of the sound's spectrum.

## Summary

The three states of sound that concern audio recording are sound as it exists in air (the physical dimensions of sound), sound as it exists in perception (the perceived parameters of sound), and the understanding of the meaning of a sound (sound as a resource for artistic expression). The physical dimensions of sound in air are transformed into neural impulses as the perceived parameters of the sound by the ear and brain. The perceived parameters of sound become understood as a resource of elements that allow for the communication and understanding of the meaning of sound (and artistic expression).

The two physical dimensions of the waveform are frequency and amplitude. They function in time and form the basis for our understanding of timbre and spatial properties. The anomalies of human hearing transform acoustic energy into our perception with marked changes to the physical dimensions: frequency becomes perceived as pitch, amplitude as loudness, time as duration, timbre as timbral characteristics, and space as perceived locations and environmental characteristics. What these changes actually are and how these changes take place are of great concern to the recording professional as they work in the many ways sound is captured, created, modified, and perceived.

Sound, as it is perceived and understood by the human mind, becomes the resource for creative and artistic expression in sound. The perceived parameters of sound become the artistic elements of sound in creating musical material and in communicating other meaningful messages. The aesthetic and artistic elements of sound in audio recording are presented in the next chapter.

# 2 The Aesthetic and Artistic Elements of Sound in Audio Recordings

The audio recording process has given creative artists the tools to very finely shape perceived sound (the perceived parameters of sound) through a direct control of the physical dimensions of sound. This control of sound is well beyond that which was available to composers and performers before recording technology. It has brought new artistic elements to music, which has led to new characteristics of sound and to a redefinition of the musician. While our discussion is focused on musical applications, it should be remembered that all aspects of these artistic elements of music also function as artistic elements in other areas of audio (such as broadcast media, film, multimedia, theater, etc.).

A new creative artist has evolved. This person uses the tools of recording technology as sound resources for the creation (or recreation) of an artistic product. This person may be a performer or composer in the traditional sense, or this person may be one of the new musicians: a producer, or sound engineer, or any of the host of other related job titles. Throughout this book, these people are referred to as *recordists*.

With this detailed control of sound, the audio-recording medium has resources for creative expression that are not possible acoustically. Sounds are created, altered, and combined in ways that are beyond the reality of live, acoustic performance. New creative ideas and new additions to our musical language have emerged as a result of recording practice and technologies. These are new and unique artistic elements.

Creative ideas are defined by these aesthetic and artistic elements. The artistic elements are the aspects of sound that comprise or characterize creative ideas (or entire works of art or pieces of music). Study of the artistic elements will allow us to understand individual musical ideas and the larger musical event, and to recognize how those ideas and sound events contribute to the entire piece of music—and how the recording shapes the music and sound. Our discussion will emphasize the artistic elements that are unique to recorded music, especially music created through the use of modern recording techniques and technologies.

As we have learned from the previous chapter, the artistic elements of sound are the interpretation of the perceived parameters of sound. Sound as it is perceived and understood becomes the resource for creative and artistic expression. The perceived parameters of sound are utilized as the artistic elements of sound to create and ensure the communication of meaningful (musical) messages.

*The Art of Recording* occurs through an understanding that the parameters of sound are a resource for artistic expression. Recording becomes an art when it is used to shape the substance of sound and music. These materials that allow for artistic expression will be understood through a study of their component parts: the artistic elements of sound.

## The States of Sound and the Aesthetic/Artistic Elements

The aesthetic/artistic elements are directly related to specific perceived parameters of sound, just as the perceived parameters of sound were directly related to specific physical dimensions of sound.

As will be remembered, sound in audio recording is in three states: physical dimensions, perceived parameters, and artistic elements.

The perceived parameters translate into the artistic elements:

**Table 2.1** The perceived parameters and the aesthetic/artistic elements of sound

| ***Perceived Parameters*** | ***Aesthetic/Artistic Elements*** |
|---|---|
| Pitch | Pitch Levels and Relationships |
| Loudness | Dynamic Levels and Relationships |
| Duration | Rhythmic Patterns and Rate of Activity |
| Timbre (perceived overall quality) | Sound Sources and Sound Quality |
| Space (perceived characteristics) | Spatial Properties |

**Table 2.2** The states of sound in audio recording

| ***Physical Dimensions*** | ***Perceived Parameters*** | ***Artistic/Aesthetic Elements*** |
|---|---|---|
| (Acoustic State) | (Psychoacoustic Conception) | (Resources for Artistic Expression) |
| Frequency | Pitch | Pitch Levels and Relationships—melodic lines, chords, register, harmony, tonal organization, range, pitch density, pitch areas, vibrato |
| Amplitude | Loudness | Dynamic Levels and Relationships—program dynamic contour, accents, tremolo, musical balance, reference dynamic level |
| Time | Duration (time perception) | Rhythmic Patterns and Rates of Activities—tempo, time, patterns of durations |
| Timbre (comprised of physical components: dynamic envelope, spectrum, and spectral envelope) | Timbre (perceived as overall quality) | Sound Sources and Sound Quality—timbral balance, pitch density, arranging, performance intensity, performance techniques |
| Space (comprised of physical components created by the interaction of the sound source and the environment, and their relationship to a microphone) | Space (perception of the sound source as it interacts with the environment, and perception of the physical relationship of the sound source and the listener) | Spatial Properties—stereo location, surround location, phantom images, moving sources, distance location, sound-stage dimensions, imaging, environmental characteristics, perceived performance environment, space within space |

The audio-production process allows for considerable variation and a very refined control of ALL of the artistic elements of sound. All of the artistic elements of sound can be accurately and precisely controlled. This high degree of control generates many different and recognizable states and/or levels in all elements—and in ways that were possible with ONLY pitch on traditional musical instruments.

## Pitch Levels and Relationships

Pitch-level relationships present most of the significant information in most music. The artistic message of most of today's music is communicated (to a large extent) by pitch relationships. Listeners have been trained, by the music heard throughout their life, to focus on this element to obtain the most significant musical information. The other artistic elements typically support pitch patterns and relationships.

Pitch is the most precisely controlled artistic element in traditional music. The use of pitch relationships and pitch levels in music is more sophisticated than the use of the other artistic elements. Complex relationships of pitch patterns and levels are common and vary with styles of music.

Information about the artistic element of pitch levels and relationships will be related to:

1. The relative dominance of certain pitch levels
2. The relative register placement of pitch levels and patterns, or
3. Pitch relationships: patterns of successive intervals, relationships of those patterns, and relationships of simultaneous intervals.

### Traditional Uses of Pitch

The aesthetic/artistic element of pitch levels and relationships is broken into the component parts: melodic lines, chords, tonal organization, register, range, pitch density, pitch areas, and tonal speech inflection.

A series of successive, related pitches creates *melodic lines.* Melodic lines are perceived as a sequence of intervals that appear in a specific ordering and that have rhythmic characteristics. The melodic line is often the primary carrier of the artistic message in most music we encounter.

The ordering of intervals, coupled with or independent from rhythm, creates patterns. *Pattern perception* is central to how humans perceive objects and events. These basic principles relate to all of the components of the artistic elements. Melodic lines are organized by patterns of intervals (short melodic ideas, riffs, or motives), supported by corresponding rhythmic patterns. The ways in which the patterns are repeated and the ways in which the patterns are modified create a complexity from the patterns. This provides the melodic line with its unique character.

Two or more simultaneously sounding pitches create *chords.* In much of our music, chords are based on superimposing, or stacking, the intervals of a third (intervals containing three and four semitones, most commonly). Chords comprised of three pitches, combining two intervals of a third, are called *triads.* Continued stacking of thirds results in seventh, ninth, eleventh, and thirteenth chords.

The movement from one chord to another, or *harmonic progression,* is the most stylized of all the components of the artistic elements. Harmonic progression is the pattern created by successive chords, as based on the lowest note (the root) of the triads (or more complex chords). These patterns of chord progressions have become established as having general principles that occur consistently in certain types of music. Certain types of music will have stylized chord progressions (progressions that occur most frequently), whereas other types of music will have quite different movement between chords, and perhaps emphasize more complex chord types. The patterns of the harmonic progression create *harmony*.

Harmony is one of the primary components that support the melodic line. The chords in the harmonic progression reinforce pitches of the melody. The speed and direction of the melodic line is often supported by the speed at which chords are changed, and the patterns created by the pacing of changing chords: *harmonic rhythm*.

The expectations of harmonic progression create a sequence of chords, which will present areas of tension and areas of repose within the musical composition. The tendencies of *harmonic motion* do much to shape the momentum of a piece of music and can greatly enhance the character of the melodic line and musical message. Performers utilize the psychological tendencies of harmonic progression, exploiting its directional and dramatic tendencies. The expectations of harmonic movement and the psychological characteristics of harmonic progression have become important aspects of musical expression and musical performance.

The melodic and harmonic pitch materials are related through *tonal organization*. Certain pitch materials are emphasized over others, in varying degrees, in nearly all music. This emphasis creates systems of tonal organization in which a hierarchy of pitch levels exists. A hierarchy will most often place one pitch in a predominant role, with all other pitches having functions of varying importance, in supporting the primary pitch. The primary pitch, or *tonal center*, becomes a reference level to which all pitch material is related, and around which pitch patterns are organized.

Many tonal organization systems exist. These systems tend to vary significantly by cultures, with most cultures using several different but related systems. The major and minor tonal organization systems of Western music are examples of different but related systems, as are the whole-tone and pentatonic systems of Eastern Asia. The reader should consult appropriate music theory texts for more detailed information on tonal organization, as necessary.

## The New Pitch Concerns of Audio Production

Certain components of pitch levels and relationships have become more prominent in musical contexts (and other areas of audio) because of the new treatments of pitch relationships in music recordings. The components of range, register, pitch density, and pitch area can be more closely controlled in recorded music than in live (unamplified) performance. These components are more important in recorded music, because in practice they have been controlled with the intention to support and enhance the musical material and the recording.

*Range* is the span of pitches of a sound source (any instrument or voice). Range is the area of pitches that encompasses the highest note possible (or present in a certain musical example) to the lowest note possible (or present) of a particular sound source.

A *register* is a portion of a sound source's range. A register will have a unique character (such as a unique timbre, or some other determining factor) that will differentiate it from all other areas of the same range. It is a small area within the source's range that is unique in some way. Ranges are often divided into many registers; registers may encompass a very small group of successive pitches, up to a considerable portion of the source's range.

A *pitch area* is a portion of any range (or of a register) that may or may not exhibit characteristics that are unique from other areas. Instead, it is a defined area between an upper and a lower pitch level, in which a specific activity or sound exists.

*Pitch density* is the relative amount and register placement of simultaneously sounding pitch material. This can be throughout the hearing range or within a specific pitch area. Pitch density is the amount and placement of pitch material in the composite musical texture (the overall sound of the piece of music), and is defined by its boundaries of highest and lowest sounding pitches.

With pitch density, sound sources are assigned (or perceived as occupying) a certain pitch area within the entire listening range (or the smaller pitch range used for a certain piece of music). In this way, certain pitch areas will have more activity than other pitch areas; certain sound sources will be present only in certain pitch areas, and other sources present only in other pitch areas; some sources may share pitch areas and cause more activity to be present in those portions of the range; some pitch areas may be void of activity. Many possible variations exist.

Pitch density is directly related to traditional concerns of orchestration and instrumentation, with many new twists. Pitch density is a much more specific concern in recorded music because it is controllable in very fine increments. Traditional orchestration is concerned, primarily, with the selection of instruments, and with the placement of the musical parts (performed by those assigned instruments and their sound qualities) against one another.

The register placement of sound sources and their interaction with the other sound sources take on many more dimensions in recorded music.

Each sound source occupies a pitch area; the acoustic energy within the pitch area of a timbre's spectrum is distributed in ways that are unique to each sound source. The spectrum of each sound source is the pitch density of an individual sound source. The pitch density of the overall program (or musical texture) is the composite of all of the simultaneous pitch information from all sound sources, and is *timbral balance*.

Sound sources and musical ideas are often delineated by the pitch area they occupy, within the timbral balance of the overall program. Sound sources are more easily perceived as being separate entities and individual ideas when they occupy their own pitch area in the overall program. This area can be large or quite small, and still be effective.

Sounds that do not have well-defined pitch quality are still perceived to occupy a *pitch area*. These types of sounds are noise-like, in that they cannot be perceived as being at a specific pitch. Such sounds do, however, have unique pitch characteristics.

Many sounds cannot be recognized as having a specific pitch, yet have a number of frequencies that dominate their spectrum. Cymbals and drums easily fall into this category. Cymbals are easily perceived as sounding higher or lower than one another, yet a specific pitch cannot be assigned to most of these instruments.

We perceive these sounds as occupying a pitch area. The sounds have a pitch-type quality based on (1) the register placement of the area of the highest concentration of pitch information (highness or lowness of the frequency band), (2) the loudness level of the primary pitch area in relation to secondary pitch areas present in the sound, and (3) the relative density (closeness of the spacing of pitch levels) of the pitch information (spectral components) within the pitch areas. We are able to identify the approximate area of pitches in which this concentration of spectral energy occurs, and are thus able to relate that area to other sounds.

Pitch areas are defined as the range spanned by the lowest and highest dominant frequencies around an area of the spectral activity. This range is called the *bandwidth* of the pitch area. Many sounds will have several pitch areas of concentrated amounts of spectral energy. In such cases, one range will dominate and the others will be less prominent. The size of the bandwidth and the density of spectral information (the number of frequencies within the bandwidth and the spacing of those frequencies) define the sound quality of pitch areas.

## Dynamic Levels and Relationships

Dynamic levels and relationships have traditionally been used in musical contexts for expressive or dramatic purposes. Expressive changes in dynamic levels and the relationships of those changes have most often been used to support the motion of melodic lines, to enhance the sense of direction in harmonic motion, or to emphasize a particular musical idea. A change of dynamic level, in and of itself, can produce a dramatic musical event and is a common musical occurrence. Changes in dynamic level can be gradual or sudden, subtle or extreme.

Dynamics have traditionally been described by analogy: louder than, softer than, very loud (fortissimo), soft (piano), medium loud (mezzo forte), etc. The artistic element of dynamics in a piece of music is judged in relation to context. Dynamic levels are gauged in relation to (1) the overall, reference dynamic level of the piece of music, (2) the sounds occurring simultaneously with a sound source in question, and (3) the sounds that immediately follow and precede a particular sound.

The components of dynamic levels and relationships in audio recording are dynamic contour (with gradual and abrupt changes in dynamic level), emphasis/de-emphasis accents (abrupt changes in dynamic level), musical balance (gradual and abrupt changes in dynamic levels between sources), and dynamic speech inflections.

### Traditional Uses of Dynamics

It is common for the most important musical idea/sound source in a piece of music to be given prominence in one way or another. Making that sound the loudest is an easy way of achieving this prominence (though not always the most elegant). Arranging sounds by relating dynamic levels to the importance of the musical part is very common. There is a very natural association of loudness and the center of one's attention.

Gradual changes in dynamic levels can be important. The crescendo (gradual increasing in loudness) can be used to support the motion of a melodic line (for instance), or it might be used on a sustained pitch as a musical gesture itself. Likewise a diminuendo or decrescendo (a gradual decrease in loudness) may be used in the same ways.

Rapid, slight alterations or changes in dynamic level for expressive purposes are often present in live performances. This is called *tremolo,* and is used primarily to add interest and substance to a sustained sound. *Tremolo* and *vibrato* are often confused. *Vibrato* is a rapid, slight variation of the pitch of a sound; it also is used to enhance the sound quality of the sound source. At times, performers may not be able to control their sound well enough to control tremolo and vibrato alterations; in these instances, tremolo and vibrato may detract from the source's sound quality, rather than contribute to it.

To support a musical idea or to create a sense of drama, musical ideas are often brought to the listener's attention by dynamic *emphasis accents* and *attenuation accents*. A shift in dynamic level that brings the listener's attention to a musical idea is an accent. Accents are most often emphasis accents, making use of increasing the dynamic level of the sound to achieve the desired result. Much more difficult to successfully achieve, de-emphasis (or attenuation) accents draw the listener's attention to a musical idea, or a sound source, by a decrease in the dynamic level of the sound. Attenuation accents are often unsuccessful because the listener has a natural tendency to move attention away from softer sounds; these accents are most easily accomplished in sparse musical textures, where little else is going on to draw the listener's attention away from the material being accented.

### New Concepts of Dynamic Levels and Relationships

Changes in dynamic levels over time comprise *dynamic contours*. Dynamic contours can be perceived for individual sounds, individual sound sources, and individual musical ideas composed of a number of sound sources, and also for the overall piece of music. Dynamic contours are perceived at many different *perspectives* (level of detail). At their extremes, they exist as the smallest changes within the spectral envelope of a single sound source, and as great changes in the overall dynamic level of a recording.

The interaction of the dynamic contours of all sound sources in a piece of music creates *musical balance*. Musical balance is the interrelationship of the dynamic levels of each sound source, to one another and to the entire musical texture. The dynamic level of a particular sound source in relation to another sound source is a comparison of two parts of the musical balance.

Dynamic contours and musical balance have been used in supportive roles in most traditional music. At times dynamic level changes have been used for their own dramatic impact on the music (as discussed with crescendo and diminuendo, above), but most often they are used to assist the effectiveness of another artistic element. The mixing process easily alters musical balance. Recordists exercise great control over this artistic element.

The dynamic levels and relationships of a performance may be significantly different in the final recording. The recording process has very precise control over the dynamic levels of a sound source in the musical balance of the final recording. An instrument may have an audible dynamic level in the musical balance of a recording that is very different from the dynamic level at which the instrument was originally performed. The timbre of the instrument will exhibit the dynamic levels at which it was performed *(perceived performance intensity)*, but its relative dynamic level in relation to the other musical parts might be significantly altered by the mix. For example, an instrument may be recorded playing a passage loudly, and end up in the final musical balance (mix) at a very soft dynamic level; the timbre of the instrument will indicate that the passage was performed very loudly, yet the actual dynamic level (its actual loudness) will be quite soft in relation to the overall musical texture, and to the other instruments of the texture.

**Listen**

to track 38

for musical balance relationships that are changed from the original performance. The drum mix has many unnatural relationships of performance intensity versus musical balance; some are over-exaggerated to provide clarity of the topic.

Many clear examples of this are found in The Beatles' recording of "Penny Lane." Listening carefully to the flutes, piccolo, and piccolo trumpet parts throughout the song, one will find many instances where the loudness levels of the performances are not reflected in the actual loudness levels of the instruments in the recording. Among many instances of conflicting levels and timbre cues, we hear moderately loud flutes that were performed softly; loudly played piccolo sounds at a soft level in the mix; and a piccolo trumpet appearing at a softer level than in the performance. Other instruments and voices in the song also have inconsistent musical balance and performance intensity information.

The reader is encouraged to take the time now to perform the musical balance and performance intensity exercise in Chapter 5 (Exercise 5.8).

The dynamic level of a sound source in relation to other sound sources (musical balance) is quite different and distinct from the perceived distance of one sound source to another. Yet these two occurrences are often confused and are the source of much common, misleading terminology used by recordists. Significant differences are present between a softly generated sound that is close to the listener and a loudly performed sound that is at a great distance to the listener, even when the two sounds have precisely the same sound pressure level (SPL) or perceived loudness level. Loudness levels within the recording process are independently controllable from the loudness level at which the sound was performed, and are independently controllable from the distance of the sound source from the original receptor and from the perceived listening location of the final recording. Dynamics must not be confused with distance. Dynamic levels, themselves, do not define distance location. A change in distance will produce a change in sound quality that is entirely independent from loudness.

## Rhythmic Patterns and Rates of Activities

Durations of sounds (the length of time during which the sound exists) combine to create musical rhythm. Rhythm is based on the perception of a steadily recurring, underlying pulse. The pulse does not need to be strongly audible to be perceived. The underlying pulse (or *metric grid*) is easily recognized by humans as the strongest common proportion of duration (note value) heard in the music.

The rate of the pulses of the metric grid is the *tempo* of a piece of music. Tempo is measured in metronome markings (pulses per minute, abbreviated "M.M."), or in some contexts as pulses per quarter note. Tempo, in a larger sense, can be the rate of activity of any large or small aspect of the piece of music (or of some other aspect of audio, for example the tempo of a dialogue).

Durations of sound are perceived proportionally in relation to the pulse of the metric grid. The human mind will organize the durations into groups of durations, or *rhythmic patterns.* In the same ways that we perceive patterns of pitches, we perceive patterns of durations. Pattern perception is transferable to all of the components of all of the artistic elements, and is the traditional way in which we perceive pitch and rhythmic relationships.

**Listen**

to tracks 45, 46, and 53

for rhythmic patterns of timbre and location created by the drum mixes.

*Rhythmic patterns* are the durations of or between soundings of any artistic element. Rhythmic patterns might be created by the pulsing of a single percussion sound; in this way rhythmic patterns would be created by the durations between the occurrences of the starts of the same sound source. Rhythmic patterns comprising the durations of successive, single pitches (perhaps including some silences) create melody. Rhythmic patterns of the durations of successive chords (groups of pitches) create harmonic rhythm. Extending this, in the same way rhythm can be transferred to ALL artistic elements. As examples, it is possible to have rhythms of sound location (as has become a common mixing technique for percussion sounds); it is likewise possible to have timbre melodies, or rhythms applied to patterns of identifiable timbres (this is often used for drum solos).

## Sound Sources and Sound Quality

The selection, modification, or creation of *sound sources* is an important aesthetic and artistic element of audio recording. The *sound quality* of the sound sources (the timbre of the source) plays a central role in the presentation of musical ideas, and has become an increasingly significant form of musical expression.

The sound quality of a sound source may cause a musical part to stand out from others, or to blend into an ensemble. Sound quality alone can convey tension or repose and give direction to a musical idea. Sound quality can add dramatic or extra-musical meaning or significance to a musical idea. Finally, the timbral quality of a sound source can, itself, be a primary musical idea, capable of conveying a meaningful musical message.

Until recently, composers used the sound quality of a sound source (1) to assist in delineating and differentiating musical ideas (making them easier to distinguish from one another), (2) to enhance the expression of a musical idea by the careful selection of the appropriate musical instrument to perform a particular musical idea, or (3) to create a composite timbre (or *texture*) of the ensemble, thereby forming a characteristic, overall sound-quality–timbral balance (also called tonal balance).

Performers have always used the characteristic timbres of their instruments or voices to enhance musical interpretation. This activity has been greatly refined by the resources of recording and sound-reinforcement technology. Performers now have greater flexibility in shaping the timbre of their instruments for creative expression. Of equally great importance, after the performance has been captured, the recording process allows for the opportunity to return to the performance for further (perhaps extensive) modifications of sound quality.

The selection of a sound source to represent (present) a particular musical idea is critical to the successful presentation of the idea. The act of selecting a sound source is among the most important decisions composers (and producers) make. The options for selecting sound sources are (1) to choose a particular instrumentation, (2) to modify the sound quality of an existing instrument or performance, or (3) to create, or synthesize, a sound source to meet the specific need of the musical idea.

The *selection of instrumentation* was once merely a matter of deciding which generic instrument of those available would perform a certain musical line. The selection of instrumentation has now become very specific and much more important. The performance that exists as a music recording may virtually live for ever and be heard by countless people. This is very different from the typical, live music performance of the past that existed for only a passing moment and was heard by only those people present (and paying attention).

Today, the selection of instrumentation is often so specific as to be a selection of a particular performer playing a particular model of an instrument. Generally, composers and producers are very much aware of the sound quality they want for a particular musical idea. The performer, the way the performer can develop a musical idea through their own personal performance techniques, and their ability to use sound quality for musical expression are all considerations in the selection of instrumentation.

Vocalists are commonly sought for the sound quality of their voice and their abilities to perform in particular singing styles. The vocal line of most songs is the focal point that carries the weight of musical expression. Vocalists make great use of performance techniques to enhance and develop their sound quality, as well as to support the drama and meaning of the text.

Performance techniques vary greatly between instruments, musical styles, performers, and functions of a musical idea. The most suitable performance techniques will be those that achieve the desired musical results when the sound sources are finally combined. One performance technique consideration that must be singled out for special attention is the intensity level of a performance.

As touched on in the discussion above with musical balance, a performance on a musical instrument will take place at a particular intensity level. This *perceived performance intensity* is comprised of loudness, energy exerted, performance technique, and the expressive qualities of the performance. Each performance at a different intensity level results in a different characteristic timbre of that instrument, at that loudness level.

The same sound source will thus have different timbres, at different loudness levels (and at different pitch levels), through performance intensity.

Along with the timbre (sound quality) and loudness level, performance intensity can communicate a sense of drama and an artistically sensitive presentation of the music to the listener. Through performance intensity, louder sounds might be more urgent, more intense; softer sounds might be cause for relaxation of musical motion. The exact reverse is equally possible. The expressive qualities of music are contained in performance intensity cues.

*Modifying a sound source* is a common way of creating a desired sound quality. Instruments, voices, or any other sound may be modified (while being recorded, or afterwards) to achieve a desired sound quality. Most often, this takes the form of making detailed modifications to a particular instrument so it best presents the musical idea. The final sound quality will still have some (perhaps many, perhaps only a few) characteristic qualities of the original sound.

The extensive modification of an existing sound source, to the point where the characteristic qualities of the original sound are lost, is actually *the creation of a sound source*. The creation of new sound qualities (or inventing timbres) has become an important feature in many types or pieces of music. The recording process easily allows for the creation of new sound sources, with new sound qualities.

Sound qualities are created by either extensively modifying an existing sound through sound-sampling technologies, or by synthesizing a waveform. Sound-synthesis techniques allow precise control over these two processes, and are having a widespread impact on recording practice and musical styles. Many specific technologies and techniques exist for synthesizing and sampling sounds; all have unique sound qualities and unique ways of allowing the user to modify or synthesize a sound source.

A new sense of the importance of sound quality to communicate, as well as to enhance, the musical message has come from this increased emphasis on sound quality and timbre. Sound quality has become a central element in a number of the primary decisions of recording music, as well as in the creation of music through the recording process. In making these basic decisions, sound quality is conceptualized as an object. The sound is thought of as a complete and individual entity, capable of being pulled out of time and out of context.

In this way, sound quality is approached as a *sound object*. This important concept will be explored in detail later in Chapter 4, *Listening and Evaluating Sound for the Aspiring Audio Professional.*

The entire, composite sound of the music may also be conceptualized as a single entity, or overall sound quality. It is composed of all the sound sources and musical ideas. This sound quality of the overall sound, or entire program, is called *timbral balance*. It is perceived and recognized as the sound-quality characteristics of the overall program, the recording's overall sound.

The overall texture of the recording is perceived as an overall character, made up of the states and activities of all sounds and musical ideas. Pitch-register placements, rate of activities, dynamic contours, and spatial properties are all potentially important factors in defining a texture by the states or activities of its component parts. This will be covered in detail in Chapter 12.

## Spatial Properties: Stereo and Surround Sound

The *spatial properties* of sound have traditionally not been used in musical contexts. Among the few exceptions are the location effects of antiphonal ensembles of certain Renaissance composers and in certain drama-related works of the nineteenth century, such as the 1837 *Requiem* by Hector Berlioz (with its brass ensembles stationed at the corners of the church, performing against the orchestra and choir on stage).

In recorded music, spatial properties can play important roles in communicating the artistic message. The potential number of roles is many. Spatial properties may be used in supportive roles to enhance the character or effectiveness of musical ideas (large and small), to differentiate one sound source from another, to provide dramatic impact, to alter reality, to reinforce reality by providing a performance space for the music, and much more. Further, spatial properties may be used as the primary idea of an artistic gesture. The spatial property of environmental characteristics even fuses with the timbre of the sound source to add a new dimension to its sound quality. Many other possibilities certainly exist.

The number and types of roles that spatial location may play in music continue to grow, with the creativity of recordists and new technologies. The continued emergence of surround formats and productions will further multiply the possibilities.

All of the components of the spatial properties are under quite precise and independent control. All of the spatial properties may be in many markedly different and fully audible states. Further, gradual and continuously variable change between those states is possible and common.

The spatial properties of sound that are of primary concern to recorded music (sound) are:

1. The *stereo location* of the sound source on the horizontal plane of the stereo array
2. The *distance* of the sound source from the listener
3. The perceived *characteristics* of the sound source's physical *environment*
4. The *surround location* of sound sources on the lateral plane 360° around the listener
5. *Perceived performance environment* of the recording.

The perceived elevation of a sound source is not consistently reproducible in widely used playback systems, and has not yet become a resource for artistic expression.

## Two-Channel Stereo

The spatial qualities of stereo playback are perceived as relationships of lateral location and distance cues and relationships of sound sources. These create a perception of a *sound stage* contained within the *perceived performance environment* of the recording.

While surround sound struggles for commercial acceptance, two-channel sound reproduction remains the standard of the music recording industry, with monophonic capabilities still considered for certain Internet, radio broadcast, and television sound applications. The two-channel array of *stereo sound* attempts to reproduce all spatial cues through two separate sound locations (loudspeakers), each with more-or-less independent content (channel). With the two channels, it is possible to create the illusion of sound location at a loudspeaker, in between the two loudspeakers, or slightly outside the boundaries of the loudspeaker array; location is limited to the area slightly beyond that covered by the stereo array, and to the horizontal plane. The characteristics of the sound source's environment and distance from the listener are created in much more subtle ways by stereo, but can be stunning nonetheless.

A setting is created by the two-channel playback format for the reproduction of a recorded or created performance (complete with spatial cues). This establishes a conceptual and physical environment within which the recording will be reproduced more or less accurately.

The reproduced recording presents an illusion of a live performance. This performance will be perceived as having existed in reality, in a real physical space, as the listener will imagine the activity in relation to his or her own physical reality. The recording will appear to be contained in a single, *perceived physical environment*. Within this perceived space is an area that comprises the *sound stage*.

### *Sound Stage and Imaging*

The sound stage encompasses the area within which all sound sources are perceived as being located. It has an apparent physical size of width and depth. The sound sources of the recording will be grouped by the mind to occupy a single area, from which the music is being played. It is possible for different sound sources to occupy significantly different locations within the sound stage but still be grouped into the illusion of a single performance.

*Imaging* is the lateral location and distance placement of the individual sound sources within the sound stage. Imaging provides depth and width to the sound stage. The perceived locations and relationships of the sound sources create imaging, as all sources appear to exist at a certain lateral and distance location within the stereo array.

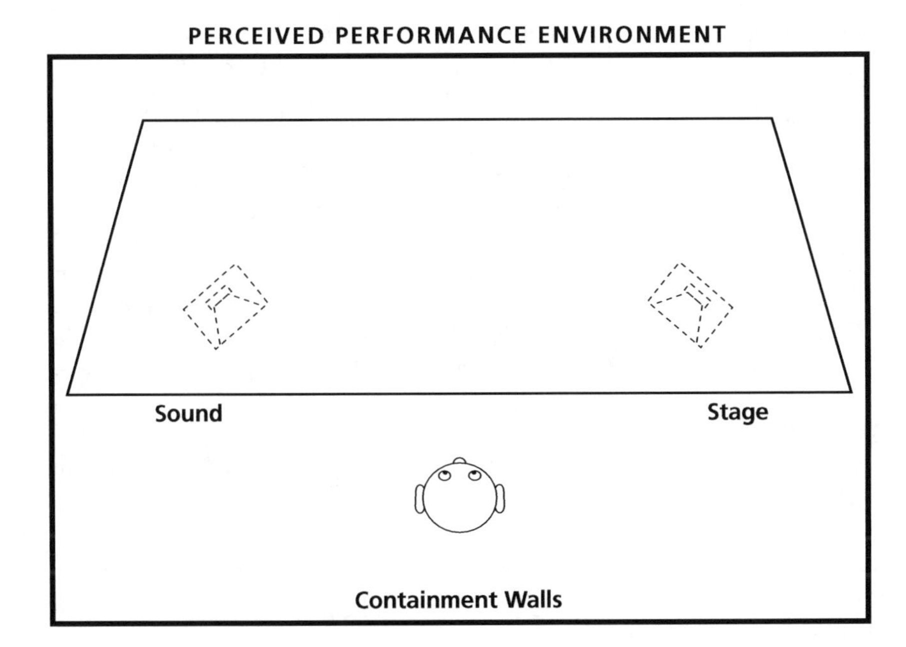


**Figure 2.1** Sound stage and the perceived performance environment.

### *Stereo Location*

The *stereo* (lateral) *location* of a sound source is the perceived placement of the sound source in relation to the stereo array. Sound sources may be perceived at any lateral location within, or slightly beyond, the stereo array.

*Phantom images* are sound sources that are perceived to be sounding at locations where a physical sound source does not exist. Imaging relies on phantom imaging to create lateral localization cues for sound sources. Through the use of phantom images, sound sources may be perceived at any physical location within the stereo loudspeaker array, and up to 15° beyond the loudspeaker array. *Stage width* (sometimes called *stereo spread*) is the width of the entire sound stage. It is the area between the extreme left and right source images and marks the sound stage boundaries.

Phantom images not only provide the illusion of the location of a sound source, but also create the illusion of the physical size (width) of the source. Two types of phantom images exist: the *spread image* and the *point source.*

A *point source* phantom image occupies a focused, precise point in the sound stage. The listener can close their eyes and point to a very precise point of little area where the source is heard to originate. Point sources exist at a specific point in space, narrow in width, and precisely located in the sound stage.

**Listen**

to tracks 42 and 43

for narrow and wide spread images of a guitar;

or to track 48

for a variety of spread image sizes of drum sounds and point source cymbal bells.

The *spread image* appears to occupy an area. It is a phantom image that has a size that extends between two audible boundaries. The potential size of the spread image varies considerably; it might be slightly wider than a point source or it may occupy the entire stereo array. The spread image is defined by its boundaries; it will be perceived to occupy an area between two points or edges. At times, a spread image may appear to have a hole in the middle, where it might occupy two more-or-less equal areas, one on either side on the stereo array.

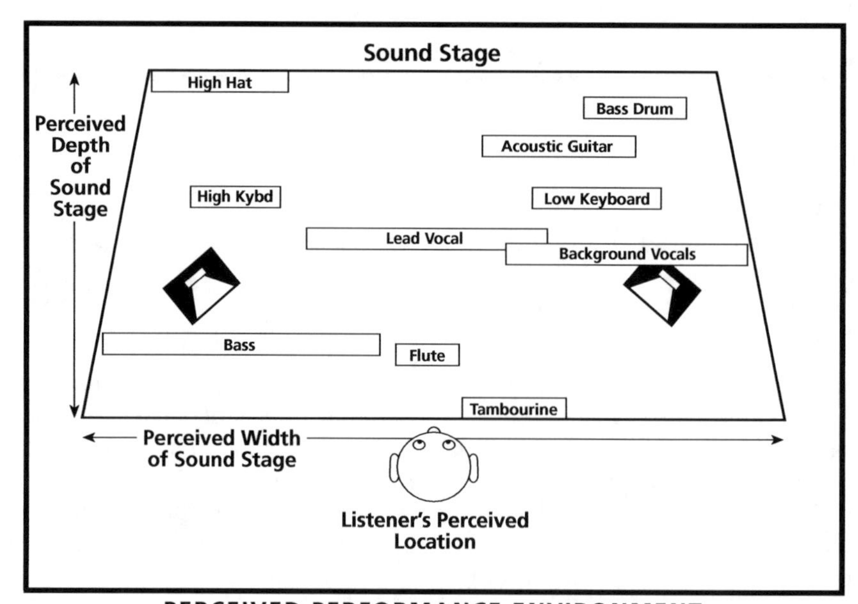


**Figure 2.2**
Sound stage and imaging, with phantom images of various sizes.

The perceived lateral location of sound sources can be altered to provide the illusion of *moving sources*. Moving sound sources may be either point sources or spread images. Point sources and narrow spread images that change location most closely resemble our real-life experiences of moving objects.

Many interesting examples of phantom images can be found on The Beatles' album *Abbey Road*. An apparent example of a spread image with a hole in the middle is the tambourine in the first chorus of "She Came In Through the Bathroom Window." The lead vocal in "You Never Give Me Your Money" begins the song as a point source. The image soon becomes a spread image that gradually grows wider, ultimately occupying a significant amount of the sound stage (this is partly due to the gradual addition and varying qualities of environmental cues, which will be discussed shortly). In the second section of the work, the new lead vocal sound gradually moves from the right to the left side of the sound stage, while maintaining a spread image of moderate size.

***Distance Location***

Two important *distance* cues shape recorded music: (1) the distance of the listener to the sound stage, and (2) the distance of each sound source from the listener.

Both of these distances rely on a perception that the entire recording emanates from a single, global environment. This *perceived performance environment* establishes a reference location of the listener, from which all judgments of distance can be calculated.

The *stage-to-listener distance* establishes the front edge of the sound stage with respect to the listener and determines the level of intimacy of the music/recording. This is the distance between the grouped sources that make up the sound stage and the perceived position of the audience/listener. This stage-to-listener distance places the sound stage within the overall environment of the recording and provides a location for the listener.

The *depth of sound stage* is the area occupied by the distance of all sound sources. The boundaries of the depth of the sound stage are the perceived nearest and the perceived furthest sound sources (fused with the depths created by their environments, discussed below). The perceived distances of sound sources within the sound stage may be extreme; they may provide the illusion of great depth and a large area, or the exact opposite.

**Listen**

to tracks 39–41

for distance changes of a single cello performance;

or to track 48

for a variety of distance locations within a single drum mix.

Stage-to-listener and depth-of-sound-stage distance cues have different levels of importance in different applications. Depth-of-sound-stage cues tend to be emphasized over stage-to-listener distance cues in many multitrack recordings; in those recordings, the cues of the distance of the source from the listener are often exploited for dramatic effect and/or to support musical ideas. In contrast, stage-to-listener distance cues are often carefully calculated in classical and some jazz recordings (especially those utilizing standardized stereo microphone techniques); in those recordings the stage-to-listener distance will not change and has been carefully selected to represent an appropriate vantage point (the ideal seat) from which to hear the music.

Turning again to *Abbey Road*, the distance cues of the various instruments of "Golden Slumbers" give the work and its companion "Carry That Weight" much space between the nearest and the furthest sources. The orchestral string and brass instruments are at some distance from the listener and give significant depth to the sound stage, while the piano establishes the front edge of the sound stage very near the listener. Remembering that timbral detail is the primary determinant of distance location will help in accurately hearing these cues.

### *Environmental Characteristics*

It has become important for music recordings to have sound sources matched with an environment with a suitable sound, and to have a suitable environment for the sound stage (the perceived performance environment). Through these, *environmental characteristics* have the potential to significantly impact music and the quality of the recording.

Environmental characteristics fuse with the sound source to create a single sonic impression. The host environment shapes the overall timbre/sound quality of each sound source; this is also true for the overall program (shaped by its perceived performance environment). Environmental characteristics contribute greatly to sound quality and also play an important role in the recording's sense of space. The characteristics provide a space for the sound sources to perform in, they supply some distance information that may be significant, and they contribute to the perceived depth of the sound stage.

The sound characteristics of the host environments of sound sources and the complete sound stage are precisely controllable. Each sound source has the potential to be assigned environmental characteristics that are different from other sound sources. The recording process allows the potential for each sound source to be given a different environment, and for the characteristics of those environments to be varied as desired. Further, each source may occupy any distance from the listener within their host environment.

The *perceived performance environment* (or the environment of the sound stage) is the overall environment where the performance (recording) is heard as taking place. This environment binds all the individual spaces together into a single performance area.

**Listen**
to tracks 45–47
for several different space-within-space examples.

The environment of the sound stage and an individual environment for each sound source (or groups of sound sources) often coexist in the same music recording. This places the individual sound sources with their individual environments within the overall, perceived performance environment of the recording. The illusion of *space within space* is thus created, with the following potential perceptions:

1. That physical spaces may exist side by side,
2. That one physical space may exist within another physical space (where often a space with the sound qualities of a physically large room may be perceived to exist within a smaller physical space), and
3. That several sounds may exist at various distances within the same host environment (space).

Any number of environments and associated stage-depth distance cues may occur simultaneously, and coexist within the same sound stage. The environments and associated distances are conceptually bound by the spatial impression of the perceived performance environment. These outer walls of the overall program establish a reference (subliminally, if not aurally) for the comparison of the sound sources.

Perhaps oddly, the overall space that serves as a reference, and that is perceived by the listener as being the space within which all activities occur, will often have the sound characteristics of an environment that is significantly smaller than the spaces it appears to contain. Such cues that send conflicting messages between our life experiences and the perceived musical occurrence are readily accepted by the listener and can be used to great artistic advantage. This is a very common space-within-space relationship.

Space within space will at times be coupled with distance cues to accentuate the different environments (spaces) of the sound sources, though often this illusion is created solely by the environmental characteristics of the different spaces of each sound source.

"Here Comes the Sun," also from *Abbey Road*, provides some clear examples of space within space. Environments clearly exist side by side from the song's opening into the first verse. The guitar has an environment all to itself in the left channel, the electronic counter melodies and Moog synthesizer glissando have similar environments distinctly different from the others, and the right channel voice has a very different third environment. The parts are held together by the notion that they all exist within a single performance space (perceived performance environment). The entry of additional instruments quickly adds numerous additional environments and enhances the sound stage. As vocal lines are added, however, they appear to be within the same environment, though at distinctly different distances from the listener. The realization of spaces within spaces is also apparent in the drum parts; the drum set seems to occupy an area, with its characteristic environment, within which low toms in a larger space are contained.

## Surround Sound

Music recordings in surround sound have become more important. Enough activity and interest exist for us to seriously explore this format now, but with some reservation. While some talented people have been working in this format and some striking recordings have been made, consistent uses of the unique sound qualities of surround are just now starting to emerge. This section will discuss prevalent aesthetic and artistic elements currently found in surround music recordings.

Listening to a stereo recording, we find ourselves observing a performance. We are viewing the activity as an outsider. And while we may get consumed by or immersed in the music, we are outside of the

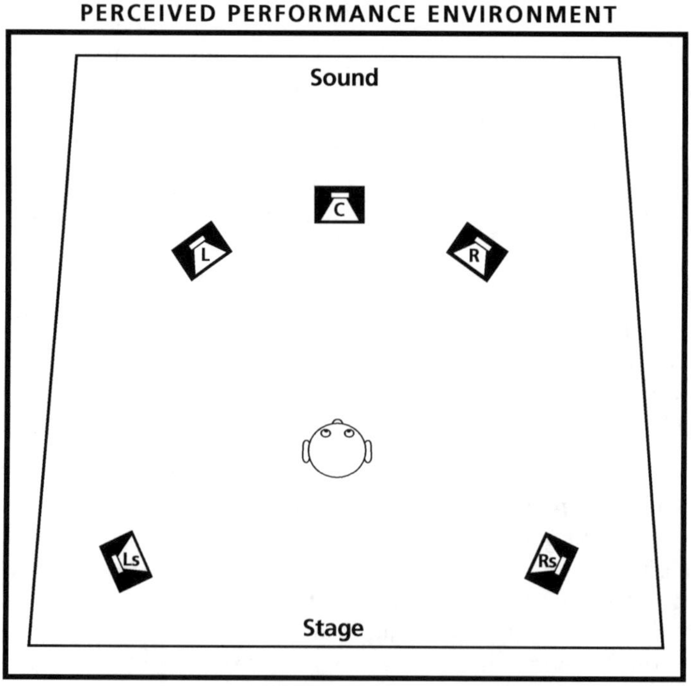

**Figure 2.3** Surround-sound stage enveloping the listener.

experience of the performance itself and are looking in as an observer, audience member. With surround sound, we can potentially find ourselves enveloped by the music. We can be surrounded by the sound, and thereby contained within the space of the recording; we are no longer outside observers, but at least inside observers if not participants (if only in our perception of the experience). Now the listener can be enveloped by the sound (and become part of the space of the recording). It is also possible to be surrounded by sound and remain detached, in the position of the observer, thereby oriented by the production techniques to observe a piece of music as a 360° panorama of sound. This aspect of surround sound has great potential for making a profound impact on music. Location and environmental characteristics will be approached differently for surround recordings, and distance cues will also take on new dimensions. These will all be explored in Chapter 11.

## Conclusion

With the recording process, it is possible for any of the artistic elements of sound to be varied in considerable detail. In so doing, all artistic elements can be shaped for artistic purposes and used to create or contribute to musical ideas. As all elements of sound can be varied by roughly equal amounts, it is possible for any element to play an important role in a piece of music. We commonly see this practice in today's music productions.

The artistic elements are used in very traditional roles in certain musical works and types of recording productions, and in very new ways in other works. These new ways the artistic elements are used tend to emphasize aspects of sound that cannot be controlled in acoustic performances. The aesthetic/artistic elements unique to audio recording (especially sound quality and spatial properties) are commonly used to support and shape musical ideas. Different musical relationships and sound properties can exist in audio recordings rather than in acoustic music. Knowing and controlling these elements gives the recordist the opportunity to contribute to the creative process and the act of making music.

The potentials of the artistic elements to convey the musical message, the musical message itself, and the characteristics and limitations of the listener are explored in the following chapter.

# 3 The Musical Message and the Listener

This chapter begins discussing the content of the musical message that appears in greater detail in Chapter 17—the last chapter of this book. There will also be initial coverage of how aesthetic and artistic elements function in communicating the message of a piece of music. These lay the groundwork for Parts Two and Three.

The listener's ability to correctly interpret the sounds of the musical message impacts their understanding of the intended artistic meaning of the music. The factors that limit the listener's ability to effectively interpret the artistic elements and understand the intended musical message (or meaning of the music) will be explored; the listener as audience member and as audio professional will be contrasted.

Today's recordist can do much to shape music and is often a key person in the creative process. This chapter is intended to be the first glimpse at what is created and shaped when crafting the mix.

## The Musical Message

Every piece of music has a message, even if only an expression. This message is related to the many purposes or functions of music. Music's message can vary widely; it can be explicitly verbal and conceptual, to fully abstract. Music's functions are equally wide, serving ceremony and celebration, entertainment and enrichment, and more. Each different purpose that music serves requires a different approach to listening. The approach will bring some aspect into the center of our attention or the center of our experience.

We listen to music with various levels of attention. This strongly impacts the listening experience. At the extremes, the listener will be focused and intent on extracting certain specific types of information (active listening); conversely their attention will be focused on some activity other than the music (eating, a conversation, a dentist's drill), with the listener perhaps not actually conscious of the music (passive listening). How the listener approaches the act of listening will largely determine what they hear in the music, and whether the music's intended message is understood. These will be covered more thoroughly later.

The purpose of a piece of music with its related musical message may serve a number of functions. These functions are most readily categorized as: (1) conceptual communication, (2) portraying an emotive state, (3) aesthetic experience, and (4) utilitarian functions. These are by no means exclusive; many pieces of music use different functions simultaneously, or at different points in time.

Music that includes a text, such as a song, will communicate other concepts. These works may tell a story, deliver the author's impressions of an experience, present a social commentary, etc. Music is used as a vehicle to deliver the tangible ideas of the author/composer. The interplay between the drama of the music and the message of the text is often an important contributor to the total experience of these works.

It is difficult for music alone (without words) to communicate specific concepts, but it is possible. *Symphonie Fantastique* by Hector Berlioz is supported by a program that brings the music to portray an involved love story. Often written works portray certain subjects without a text. The listener can associate sounds in music to their experiences with a subject if a connection can be made. The subjects of such works are often general in nature, such as Ludwig van Beethoven's "Pastoral" *Symphony No. 6*.

Certain concepts are associated with certain specific or types of musical materials (types of movie music, musical ideas associated with certain individuals or certain landscapes, etc.). These are exceptional cases where music alone can communicate specific concepts, with the aid of associations drawn from the listener's past experiences. It is easy to imagine a Western chase scene, or an impending shark attack, when listening to certain pieces of music—after one has heard the music and seen the action together enough times.

Music communicates emotions easily. One of the reasons many people listen to music is for emotional escape, relief, or a journey to another place. Music may portray a specific mood, incite a specific emotional response from the listener, or create a more general and hard to define (yet convincing) feeling or emotive impression. The composer of the music draws from the past experiences of the listener to shape their emotive reactions to the material. This is found—at least to some extent—in all music. Works of this nature may include a text, or not.

Music may be an aesthetic experience. The perception of the relationships of the musical materials alone, without the associations of concepts or emotive states, may be the vehicle for the musical message. Music has the ability to communicate on a level that is separate and distinct from the verbal (conceptual) or the emotional. Absolute music, without words, without emphasis on the emotional level, can be tremendously successful in communicating a message of great substance. Music (as all of the arts) can reach beyond the human experience; ideas that cannot be verbally defined or represented as an emotional experience can be clearly communicated. Abstract concepts may be clearly communicated; the human spirit may reach beyond reality, to experiences beyond the human condition. Some people have been so moved as to compare the aesthetic appreciation of a substantial work of art to the impressions of religious experience. Works of Johann Sebastian Bach such as the *Brandenburg Concertos* or the *Cello Suites* are excellent examples of this type of music, from the multitude that encompass hundreds of years of history and nearly all of the world's cultures.

Music serves other functions. It is used to reinforce or accompany other art forms (motion pictures, musical theater, dance, video art), to enhance the audio and visual media (television, multimedia, radio, advertisements), and to fill dead air in everyday experiences (supermarkets, elevators, etc.). In these instances, music is present to support dramatic or conceptual materials, to take the listener's attention away from some other sounds or activities (a dentist's office), or to make an environment more desirable (a restaurant, an automobile). It also marks life's activities in ceremonies, dance, worship, social interactions of many types, and more.

The complexity of the musical materials is often directly related to the function of the music. When music is the most important aspect of the listener's experience, the musical material may be more complex—as the listener will devote more effort to deciphering the materials. When music is playing a supportive role, the materials are often less sophisticated and are directly related to the primary aspect of the listener's experience. When music is being used to cover undesirable noises or to fill a void of silence that would otherwise be ignored, the musical materials are often very simple, easily recognized, and easily heard without requiring much of the listener's attention.

## Musical Form and Structure

Within the human experiences of time and space, nothing exists without shape or form. Music is no exception. Pieces of music have *form* as a global quality, as an overall concept and essence.

Pieces of music can be conceptualized as an overall quality. It is the human perception of form that provides the impression of a global quality that crystallizes the entire work into a single entity. Form is the piece of music as if perceived, in its entirety, in an instant; it is the substance and shape that is perceived from conceptualizing the whole.

Form is the global shape of a piece of music together with the fundamental concepts and expressions (emotions) it is communicating. It is the sum that is shaped from the interactions of its component parts. Form is a single concept and intrinsic quality of the piece, given shape by structure; it is comprised of component parts that are the materials of the piece of music. The materials of the music and their interrelationships provide the *structure* of the work.

The structure of a work is the architecture of its musical materials. Structure includes the characteristics of the musical materials coupled with a *hierarchy* of the interrelationships of the musical materials, as they function to shape the work. The artistic elements of sound function to provide the musical materials with their unique character, as will be later discussed. All musical materials are related by structure.

The hierarchy of musical materials provides a general framework of the materials. This brings organization by levels of detail and levels of significance. The hierarchy provides an interrelationship of the materials (with their varying levels of importance) to one another, and to the musical message as a whole. Within the hierarchy of musical structure:

- all musical materials and artistic elements will have a greater importance to the musical message than other materials or elements (except the least significant);
- all sections or ideas in the music will have greater importance over others (except the least significant);
- all musical materials are subparts of other, more significant musical materials (except the most significant materials).

Further, the hierarchy of the musical structure organizes musical materials into patterns, and patterns of patterns. In this way, relationships are established between the subparts of a work and to the work as a whole. The hierarchy is such that any time span may contain any number of smaller time spans, or be contained within any number of larger time spans; musical material at any level may be related to material at any other level of the hierarchy.

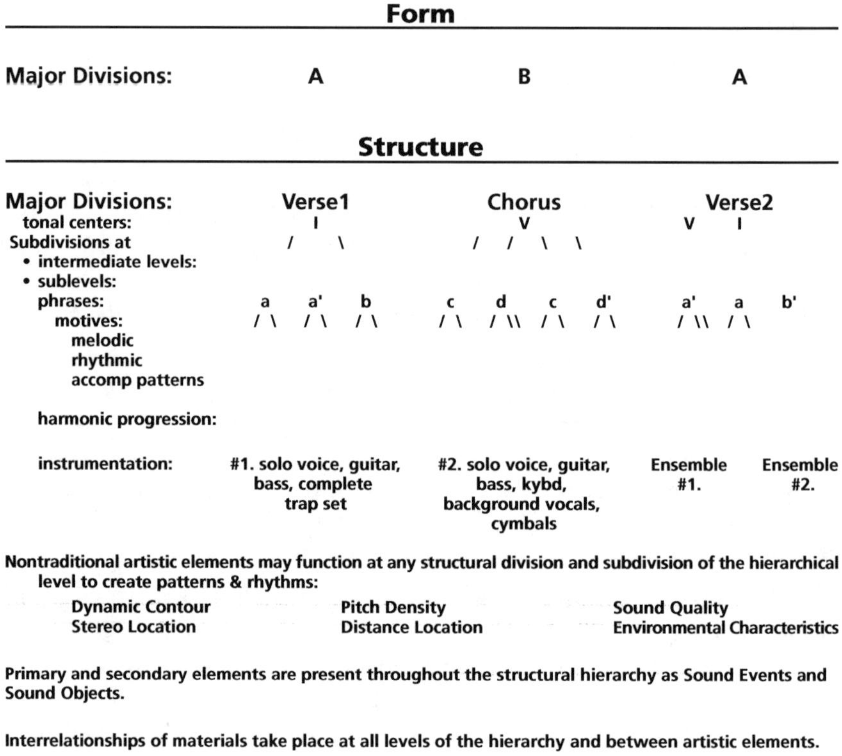

**Figure 3.1**
Form and structure in music.

**Form**

**Major Divisions:** A B A

**Structure**

**Major Divisions:** Verse1 Chorus Verse2
tonal centers: I V V I
Subdivisions at / \ / / \ \
• intermediate levels:
• sublevels:
phrases: a a' b c d c d' a' a b'
motives: / \ / \ / \ / \ / \\ / \ / \ / \\ / \
melodic
rhythmic
accomp patterns

harmonic progression:

instrumentation: #1. solo voice, guitar, bass, complete trap set | #2. solo voice, guitar, bass, kybd, background vocals, cymbals | Ensemble #1. | Ensemble #2.

Nontraditional artistic elements may function at any structural division and subdivision of the hierarchical level to create patterns & rhythms:
Dynamic Contour | Pitch Density | Sound Quality
Stereo Location | Distance Location | Environmental Characteristics

Primary and secondary elements are present throughout the structural hierarchy as Sound Events and Sound Objects.

Interrelationships of materials take place at all levels of the hierarchy and between artistic elements.

A multitude of possibilities exists for unique musical structures, supporting very similar musical forms. The innumerable popular songs that have been written during the past few decades are evidence of this great potential for variation. Most of the songs have many similarities in their structures, but they also have many significant differences. The materials that comprise the works may be very different, but the materials work toward establishing an overall shape (form) that is quite similar between songs.

Many songs share similar forms. Their overall conceptions are very similar, although the materials of the music and the interrelationships of those materials may be strikingly different. Form is an overall shape and conception that may be constructed of a multitude of materials and relationships.

Musical materials can be changed to dramatically alter the structure of a piece of music without altering its form. Many different structures can lead to the same overall design and portray the same basic artistic statement that creates form.

## Musical Materials

As music moves and unfolds through time, the mind grasps the musical message through the process of understanding the meaning and significance of the progression of sounds. During this progression of sounds, the mind is drawn to certain artistic elements (that create the characteristics of the musical materials). We perceive the *musical materials* as small patterns (small musical ideas, often called motives or gestures), and group the small patterns into related larger patterns (such as phrases and groups of phrases). The listener remembers the patterns, together with their associations to larger and smaller patterns (perceiving the structural hierarchy). In order for the listener to remember patterns, the listener must recognize some aspect of the organization of a pattern or some of the materials that comprised a pattern(s).

The use of contrast, repetition, and variation of patterns throughout the structure creates logic and coherence in the music. Several general ways in which musical materials are used and developed should make clear the multitude of possibilities. Materials are contrasted with other materials at the same and different hierarchical levels (above and below). Materials are repeated immediately or later in time, at the same or different hierarchical levels. Materials are varied by adding or deleting portions of an idea, by altering a portion of an idea, or perhaps by transposing an idea to different artistic elements (such as melodic ideas becoming rhythmic ideas; harmonic motion becoming dynamic motion).

A balance of similarities and differences within and between the musical materials is needed for successfully engaging music. A musical work will not communicate the desired message if this balance is not effectively presented to, or understood by, the listener.

The listener remembers the context in which the patterns were presented as well as the patterns themselves. Some patterns will draw the listener's attention and be perceived as being more important than other patterns—these are the *primary musical materials*. Other patterns will be perceived as being subordinate; these *secondary materials* will somehow enhance the presentation of the primary materials by their presence and activity in the music. The secondary materials that accompanied the patterns (or primary materials) are also remembered as individual entities (capable of being recognized without the primary musical idea) and as being associated with the particular musical idea (patterns).

The primary materials are traditionally: melody (with related melodic fragments or motives) and (extra-musically) any text, or lyrics of the music. The secondary materials are traditionally: accompaniment passages, bass lines, percussion rhythms, harmonic progressions, and tonal centers. Secondary materials may also be dynamic contour, pitch density, timbre development, stereo/surround location, distance location, or environmental characteristics.

The secondary materials usually function to support the primary musical ideas. It is possible to have any number of equal primary musical ideas. The potential groupings of primary and secondary musical ideas, in creating a single structural hierarchy, are limitless. Consider any number of secondary ideas (of varying degrees of importance in their support of the primary musical idea or ideas) that may coexist in a musical texture, with any number of related or unrelated primary musical ideas.

Musical materials are given their unique characters by the states and values of the aesthetic/artistic elements of sound. The artistic elements of sound function to shape and define the musical materials.

## The Relationships of Artistic Elements and Musical Materials

Musical ideas are also composed of *primary elements* and *secondary elements*. The primary elements are the aesthetic and artistic elements of sound that directly contribute to the basic shape and most significant characteristics of a musical idea. The secondary elements are those aspects of the sound that assist, enhance, or support the primary elements.

It is possible (and in fact common) to have more than one primary element and more than one secondary element contributing to the basic character of a musical idea. Primary elements provide the most significant characteristics of the musical material. The secondary elements provide support in defining or in providing movement to the primary elements.

At all levels of the structural hierarchy, musical materials (primary and secondary musical ideas) are made up of primary and secondary artistic elements. Therefore, it is possible for a certain element of sound to be a primary element on one level of the hierarchy and a secondary element on another level. This is a common situation. For example, a change in dynamic level of a drum roll might have primary significance at the hierarchical level of the individual sound source; at the same point in time but at the hierarchical level of the composite sound of the entire ensemble, changes in dynamics are insignificant to the communication of the musical message, with pitch changes being of primary importance.

All of the artistic elements of sound have the potential to function as the primary elements of the musical material. They have the potential to be the central carriers of the musical idea. Likewise, all of the artistic elements of sound have the potential to function as secondary elements of the musical material, and have the potential of functioning in supportive roles in relation to conveying the musical message. This is the concept of *equivalence* that will be thoroughly explored below.

In most music, pitch is the central element or the primary carrier of the musical message. The unique sound qualities of recording usually appear in the supportive roles of music, much more than as the primary elements. Most often, current production practice will use the unique artistic elements of recordings (such as the stereo location) to support or enhance the primary message (or perhaps to assist in making an individual sound source stand out). Rarely do the new sound resources function as the primary element of the primary musical idea, though this is entirely possible.

The new musical possibilities that make use of all artistic elements can create convincing musical ideas, even when the new elements are used as the primary carriers of the musical material. Current practice is likely to continue its gradual change toward further emphasis of these new artistic elements of sound unique to recording. It is important to recognize that the potential exists for any artistic element of sound to be the primary carrier of the musical material. The potential exists for any of the artistic elements of sound to function in support of any component of the musical idea. All of the artistic elements are equally capable of change, and that change can be perceived almost equally well in all of the artistic elements.

Traditionally, a breakdown of primary and secondary elements of a piece of music (with associated musical materials identified) would commonly appear similarly to Table 3.1. Pitch is the primary element and is supported by rhythm and dynamics. The musical materials are differentiated by sound-quality differences of instrumentation.

**Table 3.1** Traditional hierarchy of artistic elements

| *Primary Elements* | *Secondary Elements* |
|---|---|
| Pitch—melodic line #1 | Pitch—harmony |
| Pitch—melodic line #2 | Pitch—accompaniment patterns |
| | Dynamics—contour for expression |
| | Rhythm—supporting melody |
| | Sound Quality—instrument selection |

In many current recordings, a similar (and equally common) outline might appear, as Table 3.2. While pitch remains a primary element, rhythm and sound-quality changes are equally important in delivering the musical message. Sound quality, in particular, has become more

**Table 3.2** Common hierarchy in current music productions

| *Primary Elements* | *Secondary Elements* |
| --- | --- |
| Pitch—melodic line #1 | Pitch—harmony |
| Pitch—melodic line #2 | Pitch—accompaniment patterns |
| Rhythm—recurring patterns | Dynamics—contour changes without changes in timbre; accents; contour for expression |
| Sound quality—changing texture | Rhythm—supporting melody |
| | Sound quality—instrument selection; expression changes without dynamic changes |
| | Spatial properties—diverse host environments for each instrument; rhythmic pulses in different stereo locations; sound-stage location of instruments widely varied |

important with recording technology. Pitch, dynamics, and rhythm still play supportive roles, with spatial properties assisting sound quality in differentiating the musical ideas.

## Equivalence and the Expression of Musical Ideas

Pitch has functioned as the central element in nearly all music that has descended from or has been significantly influenced by the European tradition. Pitch is the primary artistic element in much of the music we know and is the perceived parameter that contains most of the information that is significant to the communication of the message of a piece of music.

Still, Western music might have developed differently. While pitch relationships have been used as the primary generator of musical materials much more than other artistic elements, this need not have been so. Indeed, some musics of other cultures use the artistic elements of music in significantly different ways. Some cultures emphasize other artistic elements, such as rhythm, and many incorporate very different types of pitch relationships.

It is difficult to justify pitch's traditional prominence in the expression of musical ideas. While it is true (1) that pitch is the perception of a primary attribute of the waveform (frequency, with the other attribute being amplitude), (2) that of all the elements it is the most easily detected in many states and values, and (3) that pitch is the only artistic element that can be readily perceived as multiples of itself (the octave repetition of pitch levels, and the perceptions of real and tonal transposition of pitch patterns), these factors do not cause pitch to be a more prominent element than the others.

Our ability to perceive pitch is not significantly more refined (if at all) than our abilities to perceive other parameters of sound. This is especially true of those parameters that utilize less precise pitch-related percepts (timbre, environmental characteristics, texture, and pitch density).

It follows that the artistic elements of sound other than pitch are equally capable of contributing to the communication of musical ideas.

This capability is being realized in the recordings of today's creative artists. In fact, this has been going on for quite some time, as the examples of recordings by The Beatles and others should indicate. This is occurring without conscious planning, but rather as a natural exploration of available sound relationships. Musicians (recordists, performers, composers, and producers) instinctively find roles for the unique artistic elements of recordings. The artistic elements that were not available, or that were underutilized, in traditional music performed in traditional contexts are now functioning in significant ways in modern music productions.

The concept that all of the artistic elements of sound have an equal potential to carry the most significant musical information is *equivalence*.

The states of the various components of the aesthetic and artistic elements of sound will make up the musical material. As such, they will function in primary or secondary roles of importance in the communication of the musical message. It is possible for any artistic element to function in any of the primary and secondary roles of shaping musical materials, and of generating the communication of the musical message. These are under the control of the recordist and have been commonly used to shape the musicality of music productions for years. Bringing these musical ideas into acceptance by the listener will require finesse and control of craft by the recordist.

Equivalence is also a framework for listening. It is a point of departure that reminds the recordist that any element or aspect of sound can change or can demand attention. Any change in the sound must be detected by the recordist and understood—no matter its significance to the music. Any aspect of sound might require the attention of the recordist at any moment during the listening experience. Equivalence provides this guidance and awareness.

## Text as Song Lyrics

When a text is present in a piece of music, it is a significant addition to the musical experience. Through the text, language communicates a concept or describes a drama within the work. Further, the sound resources of the language will be exploited to enhance the aesthetic experience of the music. Songs are often relatively short musical pieces that contain a text (usually a single, rather short text). The song is the most common form of music today. It is often said: The song is about the story.

The text, or lyrics, of a song is a poem set to music. The text's elements are arranged in some sort of structure (similar to the structural construction of music). The individual concepts of the text will create structural areas. These are then conceived as establishing a single entity, an overall idea and meaning of the text.

The lyrics of songs are constructed in many of the same ways as traditional poetry, written for its own sake and not intended to be set to music. The primary differences between the traditional poetry and poetry as song lyrics lie (1) in the repetitions of certain stanzas or phrases of the poem (unaltered or with slight changes), (2) in the careful crafting of the meters of the text, the rhythms of the lines, and the timing of the conceptual ideas of the text often found in song lyrics, and (3) in the sound qualities of the words that can be chosen to enhance the musical setting. This will be covered in greater detail in Chapter 17.

### Literary Meaning

The *literary meaning* of the text brings the dimensions of verbal communication of ideas and concepts to the musical experience. Songs have been written on a multitude of subjects, from common, everyday small occurrences, to the highest of human ideals. The lyrics of a song might present a storyline, or it might be a description of an event or the author's feelings about some aspect of the world around them. The text might be a presentation of the social-political philosophies of the author, or it might be a love song. The potential subjects for a song are perhaps limitless.

The text's literary meaning is often enhanced by subordinate phrases of text segments that create new dimensions in the lyrics. These subordinate ideas provide the turns of phrase or concepts that enrich the meaning of the text as a whole. The turning of the phrase allows for different interpretations of the meaning of certain ideas, at times different meanings to different individuals (or groups of people), depending on the experiences of the audience.

The potential for different interpretations allows for some (or much) ambiguity and intrigue in the text. The ambiguity may be clarified with a study of the central concepts of the song lyrics. Reevaluating a well-crafted text will often allow the listener to find new relationships of ideas or meanings of materials that enhance the experience of the song for the listener (or recordist). This is common in songs from many styles of music and lends a considerable dimension to the musical experience.

The concepts used to enhance the literary meaning of the text may or may not be directly related to its central ideas. These ancillary concepts embellish or otherwise enhance the text. They may take many forms and are important in shaping the presentation of the communicative aspects of poetry. A study of poetry, or of the setting of texts to music, may be very appropriate for the individual recordist, but is out of the scope of this writing. Some general observations are instead offered.

## Structure and Form of Song Lyrics

The *structure* of the text exists on many levels, similar to the hierarchy of musical structure. The conceptual meanings of the text and the sounds and rhythms of the text bridge the division between the structural aspects of the text and the form-related aspects of the text. An examination of the structure of the text should address the sound qualities of the text and its organization of component parts. The *form* of the text should address the conceptual, often with a recurring concept or theme, a refrain (as the song's chorus).

Some crossover will occur between the two areas: (1) the structure of the text's presentation may alter the statement's meaning, and (2) concepts can, at times, function as structural subparts. These are the result of the ways we conceptualize in verbal communication, and the previous experiences and social-cultural conditioning of the individual.

The components of the structure of a text will be major divisions of the materials of the text, and the subdivisions they contain. The materials that comprise the components are words, with all of their associated meanings, and the thoughts and feelings they invoke from within the individual. Words will be related by their sound qualities, rhyme schemes, rhythms and meters of groups of words, repetitions of words and words sounds, and by tonal and dynamic vocal inflections. Meanings of the words, repetitions of words with different associated meanings, phrases created by the concepts (sentences), and groupings of phrases by subject matter or concepts are also used.

This format will not necessarily be directly transferable to all text settings, but these concepts can provide a meaningful point of departure.

**Figure 3.2**
Form and structure of song lyrics.

**Form**

| | | | | | | | | | |
|---|---|---|---|---|---|---|---|---|---|
| **Major Divisions:** | **A** | | | **B** | | | | **A** | | |
| **Subdivisions** | | | | | | | | | | |
| **phrases:** | a | b | c | d | e | d | f | g | h | i |
| **concepts:** | | | | | | | | | | |

**Structure**

| | | | |
|---|---|---|---|
| **Major Divisions:** | **Stanza #1** | **Refrain** | **Stanza #2** |
| **Subgrouping by function:** | beginning of plot | author's impressions | plot continued |

**Subdivisions**

- **groupings of phrases by:**
  - rhythm
  - rhyme scheme
  - word usage
  - sound quality
- **lines by:**
  - rhythm
  - rhyme scheme
  - word usage
  - sound quality
- **words by:**
  - repetition
  - varied meaning
  - tonal inflection
  - dynamic inflection

## Texts and Music in Combination

The structure of the text and the structure of the music interact in the overall perception of the song. They are perceived as being interrelated. They serve to enhance each other. The structures may complement one another, or they may serve as areas of contrast, with the text and the music grouped in overlapping segments, unfolding over time.

Both complementary and contrasting relationships of the structural elements of the text and the music exist in most works. The two play off one another, creating a sense of drama between the text and the music.

The relationships of structures create our impression of form: our conceptualization of grasping the essence of the entire work in an instant of realization. Within our impression of form as the overall conception of the work, we conceptualize points of climax and points of repose; we conceptualize the characteristics of design and shape of the materials that create the movement from one important event, or moment, to the next.

We recognize the shape and design of the work as it is represented in our perception of the significant moments of the work, and in the movement between the moments as they unfold over time.

The relationships of the musical materials create structure in a piece of music. Our perception of the design of structure is our conception of form. The structure of a piece of music may be altered significantly without altering its form. Even when the primary musical materials and the structure of a work are significantly altered, two very different interpretations of the same piece of music will be perceived as being similar when the forms (or overall conception) of both performances are similar.

Contrast, for example, the following two performances (recordings) of the song, "Every Little Thing," the original by The Beatles (*Beatles for Sale*, 1964) and a cover by Yes (*Yes*, 1969). The overall shape of the piece is not dramatically altered, but the structures of the two performances are quite different. Great differences exist between the lengths of sections, as well as the treatments of the basic musical materials and how they are organized. Few people would argue that both performances are of the same piece of music. Few people could not perceive dramatic changes in the structure and materials between the two different versions.

The reader is encouraged to perform the exercise in identifying the structure of a song found in Chapter 5 (Exercise 5.5), to compile a timeline similar to those in Figure 3.3.

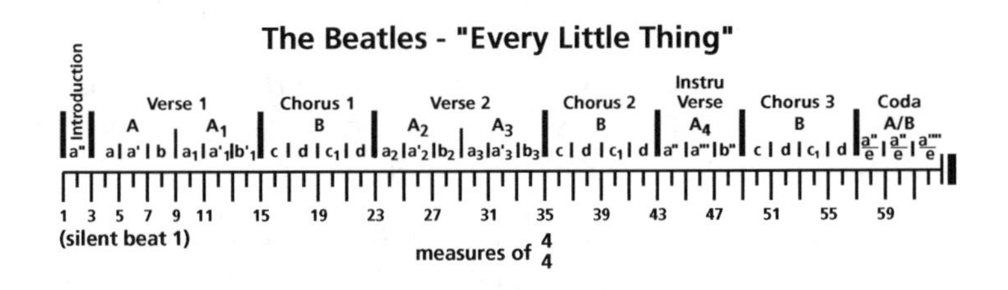


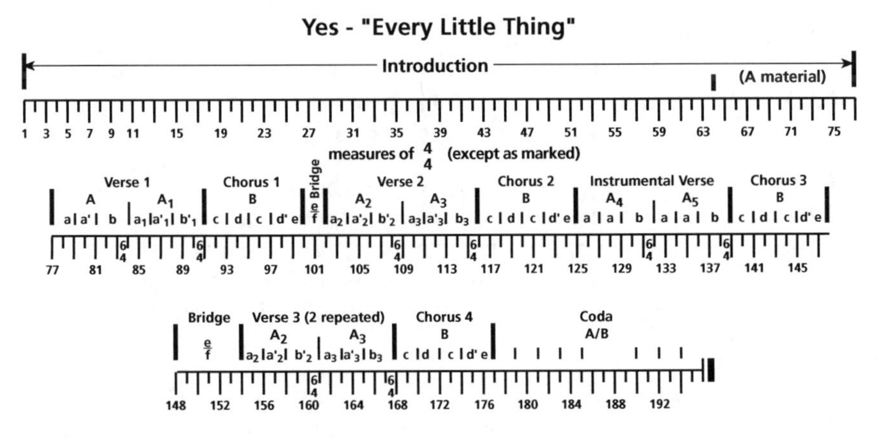


**Figure 3.3** "Every Little Thing" as performed by The Beatles and by Yes.

## The Listener

Typical audience members will have very different levels of listening expertise and usually dramatically different purposes for listening than professionals in the audio industry.

The recordist will have knowledge of the recording process (that which appears in Part Four, and more), the states of sound in audio recording, the materials of music, and of the hearing mechanism (previously discussed). Further, the recordist will have spent considerable time acquiring the listening skills for the evaluation of recorded and reproduced music, and sound that will be covered in Parts Two and Three. The recordist is often equally skilled at evaluating the technical integrity of the audio signal (perceived parameters of sound), and at evaluating the artistic elements of sound and the materials of the musical message, also to be covered in Parts Two and Three.

The *lay listener* is the audience for a recording. The lay listener will listen by relying primarily on their previous listening experiences. They typically have little or no formal training. The lay listener may be listening for some meaning in the music and be concerned with the relationship of the musical (or literary) message to their personal preferences of musical style, and musical and dramatic meanings. They may be listening for the sensual aspects of the music or be listening for the aesthetic experience.

The listener will likely be listening for pleasure and be concerned about enjoyment. People most often listen for entertainment, and perhaps for escape or enrichment.

While an album might be carefully crafted as a complete experience from beginning to end, most people will not sit in a position equidistant from two loudspeakers for about an hour listening from beginning to end. Most people do not dedicate time to focus their attention to music listening, or to sitting in one place while listening (unless they are driving, and then we hope they are not listening too intently). This is simply not the normal listening practice of people today.

Whatever the purpose for listening to music, the listener will not listen in the same way or for the same sound qualities as the recordist. Nor should they. The audience member should not be expected to listen in the same way as the recordist. Their purpose for listening is very different. It is necessary, however, for the recordist to deliver the recording in such a way that it can be understood.

The receiver and the quality of the communication limit the success of any communication. The receiver (listener) must be able to accurately process information (recording/music) for communication to occur.

Humans are limited in their abilities to understand the content and/or meaning of what they perceive. These limitations are primarily the result of the listener's experience and knowledge, but are also dependent upon the listener's degree of interest in the material, intellect, and physical condition. The same material (music recording) will yield different information to different listeners (or to the same listener on different hearings), depending on knowledge, experience, analytic reasoning, social-cultural conditioning, expectations of context, attentiveness, and the condition of the hearing mechanism.

In crafting recordings for an audience, the recordist might need/wish to directly consider the listener (audience). An examination of these factors will provide a realistic assessment of a target audience and perhaps allow the recordist to reach them more readily. These are the conditions that shape listeners.

### Knowledge

The listener's accumulated information related to what is being heard, as well as of all subjects related to their existence, plays a substantial role in the understanding of music and sound. *Knowledge* allows the listener to understand a sound, or a musical passage, by relating the experienced sound material to a body of known information. When the listener has a body of known information and/or possible circumstances, the music can be matched against those possibilities. With a match, listeners can then comprehend (and potentially reason) the meaning of the material.

Knowledge is the amassed body of learned information, or known truths. The listener can draw on their knowledge to make evaluations and judgments on what is being heard. Knowledge areas related to the understanding of sound (and music) would include acoustics, psychoacoustics, music theory, music history and literature, language, audio-recording theory and practice, mathematics, physics, engineering, computer science, communications, and more. The listener can formally and consciously know these subjects, or they might be more or less intuitively learned through sensitivity to life's experiences.

### Experience

The listener's past life *experiences* are directly related to knowledge. Sound is experienced. In its conceptualized state, sound becomes experienced information. A personal knowledge, or experience, of the sound is the result of the listening process. Prior listening experiences are a resource that can be drawn from to recognize certain sound events or relationships. Sounds are mapped into the memory. The listener is better able to retain sound events in memory when a sound is the same as, or similar to, a sound that has been previously experienced. The act of listening is itself an experience, involving the learning of new information from what is going on in the listener's "present." New information is recognized and understood by comparing it with what has been previously experienced.

The type and quantity of listening experiences, and the personal knowledge gained, will vary significantly between individuals. These listening experiences are significant factors in understanding the messages of music. Different types of music will communicate different messages and may communicate the message through different musical styles. Difficulties people experience in understanding or appreciating different types of music can often be attributed to a limited experience with a certain type of music. An individual's listening experiences may have limited their ability to understand the materials (language) of the music, or to appreciate what the music is trying to communicate. Increased knowledge of a type of music, and/or an increased number of experiences in listening to the type of music, will increase the listener's ability to understand or appreciate the type of music.

Listening experiences are greatly influenced by the life environment of the individual. The social and cultural environment(s) in which the individual lives, and has experienced, provides opportunities for a certain finite number of listening experiences. Within any environment, certain experiences will occur much more frequently than others. Certain types of listening experiences will be very common, and certain types of listening experiences will never occur or occur only rarely.

*Social-cultural conditioning* will predispose the listener to a certain set of available previous experiences. People are conditioned by their environment (social and cultural) to apply meanings to sounds, and to understand stylized musical relationships. We learn to listen for certain relationships in musical materials and the artistic elements of sound. For example, the music of India uses pitch and rhythm in significantly different ways than American popular music. Individuals from either culture will not readily understand the meaning or appreciate the subtleties of the music of the other culture, upon initial hearings.

The application of meanings to sounds is the basis for language. Sounds have meaning, and can represent ideas. In this manner, a series of short sounds as narrowly defined, isolated ideas can combine (in a prescribed ordering) to create a complex concept. Communication of simple ideas to complex thoughts is thus accomplished by language. As we well know, different cultures have strikingly different languages. Some languages have common elements to other languages, and certain languages have elements that are largely unique.

Social context also plays a significant role in defining language sounds and meaning. Quite different meanings may be associated with a single sound, in the same language, by people of the same culture/society. This most often occurs between different social groups (ethnic origins, religious beliefs, age groups, etc.), groups of different economic status, and between geographic locations.

Sounds have meanings associated with their source. A sound produced by a car horn will invoke in the listener the thought of an automobile, not of the horn itself. Such referential listening only occurs when the listener has a certain set of life/listening experiences. Associations between sounds and their sources are largely dependent upon the listener's set of life/listening experiences, as provided by social-cultural environments. One can imagine living conditions under which an individual might never have experienced the sound of a car horn (perhaps the nineteenth century). The sound would not elicit the same response from this person as it would from a modern urbanite.

The meanings of musical sounds transfer between cultural and social groups in very similar ways to language sounds.

Social-cultural conditioning creates expectations as to the function of music. People are conditioned to relate various functions and applications to certain types of music. Dance, celebration, worship, ceremony, accompaniment to visual media, and aesthetic listening are but a few of the functions that music serves in various societies. Each function carries with it certain expectations for musical style. These expectations are defined differently in different cultures and societies.

The listener's life/listening experiences make available the information needed to understand sounds. Social-cultural environment conditions the listener through (1) providing a predominance of certain listening experiences, (2) providing certain expectations as to the content of musical materials (the applications of the artistic elements), (3) providing certain expectations as to the context within which certain types of music will be heard (in church, in a club, in the street, etc.), (4) providing meanings of association for certain significant sounds (significant sounds being perhaps a siren, or a falling tree), and (5) providing associations of group activity for certain types of music (ceremony, dance, group experiences).

While broadcast media have broadened the number of common elements between social and cultural groups throughout the world, great diversity still exists among human cultures and societal groups. Social-cultural conditioning remains a significant factor in our realization of the limitations of the listener. For example, it might be unrealistic to expect the lay person from China to understand the musical nuance and message of rap music, just as it might be unrealistic to expect the typical American, suburban 16-year-old to understand the meaning and significance, or to appreciate the aesthetic qualities, of Tibetan chant.

## Expectation

Knowledge, experience, and social-cultural conditioning create *expectations* for the listener. The listener will expect to hear certain sounds (or sequences of sounds) under certain circumstances. They will expect certain types of sounds to follow what has already been heard. They can expect to hear materials in certain relationships (melody with certain harmony), and to hear certain sounds within a given physical environment (one would not expect to hear a lion sound on a city street). The listener will likewise expect certain sounds in a given musical context (an operatic vocal technique would be unexpected in a reggae work) and expect to hear certain kinds of music in certain social-cultural contexts (the listener will expect to hear different music in church, movies, dance clubs, etc.).

When listeners are presented with something that is not expected, they may be surprised if they are able to recognize the material enough to understand it and its context, or may be confused if they cannot recognize the sound or relate the sound to its context. An unexpected sound might intrigue the listener as a unique turn of a musical idea or as a sound slightly out of context. Conversely, if unexpected sounds that are also unfamiliar to the listener are present, they will not be able to understand the sound, they will not receive the message of the material, and may likely be dissatisfied or frustrated by the listening experience. Among other possibilities, the listener might have a dislike for the original context of this unexpected sound, and thus cause this new experience to be unenjoyable.

Expected and unexpected sounds and relationships are balanced within all musical styles. A musical style is a set of expectations. Certain types of musical events and relationships are present that provide a musical style with consistency and a unique character. And it must be remembered: The unexpected in music is necessary to keep the music interesting and the listener engaged.

## Analytical Reasoning

The listener's knowledge, experience, and *analytical reasoning* play important roles in understanding the musical messages. Too many unexpected sounds or situations will result in confusion and frustration on the part of the listener. If expectations are filled in predictable ways, the listener will become bored with the material. They perceive logic and coherence of the musical materials through a fulfillment of expectations in the characteristics and functions of the musical materials, coupled with enough unexpected activity to maintain interest.

Listeners use analytical reasoning to extract the meaning of musical materials when they are unable to identify the material. Analytical reasoning, in music listening, is the ability to relate immediate listening experience to knowledge, in a manner capable of deducing meaningful observations and information. The ability to perform this type of listening activity is dependent upon intellect, the amount of knowledge the listener is able to draw from, the listener's previous experience in performing analytic reasoning exercises, and the listener's knowledge of the types of information to extract from the listening experience. This method of listening works similar skills as the critical- and analytical-listening skills addressed in Parts Two and Three.

## Active and Passive Listening

The listener's level of attentiveness plays an important role in understanding the musical message. Listener attentiveness and musical understanding are related to active and passive listening, and to the listener's interest in the music.

The difference between active and passive listening is the listener's attention and involvement. *Passive listening* occurs when the listener is not focused on the listening process, or on the music itself within the listening experience. Passive listening might find listeners otherwise occupied and listening to music as a background activity (reading a book, for example). Alternatively, listeners might be listening to music for reasons other than understanding. They may be listening for relaxation purposes. Other types of passive listening include approaching music for its emotive state, or feeling, or listening to music for its pulse only, such as an accompaniment to dancing. In all of these cases, the listener is not listening to the musical materials themselves, and they might not be aware of the music during certain periods of time. In passive listening the music itself is not the center of the listener's attention.

Music is at the center of the listener's attention in *active listening.* Various levels of detail can be extracted during the active listening process. Among many possible states, active listening might take the form of listening to the text and primary melodic lines of a work. It may take the form of following the intricacies of motivic development in a Beethoven string quartet, or of evaluating the characteristics of a sound system. In all cases, only listening has the listener's attention, whether listening for musical materials or sound quality.

The listener is most likely to be an active listener if they are interested in the music or have a specific reason to be listening carefully. The listener's interest in the music may be determined by mood or energy level at the time of the listening experience but is most often associated with listening preferences (and the previous experiences that have shaped those preferences). These preferences lead to the types of music the listener listens to most often and what they prefer to hear.

### Hearing Mechanism Condition

The final variable between individual listeners is the *condition of the hearing mechanism*. Some individuals have impaired hearing; some have knowledge of their condition and others do not. The hearing of the individual might vary from the norm because of a defect at birth, from accidental damage from physical trauma or prolonged exposure to high sound-pressure levels, or from the natural deterioration caused by the aging process. The recordist cannot anticipate hearing impairment of the listener, nor create recordings that can be heard well by those with diminished hearing. It is quite important, however, that the recordist know the condition of their hearing.

Variation of the individual's hearing characteristics from normal human hearing is of great importance to the recordist. The recordist must have knowledge of their hearing and make use of that information in evaluating sound in their job function. Significant hearing problems may make a person poorly suited for certain positions in the recording industry. Normally, a recordist might find they are less sensitive to sound in certain frequency ranges or that the two ears have different frequency and amplitude sensitivities. This information will serve the recording professional well in evaluating sound, as it will allow them to make adjustments in their work by knowing how their personal perception is different from the existing sound.

### Target Audience

The typical listener envisioned for a specific recording project, or piece of music, is often called a *target audience*. The target audience for a piece of music is often determined to help focus a project and to seek a way of predicting the success of the music in communicating its message. The knowledge, musical and sociological expectations, and the listening experience of a typical audience member will define the target audience. The music can then be shaped to conform to the abilities and expectations of the typical member, thereby increasing the chance it will successfully communicate its musical message. The goal is to create a recording/song that this defined audience will find engaging and that will be commercially successful.

## Conclusion

Recording professionals should not expect people to listen to their recordings with undivided attention or with the same level of accomplishment they have attained. At the same time the recordist must not underestimate or undervalue the listener. Listeners are often passionate about the music we record and the music they listen to.

Music audiences feel strongly about "their" music. They are often very possessive about the type of music they enjoy and the performers they follow. Music can speak deeply to people and bring people to identify with music on a very fundamental and personal level. Commentaries about their music can be perceived as reflections on themselves. While the listener may not know much about music or recording, they know what they like—and usually are willing to tell you about it.

Similarly, listeners are often quick to identify the quality of productions. Well-crafted and successful recordings are easily identified by listeners, not for their quality but because they present the musical material in a way that communicates well and directly to the listener. Further, sound qualities of the recordings of one type of music will be different from others, and will draw the listener, or not. The listener may not recognize technical integrity, but any signal problems will detract from the recording and the listening experience. Listeners will not miss this. They may not be able to tell what is wrong but they will recognize that it is not right.

The recordist is in the position to play a central role in the creation of music. They may use the recording process to shape music performances and the music itself. This new role for the recordist has been widely recognized since the early 1960s. While perhaps it is new when we think back over hundreds of years of music, recordists are currently very much a part of the creative process of nearly all recordings. With nearly 50 years of sophisticated practice in crafting sounds through multitrack production and stereo reproduction, the recordist as an artist is no longer something new.

The more the recordist understands music and the listener, the more likely it is that they will be in a position to assist the artist in delivering a performance and recording of a piece of music that will be successful. To bring the reader to appreciate some of what is involved with this feat was the goal of this chapter. It is wished that the recordist would want the listener to find enjoyment in what is recorded—for the artist's sake and for the music.

# Part Two

---

## Learning to Listen, Beginning to Hear:

### Acquiring Fundamental Listening Skills and Establishing an Effective Approach to Listening

# 4 Listening and Evaluating Sound for the Aspiring Audio Professional

People in the audio industry need to listen to and evaluate sound. Carefully evaluating sound, for one reason or another, is an integral part of most positions in the audio industry. Sound must be evaluated in all areas of audio production, manufacturing, and support. These areas are very diverse. They may be equipment performance or microphone placements, music mixes or the technical quality of the signal, or any one of many other possibilities.

Sound is being evaluated by the audio professional in all these cases and more.

Saying "sound is central to audio" is obvious to the point of sounding trivial. It is equally ironic that the audio and music community has not developed a way to clearly communicate meaningful information about sound. No language or vocabulary exists for qualities of sound. Part Two begins the creation of a means and a vocabulary to communicate about sound.

While this book is focused on the artistic roles of the recording professional, sound evaluation is important to everyone in the industry who listens to, evaluates, and talks about sound. Parts Two and Three of the book can and should be used by anyone in need of developing the ability to understand, evaluate, and communicate about sound. It should be a primary objective of all people in the audio industry to be more sensitive and reliable in their evaluations of sound. While the term "recordist" will still be used in here, it should be interpreted to mean "any audio professional" during discussions of sound evaluation.

The sequence of chapters in Part Three will present a system for understanding and evaluating sound that will substantially develop the reader's ability when mastered. Here in Part Two will be information on how to listen and what to listen for, exercises to acquire basic listening skills that will be the foundation for Part Three, and a general overview of the system for evaluating sound and its use.

It is necessary for all people related to the audio industry to be accurate and consistent in their evaluations of the quality and content of sound and audio. As we have seen, the previous experience, knowledge, cultural conditioning, and expectations of the listener (in this case the audio professional) have a direct impact on the level of proficiency at which the listener is able to evaluate sound. With increased experience in evaluating sound comes increased skill and accuracy.

The act of listening and the process of evaluating sound can be learned and greatly refined. The following is a presentation of the need for sound evaluation and the listening process, leading to a discussion about how we talk about sound, and the development of listening and sound-evaluation skills.

## Why Audio Professionals Need to Evaluate Sound

Audio professionals need to evaluate sound to define what they hear, to understand what they hear, and to communicate with one another about sound. These are important aspects of the job functions for almost all people in audio.

Recording engineers and producers, obviously, must have well-developed listening skills because evaluating sound is one of the most important things they do in their work. The need for highly refined skills obviously holds true for composers and performing musicians, especially those involved in the audio-recording processes. All audio professionals who listen to sound share a similar need for these skills. The technical people of the industry, those involved in artistic roles, and those in manufacturing or facility design, or product sales and many others, all must share observations and information about sound.

There are other reasons audio professionals need to evaluate sound in addition to talking about sound in precise and meaningful terms. The recording's sound qualities need to be observed, recognized, and understood to perform a great many jobs in the industry. Nearly all positions approach sound evaluation in a somewhat unique way. In fact, there might be as many reasons (significantly or slightly unique) for evaluating sound as there are job functions within the multitude of positions in the audio industry.

For the recordist, there are additional benefits to sound evaluation, and some will be discussed in detail in later chapters. These include ways to (1) keep track of one's work so that the audio professional can return to those thoughts/activities in the future, (2) plan recording projects out of the studio, (3) understand the work and ideas of others, (4) recreate sounds and musical styles, and many more.

Nearly all people in audio work directly with some aspect of sound. These aspects might be vastly different, yet these people must communicate directly and accurately to share information. In order to share information, sound must first be evaluated and understood by the listener.

Understanding sound begins with perceiving the sound through active attention. One can then recognize what is happening in the sound or recognize the nature of the sound, provided the listener has sufficient knowledge and experience. The listener must know what to listen for (i.e., the artistic elements of sound) and where to find that information (perhaps a particular musical part). This recognition can lead to understanding, given sufficient information. What is understood can be communicated, with the presence of a vocabulary to exchange meaningful information that is based on a common experience.

## Talking About Sound

People in the audio industry, as in all industries, work together toward common goals. In order to achieve those goals, people must communicate clearly and effectively. A vocabulary for communicating specific, pertinent information about sound quality does not currently exist. People have been talking about sound for hundreds of years without a vocabulary to describe their actual perceptions and experiences. Instead, people have used imprecise terms to associate other perceptions and experiences to sound—unsuccessfully and inaccurately.

Describing the characteristics of sound quality through associations with the other senses (through terminology such as "dark," "crisp," or "bright" sounds) is of little use in communicating precise and meaningful information about the sound source. "Bright" to one person may be associated with a narrow, prominent band of spectral activity around 15 kHz throughout the sound source's duration. To another person the term may be associated with fast transient response in a broader frequency band around 8 kHz, and present only for the initial third of the sound's duration. A third person might easily provide a different, yet an equally valid, definition of "bright" within the context of the same sound. The three people would be using different criteria of evaluation and would be identifying markedly different characteristics of the sound source, yet the three people would be calling three potentially quite different sounds the same thing—"bright." This terminology will not communicate specific information about the sound and will not be universally understood. It will not have the same meaning to all people.

Analogies such as "metallic," "violin-like," "buzzing," or "percussive" might appear to supply more useful information about the sound than the intersensory approach. This is not so. Analogies are, by nature, imprecise. They compare a given sound quality to a sound the individuals already know. A common reference between the individuals attempting to communicate is often absent. Sounds have many possible states of sound quality.

"Violin-like" to one person may actually be quite different to another person. One person's reference experience of a "violin" sound may be an historic instrument built by Stradivarius and performed by a leading classical artist at Carnegie Hall. Another person may use the sound of a bluegrass fiddler, performing on a locally crafted instrument in the open air, as their reference for defining the sound quality of a "violin." The sound references are equally valid for the individuals involved, but the references are far from consistent and will not generate much common ground for communication. The sound qualities of the two sounds are strikingly different. The two people will be referencing different sound characteristics, while using the same term. Certainly there will be strong similarities between these two instrument sounds, but there will be great differences in the subtle details of the sound qualities; it is in these subtle details that quality recordings are created, and where the skills of the recordist must be drawn. An accurate exchange of important information will not occur without a clear communication of this detail.

The imprecision of terminology related to sound quality is at its most extreme when sounds are categorized by mood connotations. Sound qualities are sometimes described in relation to the emotive response they invoke in the listener. The communication of sound quality through terminology such as "somber," for example, will mean very different things to different people. Such terminology is so imprecise it is useless in communicating meaningful information about sound.

People can only communicate effectively through the use of common experiences or knowledge. The sound source itself, as it exists in its physical dimensions in air, is presently the only common experience between two or more humans.

As we hear sounds, we make many individualized interpretations and personal experiences. These individual interpretations and impressions are present within the human perceptual functions of hearing and evaluating sound. They cause individualized changes of the meaning and content of the sound. Therefore, our interpretations and impressions are of little use in communicating about sound. Humans have few listening experiences that are common between individuals and that are available to function as the reference necessary for a meaningful exchange of information (communication).

This absence of reference experiences and knowledge makes it necessary for the sound source itself to be described. Meaningful communication about sound will not be precise and relevant without such a description.

The states and activities of the physical characteristics of the sound will be described directly in our communications about sound. This approach to evaluating sound requires knowledge of the physical dimensions of sound and how they are transformed by perception. Meaningful communication between individuals is possible when the actual, physical dimensions of sound are described through defining the activities of its component parts.

By describing the states and activities of the physical components of a sound, people may communicate precise, detailed, and meaningful information. The information must be communicated clearly and objectively. All of the listener's subjective impressions about the sound, and all subjective descriptions in relation to comparing the sound to other sounds, must be avoided for meaningful communication to occur.

Subjective information does not transfer to another individual. As people attempt to exchange their unique, personal impressions, the lack of a common reference does not allow for the ideas to be accurately exchanged.

Meaningful communication about sound can be accomplished through describing the values and activities of the physical states of sound. Sounds will be described by the characteristics that make them unique. Meaningful information about sound can be communicated through verbally describing the values and activities of the physical states of sound in a general way. Information is communicated in a more detailed and precise manner through graphing the activity, as will be explored and described in the following chapters.

A vocabulary for sound is essential for audio professionals to recognize and understand their perceptions, as well as to accurately convey to others what they hear.

## The Listening Process

Recording engineers and other industry professionals must learn to listen in very exacting ways. The profile of the listener discussed in Part One assisted us in identifying how the recordist has different purposes for listening and needs a much higher skill level. It is necessary for audio professionals to be accurate and consistent in the listening process and its observations. Likely the most difficult job of the recordist is listening and paying attention.

Listening skills need to be developed for the recordist to function in their job. They will be focused in their attention and ultimately become systematic in how they listen to hear detail in sound. The recordist will not be listening passively, but rather will be actively engaged in seeking out information with each passing sound. They will be concerned about a multitude of things, from the quality of a performance, to its technical accuracy; from the quality of a microphone selection, to its appropriate placement; from the quality of the signal path, to the inherent sound quality of a signal processor. All of these things and many more might pass through the thoughts of the recordist frequently and regularly throughout any work session. The listening experience of the audio professional will be multidimensional in many ways. All of their work comes back to learning how to listen.

The recordist must acquire a systematic approach to listening that will involve quickly switching between critical and analytical listening information. It will involve quickly switching between levels of detail, or perspective, and focus on various artistic elements and musical materials.

In many ways the recordist's listening process is like a scanner—always moving between types of information and between levels of detail.

### Critical Listening versus Analytical Listening

Audio professionals evaluate sound in two ways: *critical listening* and *analytical listening*. Critical listening and analytical listening seek different information from the same sound. Analytical listening evaluates the artistic elements of sound, and critical listening evaluates the perceived parameters. A different understanding of the sound is achieved in each case.

The artistic elements are the functions of the physical dimensions of sound, applied to the artistic message of the recording. We recognize the physical dimensions of sound through our perception, as perceived parameters. This allows understanding of the technical integrity of sound quality to be contrasted with musical meaning and relationships of the artistic elements.

The same aspects of sound quality may provide two different sets of information. This is entirely dependent upon the way we listen to the sound material, evaluating the sound for its own content (critical listening) or evaluating sound for its relationships to context (analytical listening). The recordist must understand how the components of sound function in relation to the musical ideas of a piece of music and the message of the piece itself. These are analytical-listening tasks. The audio professional must also understand how the components of sound function to create the impression of a single sound quality, and how they function in relation to the technical quality of the audio signal. These aspects are critical-listening tasks.

Analytical listening is the evaluation of the content and the function of the sound in relation to the musical or communication context in which it exists. Analytical listening seeks to define the function (or significance) of the musical material (or sound) to the other musical materials in the structural hierarchy. This type of listening is a detailed observation of the interrelationships of all musical materials, and of any text (lyrics). It will enhance the recordist's understanding of the music being recorded, and will allow the recordist to conceive of the artistic elements as musical materials that interact with traditional aspects of music.

Critical listening is the evaluation of the characteristics of the sound itself. It is the evaluation of the quality of the audio signal (technical integrity) through human perception, and it can be used for the evaluation of sound quality out of the context of a piece of music. Critical listening is the process of evaluating the dimensions of the artistic elements of sound as perceived parameters—out of the context of the music. In critical listening, the states and values of the artistic elements function as subparts of the perceived parameters of sound. These aspects of sound are perceived in relation to their contribution to the characteristics of the sound, or sound quality.

Critical listening seeks to define the perception of the physical dimensions of sound, as the dimensions appear throughout the recording process. It is concerned with making evaluations of the characteristics of the sound itself, without relation to the material surrounding the sound, or to the meaning of the sound. Critical listening must take place at all levels of listening *perspective* (see below), from the overall program to the minutest aspects of sound.

## The Sound Event and Sound Object

The concepts of the *sound event* and the *sound object* assist in understanding how the musical materials (analytical listening) and sound quality (critical listening) are shaped by the artistic elements. A sound event is the shape or design of the musical idea (or abstract sound) as it is experienced over time. The sound object is the perception of the whole musical idea (or abstract sound) at an instant, out of time.

The *sound event* is a complete musical idea (at any hierarchical level) that is perceived by the states and values of the artistic elements of sound. The term designates a musical event that is perceived as being extended over time, and has significance to the meaning of the work. The sound event is a musical idea perceived by its various dimensions, as shaped by the artistic elements of sound. It is a perception of how the artistic elements of sound are used to provide the musical section with its unique character. The sound event is understood as unfolding and evolving over time, and is used in analytical listening observations.

*Sound object* refers to sound material out of its original musical context. For example, in a discussion of the sound quality of George Harrison's Gibson J-200 on "Here Comes The Sun" compared to its sound on "While My Guitar Gently Weeps," the two sound qualities of the instrument would be thought of as sound objects during that evaluation and comparison process. A sound object is a conceptualization of a sound as existing out of time, and without relationship to another sound (except its possible direct comparison with another sound object).

The concepts, sound object and sound event, are contrasted at any hierarchical level. They allow analytical-listening and critical-listening evaluations to be performed, interchangeably and/or simultaneously, on the same sound materials.

These concepts are able to provide an evaluation of the music's use of the artistic elements of sound, in ways that are not necessarily related to the importance or function of the musical materials. Rather, these concepts seek to determine information on the artistic elements (or perceived parameters) themselves, as they exist as singular and unique entities (sound objects), and as they change over time (sound events).

## Perspective and Focus

For sound-evaluation purposes, the audio professional must be able to understand the artistic elements of sound, how those elements relate to the perceived parameters of sound, and how those two conceptions of sound are used with *perspective* and with *focus*. The concepts of perspective and focus are central to the listening process and evaluating sound. The audience will go through this process in a general and intuitive manner. The audio professional must be thorough and systematized in approaching this aspect of the listening process.

In order for the message carried by the artistic elements to be perceived, the listener (audience or audio professional) must recognize that important information is being communicated in a certain artistic element. The listener must then decipher the information to understand the message, or recognize the qualities of the sound. The listener will identify the artistic elements that are conveying the important information by scanning the sound material at different perspectives, while focusing attention on the various artistic elements at the various levels of perspective.

*Focus* is the act of bringing some aspect of sound to the center of one's attention. The listener needs to identify the appropriate, perceived parameter of sound that will become the center (focus) of attention in deciphering the sound information. Further, the listener needs to determine a specific level of detail on which to focus attention.

The *perspective* of the listener determines the level of detail at which the sound material will be perceived. Perspective is the perception of the piece of music (or of sound quality) at a specific level of the structural hierarchy. The content of a hierarchy is entirely dependent upon the nature of the music or program material, at any specific time.

In a musical context, the detail might break down as in Table 4.1. Each level of detail represents a unique perspective from which the material can be perceived. Each perspective will allow the listener to observe

**Table 4.1** Examples of hierarchical levels of perspective

| | |
|---|---|
| Level 1 | Overall musical texture and form |
| Level 2 | Text (lyrics) |
| Level 3 | Program dynamic contour, timbral balance, sound stage |
| Level 4 | Individual musical parts (melody, harmony, etc.) |
| Level 5 | Groupings of instruments and voices |
| Level 6 | Individual sound sources (instruments and voices) |
| Level 7 | Dynamic relationships of instruments (musical balance) |
| Level 8 | Composite sound of individual sources (timbre and spatial qualities) |
| Level 9 | Pitch, duration, loudness, timbre, space, and duration elements of an individual sound of a specific sound source |
| Level 10 | Dynamic contour; definition of important components of timbre and space of an individual sound of a specific sound source |
| Level 11 | The spectral content (harmonics and overtones) of the individual sound of a specific sound source |
| Level 12 | The spectral envelope (dynamic envelopes of the overtones and harmonics) of that specific sound |

different dimensions and activities of the sound material. A perspective might be thought of as a type of distance of the listener from the sound material; the nearer the listener to the material, the more detail the listener is able to perceive. Perspective is the level of detail at which one is listening.

The listener may approach any perspective to extract analytical-listening information (pertaining to the function of the musical materials and artistic elements at that level of the structural hierarchy) or to extract critical-listening information (pertaining to defining the characteristics of the sound itself). Focus, again, is the act of bringing one's attention to the activity and information occurring at a specific perspective of the structural hierarchy, and/or within a particular artistic element or perceived parameter.

Attention to focus and perspective is needed in both critical-listening and analytical-listening activities, and should be considered before starting any listening session. It is important for the recording professional to define the focus and level of perspective of the listening experience before the sound material begins, as they can shape the listening experience in strikingly different ways for different situations. In many listening situations, all parameters of sound will need to be continually scanned to determine their influence on the integrity of the audio signal, and all artistic elements will need to be scanned to determine their importance as carriers and shapers of the musical message. In other listening situations, the recordist might need to carefully follow a specific artistic element at a specific level of perspective throughout the listening experience. Different situations will require a different approach to listening. It is important that the recording professional have a clear idea of what needs to be the focus of their attention and the level of detail required (perspective) before beginning to listen—or of the need for continually shifting focus and levels of perspective.

In beginning studies it is very important for the listener to have a clear purpose for each listening experience. This will greatly assist the learning process, and will make each listening session more productive and successful. They should be focused on a specific level of perspective and on a specific aspect of the sound, and should seek to ignore other aspects of sound. They will listen to the material repeatedly with a focus on a new aspect of sound at each repetition. With practice, one will be able to listen to (and recognize and understand what is happening to) many elements "at once."

## Multidimensional Listening Skills

Equal attention must be given to all aspects of sound as, depending on the sound material and purpose of the listening, any perceived parameter of sound or any artistic element may be the correct focus of the listener's attention. An incorrect focus will cause important information to go unperceived and will cause unimportant information to incorrectly skew the listener's perception of the material. The recording professional will often face the possibility that a change might happen in any of the dimensions of sound, at any point in time, at any level of perspective. It is necessary that recording professionals hear, recognize, and understand the character of the sound and any changes that might occur. This awareness needs to be cultivated, as it is counter to our learned listening tendencies.

Audio professionals must develop their listening skills to be multidimensional. The listening process involves the potential need to listen to many things "simultaneously." Though on one hand impossible, this is in practice often necessary. The actual process switches attention very quickly between materials. To accurately evaluate sound, they must learn to:

1. Shift perspective between all levels of detail
2. Focus on appropriate elements and parameters at all levels of perspective (and not allow their attention to be pulled away to activity in another element or level of perspective), and
3. Shift between analytical listening (for the qualities and relationships of musical material) and critical listening (for the characteristics of the sound itself) to allow the evaluation of sound.

## Distractions

It is often difficult for the recordist to keep from being distracted. Maintaining focus on the purpose and intent of the particular listening experience is very important. Common distractions include becoming preoccupied with the music, being drawn to sounds and sound qualities other than those under evaluation, and being curious about how a sound quality was created (as opposed to character of the sound).

Most of us are drawn to a career in recording because of a love of music. When working on a recording, we can lose our focus by becoming engaged with the musicality of the material. This focus is similar to listening for entertainment. However, there is a time and place to listen for entertainment. Most often recordists listen to qualities that are more precise and exacting. Even when listening within musical contexts, working directly with musical materials, and thinking about the musicality of the recording, the audio professional will be working at a level of perspective that is far removed from the passive music listening experience enjoyed by most people.

While focused on listening to the characteristics of one element of sound, the sound qualities of another element can draw the listener's attention. It is very important that the listener remain mindful of the purpose of the listening experience. For example, if the listening activity is intended to determine the musical balance of the snare drum against the toms, one should not allow oneself to get distracted by the sound quality of the piano.

In evaluating sound, audio professionals must remember that they are seeking to understand the sound that is present. It is possible for the listener to become distracted from listening by their own knowledge of the recording process or by their wanting to learn more about the recording process. At times people are drawn to thinking about how sound qualities were created—equipment, recording techniques, etc. Bringing production concerns into the process of evaluating sound is counterproductive, unless the recordist is specifically trying to identify equipment choices and production techniques, but this is a very different matter.

Listening sessions should have a clearly defined function. If the recordist is listening to determine equipment that may have been used in a recording, then that is the purpose of the session. If the recordist is listening to understand the sound quality of a certain environmental characteristic, then they should be listening to the various components of that sound and not be concerned about identifying the manufacturer or model number or the settings of the device that created the environment.

## Personal Development for Listening and Sound Evaluation

The skills and thought processes required for listening and sound evaluation must be learned. The development of any skill requires regular, focused, and attentive practice. Patience is required to work through the many repetitions that will be needed to master all of the skills necessary to accurately evaluate sound. Each individual will develop at his or her own, separate pace, as with any other learning.

### Memory Development

The recordist will evaluate sound more quickly and accurately with the development of their auditory memory. This will often be accomplished through their ability to recognize the unique characteristics of the various aspects of sound, and the patterns those characteristics might create. The listener must be conscious of the memory of the sound event, and they must seek to develop their memory to sustain an impression of the sound long enough to describe, annotate, or graph certain characteristics about the sound event.

Auditory memory can be developed. As one learns what to listen for, and as one understands more about sound and how it is used, the listener's ability to remember material increases proportionally. This is similar to the process of learning to perform pieces of music through listening to recordings of performances and mimicking the performances. With repetition, this seemingly impossible task becomes a skill that is much easier to perform. Listeners often remember more than their confidence allows them to recognize. The listener must learn to explore their memory and immediately check their evaluations to confirm the information. Accept now that mistakes will happen, and that mistakes are necessary for learning to take place.

The human mind seeks to organize objects into patterns. Sound events have states or levels of activity of their component parts that will often tend to fall into an organized pattern. The listener must become sensitive to the possibility of patterns forming in all aspects of the sound event, to allow greater ease in the process of evaluating sound. Recognizing patterns will assist in understanding sound and sequences of sounds, and will make remembering them more possible.

Developing memory is very possible and very important.

Sound is a memory. Sound is an experience that is understood backwards in time. Considering sound takes place over time and can only exist by atmospheric changes over time, it should be understood that sound is processed in memory. Sound is perceived after it has passed, using memory. Sound does not happen now (at a specific moment), but rather it happened then (over a span of time). It can start or stop "now," but it exists over a stretch of time (duration).

The reader is encouraged to embrace fully the Musical Memory Development Exercise (Exercise 5.6) in the next chapter and to return to that exercise regularly during the course of their work in listening skill development.

### Success and Improvement

With increased experience in evaluating sound, comes increased skill and accuracy. The act of listening and the process of evaluating sound can be learned and become greatly refined.

The reader will continue to become more accurate and consistent in evaluating sound the more they practice the skills and follow the exercises in the following chapters. The development of these skills must be viewed as a long-term undertaking. Some of the skills might seem difficult, or impossible, during the first attempts. The reader must remember their previous experiences might not have prepared them for certain tasks. The skills are, however, very obtainable. Further, the skills are desirable and indispensible, as the individual will function at a much higher level of proficiency in the audio industry after they have obtained these evaluation and listening skills.

The mastery of the skills of sound evaluation is a lifelong process, one that should be consistently practiced and itself evaluated. New controls of sound are continually being developed and incorporated into the recording process. These new controls create new challenges to the listening abilities of those in the audio industry—even seasoned professionals.

### Discovering Sound

Things are present in recorded music that are subtle and difficult to hear. Most people have never really consciously experienced a good number of these subtle dimensions of recordings. When something has never been experienced or perceived, one does not know it exists. It is possible for people to simply not hear some aspect of sound, simply because they do not have an awareness of or sensitivity to that dimension. Once that awareness and sensitivity is developed, those sounds are heard as easily as any other.

In *Personal Knowledge,* Michael Polanyi conveys the experience of a medical student attending a course in the X-ray diagnosis of pulmonary diseases. The student watches dark shadows on a fluorescent screen against a patient's chest while listening to the radiologist describe the significance of those shadows in detailed and specific terms. At first the student is puzzled and can only see the shadows of the heart and ribs, with some spidery blotches between them. The student does not see what is being discussed. It appears to be a figment of the radiologist's imagination. As a few weeks progress, and the student continues to look carefully at the X-rays of new cases and listen to the radiologist, a tentative understanding begins to dawn on the student. Gradually the student begins to forget about the ribs and the heart, and starts to see the lungs. With perseverance in maintaining intellectual involvement, the student ultimately perceives many significant details, and a rich panorama is revealed. The student has entered a new world of perception and understanding. The student may still see only a fraction of what the seasoned radiologist sees, but the pictures now make definite sense, as do most of the comments made by the instructor.

Many readers will likely discover a new world of sound. Dimensions of sound exist that are out of normal listening experience. We are not aware of those sounds until we learn what they are, and learn to bring the focus of our attention to those elements. Only then can we discover them and begin to understand them.

We have already learned to focus our attention on certain aspects of sound. In music, we have learned that pitch relationships will give us the most important information. In speech, we know that the sound qualities of words make up language, and the sound qualities of the speaker will inform us who is talking. We know dynamics will simply enhance the message of these two communications, and we listen to them in that way. We have been taught that *where* a musical instrument is playing is not important (and therefore not worth the effort of recognizing the sound characteristics of location and environment), but *what* pitches they are playing *is* important (and worthy of attention).

The reader will now be asked to perform listening exercises and to evaluate sound in ways that work *against* these learned (and perhaps natural) listening tendencies. This requires conscious effort, focused attention, patience, and diligence. With the knowledge that the listener is working against natural tendencies, it will make sense that certain things are difficult. This does not mean they are impossible; many people accomplish them daily. Nor does it mean that this way of listening should not take place. This way of listening is necessary to evaluate and understand many aspects of recorded sound that are simply not normally at the center of one's attention. As we know, the audio professional needs to listen in ways and for things that are not part of a lay person's normal listening experiences.

Above, the student took the leap of faith that is necessary in learning. The student believed in the radiologist and continued to try to understand, initially perhaps considering the material was an illusion, not really present, a creation of the radiologist's imagination. The student ultimately reached a moment of revelation when suddenly an image was perceived. It was always there. The student was now able to see it because of increased sensitivity to the possibility of its existence and an understanding of what that existence might be.

If the reader can commit to a similar leap of faith, they may be rewarded with the discovery of a remarkable new world of sound.

## Summary

Understanding sound must begin with perceiving sound. This requires active attention, and sufficient knowledge and experience to know what to listen for. One can then recognize what is happening in the sound or recognize the nature of the sound. This recognition can lead to understanding. What is understood can be communicated, given a vocabulary to exchange meaningful information that is based on a common experience.

In the next chapter the basic skills that form groundwork for the method for listening will be presented. Exercises for acquiring these skills will be offered, as well as techniques to practice for their development. This is enhanced by guidance in how to shape your listening process to best utilize these skills as an aspiring recordist.

# 5 Fundamental Listening Skills

## Introduction

This chapter will present some fundamental listening skills for the recordist. These skills will allow more advanced audio-related listening skills to be more readily understood and acquired, and will form the basis for many future exercises and skills. Some of this chapter's material may seem unusually detailed, may seem somehow irrelevant, might at first be extraordinarily difficult, or any number of other impressions might arise (perhaps even "this is easy"). Within these impressions the reader will need to begin their leap of faith, and dig into the materials, remembering that becoming confused is necessary for learning, acquiring new skills will require making some mistakes, and that by embracing not knowing, one can move toward knowing.

Also included here are some more "general" exercises to pull the reader into an overall impression of the sound relationships of the mix, without the burden of exacting detail, or precise observations. These exercises are designed to allow the reader to engage some of the core aspects of the mix and the sound stage, to examine some of the more readily apparent general qualities, before encountering the detailed studies of later chapters.

The topics and exercises of this chapter are not in a set, unalterable sequence. These areas are interrelated, but not sequential. While there is a logical progression of information and skill types presented here, the reader may move ahead or skip between skills before any single one is mastered. The reader should expect they will acquire some knowledge/skills more readily than others, and that every reader will be unique in this respect. Progressing from the beginning of the chapter to the end, step-by-step through each exercise in order, is not necessary, but may be most beneficial (and easiest to follow).

Exercises are separated from the body of the text and appear at the end of the chapter. Throughout this book, exercises will continue to be separated out for clarity and easier reference.

You are about to find frequency/pitch and time/rhythm approached in new ways, but in ways common in audio. With this chapter we begin to develop new ways of hearing them and thinking about them, that lead us to new ways of listening to and thinking about dynamics/loudness, timbre, and space in recordings.

## Background Knowledge and Preparation

The reader should memorize the materials in this section. These items will be of use to the reader throughout the remainder of this book, but—most importantly—they will be of use throughout one's career in audio. The reader is encouraged to gain control of these quickly, so the following exercises can be learned more deeply and directly.

Certain materials will need to be memorized to make learning certain listening skills easier, and more meaningful. The following will inform your listening process, and allow you to access more information faster and more meaningfully.

### Learning Frequencies of Pitches, Pitches of Frequencies

Recordists often work in frequencies as designated on equipment and software, and they often work with pitches when recording music and interacting with musicians. It is very helpful to know the equivalents of each, to cross between those two worlds. Knowing what pitches are being altered by a piece of equipment allows the recordist to know directly how the music is being altered; in reverse, knowing the frequencies of the music or the instrument being recorded allows the recordist to select the appropriate frequencies to alter and effect the music or the instrument's sound as desired.

Table 5.1 provides select reference pitches and frequency equivalents. These are some important ones, as they represent very common pitches in modern music and very common frequencies incorporated into recording devices and software. The reader should study this list and commit it to memory. These are not the only pitches and frequencies the reader will find useful in their studies and in their career, but they represent a reasonable starting point for acquiring this skill of transferring

**Table 5.1** Frequencies of select common pitches, and pitches of select common frequencies

| Pitch | Equivalent Frequency (Hz) | Frequency | Equivalent Pitch (with octave) |
|---|---|---|---|
| $E_2$ | 82.4 | 60 Hz | Between $A^{\sharp}_1$ and $B_1$ |
| $G_2$ | 98 | 100 Hz | Sharp $G_2$ |
| $A_3$ | 220 | 250 Hz | Sharp $B_3$ |
| $C_4$ | 261.63 | 500 Hz | Sharp $B_4$ |
| $E_4$ | 329.63 | 1 kHz | Sharp $B_5$ |
| $A_4$ | 440 | 2.5 kHz | Sharp $D^{\sharp}_7$ |
| $B^{\flat}_4$ | 466.16 | 4 kHz | Sharp $B_7$ |
| $F_5$ | 698.46 | 8 kHz | Sharp $B_8$ |

pitch into frequency, and the reverse. The number following the letter names of pitches is the octave in which the pitch is sounding. Figure 5.2 provides the relationship of these pitches to the keyboard, to help clarify octave placement.

The reader is encouraged to extend this table by creating a list of other, additional frequencies and pitches (and their equivalents in pitch or frequency) that are relevant to their own musical experiences or situation, and/or relevant to their particular recording system and devices.

## Harmonic Series

The harmonic series provides a template from which we can identify the pitch/frequency content of sounds. This will be explored in great

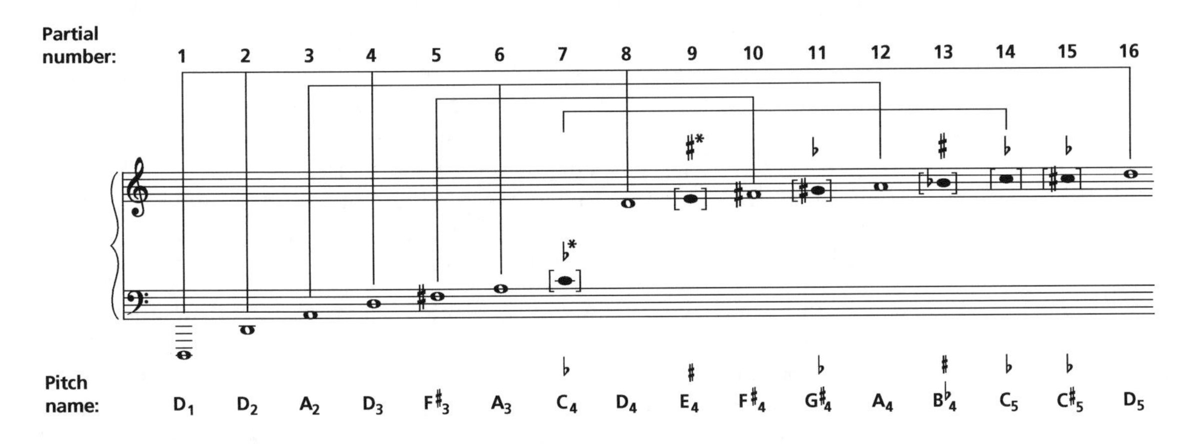


**Figure 5.1** The sequence of pitches and interval content of the harmonic series, through 16 partials.

depth as the book progresses. The reader is encouraged to memorize the sequence of pitches/frequencies that comprise the harmonic series through 16 partials, and also to memorize the sequence of intervals that separate the pitches. Later we will bring our attention to the "sound" of the harmonic series.

Notice the pitches/frequencies that repeat, as octaves progress. Observe how many times the fundamental frequency reappears as partial 1, 2, 4, 8, 16. The first fifth above the fundamental returns: 3, 6, 12; the 5th partial returns as the 10th. This leaves only the 7th and its octave repetition at 14, then the 9th, 11th, 13th, and 15th to remember as individual pitches/frequencies; notice these, and their interval spacing with surrounding partials.

### Pitch Registers

Pitch registers have been devised for dividing the hearing range into recognizable frequency/pitch bands. The ranges of the pitch registers will provide the reader with important points of reference to develop their skill in recognizing pitches and frequencies. They will be used in upcoming exercises extensively. At this time the reader should begin memorizing the boundary frequencies/pitches of these ranges.

Notice the boundaries of the pitch registers. The boundaries are defined by a group of pitches, to bring the reader to a general sensation of pitch level now, at the beginning of our work. The pitches/frequencies of the boundaries represent readily distinguishable and recognizable pitch areas, that allow physical sensation, of sorts, to be used to aid the cognitive learning of the pitch/frequency materials.

Figure 5.3 will assist the reader to commit these boundaries to memory. Notice the progression of pitches separating the registers, and the intervals of the boundaries that separate the registers. Note and memorize the frequencies of the pitches; both are equally important. By memorizing these pitches and frequencies you are also increasing the number of pitches you know by frequency, and frequencies you know by pitch. Bringing together these observations will allow you to more directly engage these topics in the future, especially as we begin to recognize pitch levels and frequencies.

Figure 5.4 provides a view of the pitch registers related to the grand staff. Here, for example, we can observe that the boundary of A-440/B-494 that separates the *mid* and *mid-upper* registers sits on the treble clef, almost in the center. We also notice that the *low* register is entirely below the bass clef, and that the *very high* register begins over two octaves above the treble clef; referring to Figure 5.2, it is apparent that the *very high* register begins where the piano keyboard ends.

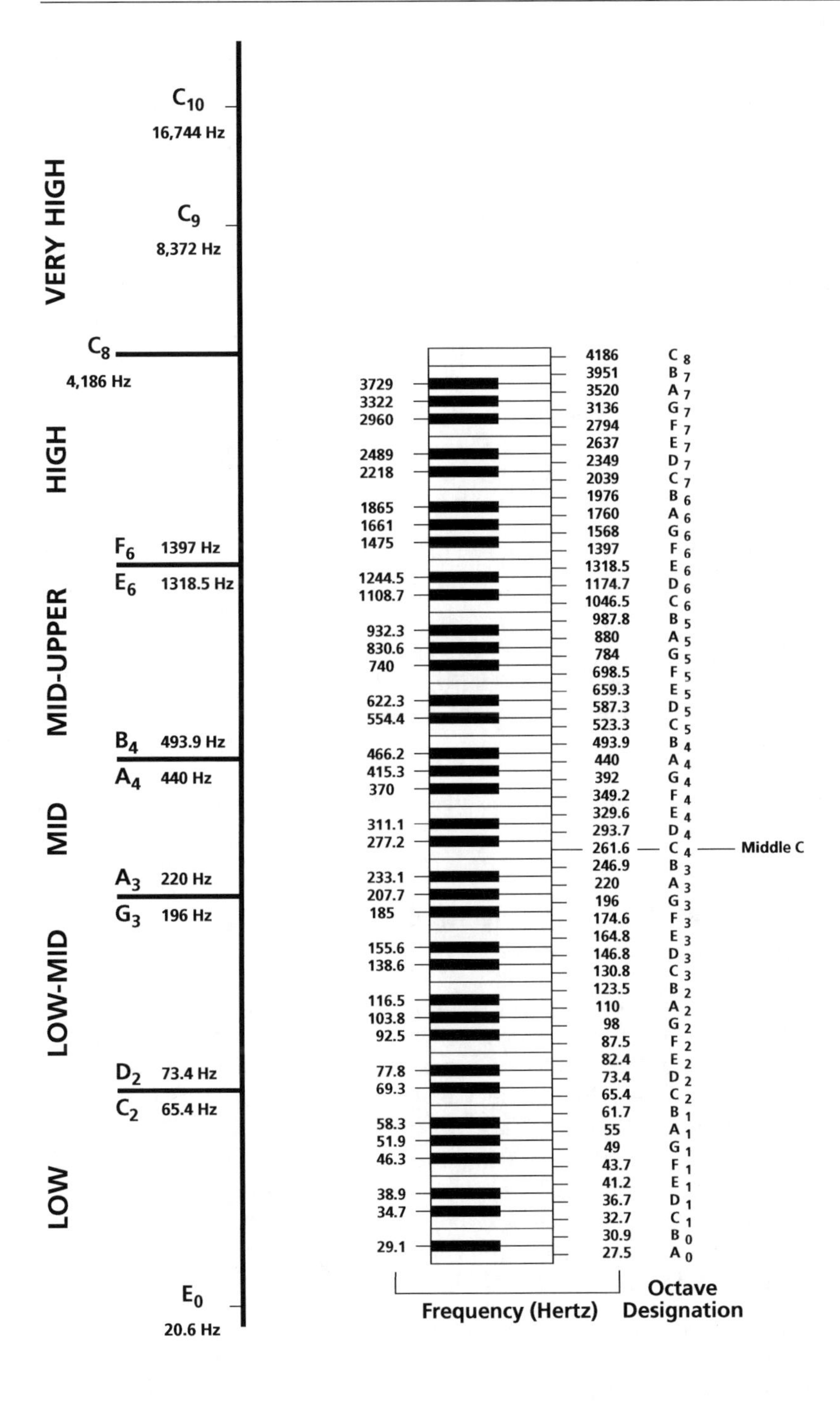


**Figure 5.2** Pitch/frequency registers in relation to keyboard with pitches, octave designations, and equivalent frequencies.

**Figure 5.3** Pitch and frequency ranges of registers.

| Register | Pitch Range | Frequency Range |
|---|---|---|
| LOW | up to $C_2$ | up to 65.41 Hz |
| LOW-MID | $D_2$ to $G_3$ | 73.42 to 196 Hz |
| MID | $A_3$ to $A_4$ | 220 to 440 Hz |
| MID-UPPER | $B_4$ to $E_6$ | 493.88 to 1,318.51 Hz |
| HIGH | $F_6$ to $C_8$ | 1,396.91 to 4,186.01 Hz |
| VERY HIGH | $C_8$ and above | 4,186.01 and above |

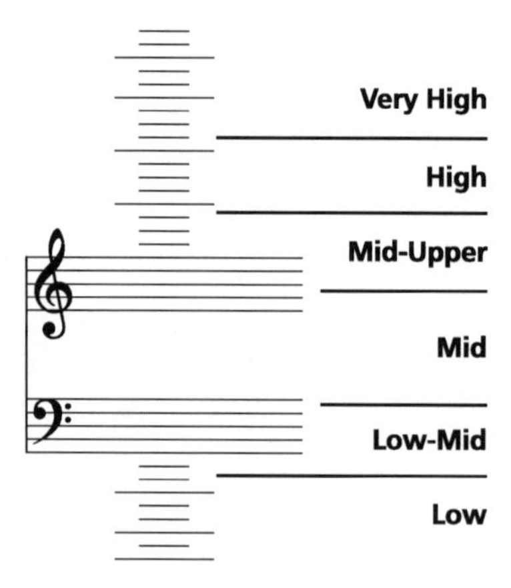


**Figure 5.4** Pitch registers against the grand staff of music notation.

## Self-Discovery and Realization: "What is Sound to You?"

We all hear sound differently. We all process sound, or make sense of sound, in our own unique way, as well. In a way, each of us has our own relationship with sound; we understand it somewhat uniquely and related to who we are. In a sense, we also each feel sound (as vibrations) in our own way. We each have an intimate connection with sound, internalizing the external, hearing the vibrations around us, and also feeling their presence in our bodies (though very often not within our awareness). This makes sense when we consider that we each have our own, unique physical shape, and we each have our own set of experiences, knowledge, upbringing, and more.

We will tap into this personal relationship with sound, to bring us to develop our *own* listening skills, in a way that is most meaningful and tangible for us as individuals.

### Realizing a Sense of Pitch

Everyone has an *internal pitch reference*. This is a sense of pitch level that is present, usually unconsciously, within each individual. Though it is often undeveloped, this reference can be brought to the consciousness of the recording professional. This will, however, require focused attention and concerted efforts over a period of time. The skill acquired will be well worth the time and effort involved.

Every individual is different and has a unique internal pitch reference. The things that make us unique human beings likely also contribute to our unique sense of pitch/frequency. What goes into making this reference is our own complex set of experiences, which vary markedly between individuals. It is often related to the timbre of certain sounds, and uses our exceptional ability of remembering the "correct" sounds of a particular sound source. Therefore, the reference itself is often related to an instrument one has played for a considerable length of time or to the sound and/or sensation of one's own voice. Some people identify strongly with certain pieces of music and can use their memory of the pitch level (tonal center) of that piece of music as a reliable reference. Other experiences or characteristics of an individual can generate or hold this pitch reference impression.

The process of realizing and then developing one's unique sense of pitch can seem perplexing, daunting, or simply impossible. This is something that may at first seem beyond human capability, simply because it is beyond our own experience. Remember: To believe something is possible is the first step toward making it so. One must become confused and grapple with the confusion in order to learn.

**Listen**

to tracks 4–13

for potential reference frequencies and potential reference pitches.

The reader should work through the Pitch Reference Exercise (Exercise 5.1) at the end of the chapter over a period of several weeks. Daily attention with a number of 5-, 10-, or 15-minute work sessions will yield results quickly. The reader should always try to find a quiet location where they will not be distracted.

This exercise will bring the reader to develop a consistent and reliable sense of *relative pitch*. This relative pitch may change by 5 percent to 10 percent, depending on mood, energy level, distractions, or countless other factors. Still the core of the individual's sense of pitch will be present and can be relied upon for specific tasks or for general use. For tasks that require precision, periodic checks of the reference level for accuracy may be necessary—especially in the beginning. As with any skill, the more it is used, the greater it will be refined. Once the skill is acquired, refining and maintaining the skill of remembering a reference pitch level can become intuitive and readily accomplished.

Recordists have cultivated this sense of pitch to a reliable reference for many practical uses. As examples, it is commonly used to identify frequency levels (such as what is required to immediately determine an appropriate equalization [EQ] setting). It is also used to hear the tuning of performances or to keep tuning consistent throughout a project, especially for an ensemble without a keyboard. All of the judgments the recording professional makes related to pitch and frequency will be enhanced with a stable sense of pitch/frequency gained through understanding one's own internal pitch reference.

### Recognizing Pitch Register Boundaries and Pitch Levels

The internal pitch reference is an important aid in recognizing frequencies and pitches throughout the hearing range. This reference will now be used to help identify the general locations of frequencies and pitches in relation to carefully devised pitch registers.

Critical listening and analytical listening define perceived pitch differently. Frequency estimation through pitch perception allows for critical-listening observations, while the same sound will need to be thought of as pitch relationships to understand analytical observations. The *pitch/frequency registers* (Figures 5.2, 5.3, and 5.4) will be used to estimate the relative level of the pitch material and to allow the information to be directly transferred between these two contexts.

**Listen**

to tracks 14–18

for pitch register boundaries performed on a piano.

Pitch and frequency estimation are fundamental skills that must be developed by the audio professional. The use of pitch/frequency registers will assist the reader in developing this skill of identifying perceived pitch and frequency levels. These registers will serve as reference areas and will provide a basis for a general description of perceived frequency and pitch levels. With the registers already committed to memory, we will now put them to use.

It will be helpful to relate this material to actual sound sources. The ranges of human singing voices stretch from the *low-mid* and *mid* registers (male voices) to the *mid* and *mid-upper* registers (female voices). Most musical activity occurs in the *mid* and *mid-upper* registers. This is where many instruments sound their fundamental frequencies, and where most melodic lines and most closely spaced chords are placed in musical practice. Take a moment to again notice the frequency ranges spanned by these registers.

The sibilant sounds of the human voice occur primarily in the *high* register; typically emphasized around 2 to 3 kHz, they extend to 8 kHz depending on the sound and the voice. Within the *very high* register, humans have the ability to hear nearly two and one half octaves. While this register is not playable by acoustic instruments or singable by human voices, much spectral information is often in this register.

The reader should work through Exercise 5.2 (at the end of the chapter) to begin developing skill at estimating pitch levels and octave placements. This is designed to take the reader from making general judgments to identifying precise pitch levels. The boundaries between registers are purposefully large to give the reader a sense of moving from the general to the specific and to acquire the skill with meaningful successes along the way.

First learn the sounds of the boundaries themselves. Get a sense of where the boundaries start and stop, of the small area that spans the boundaries. Once you have a sense of these boundaries, seek more detail by seeking pitches/frequencies within the pitch registers. We are beginning by "estimating," or coming close to the answer by using our slowly developing skill coupled with what we have learned with the pitch registers and their sound qualities. Our goal is to "recognize" and "identify" specific placements of pitches/frequencies anywhere in the hearing range; the pitch registers will help us develop this skill.

This skill is central to many of the evaluations commonly performed by all people in audio, and will be continually developed throughout the remainder of this book. The listener can easily transfer perceived pitch level into frequency estimation through using the above registers. The reader should continue practicing transferring various pitch levels to frequency and the reverse.

It is possible for the experienced listener to quite quickly learn to consistently estimate pitch/frequency level to within an interval of a minor third (one quarter of an octave). After considerable practice and experience, and gaining a clearer sense of an internal reference pitch, significant accuracy is possible. Within several weeks of thoughtful effort, the reader should be consistent within a perfect fourth (a bit less than one half an octave), and accuracy will continue to increase at a rapid pace with regular study. Once acquired, this skill will be invaluable.

## Recognizing Clock Time

We can use our ability to track time with the metric grid to our advantage in perceiving clock time. We can recognize the tempo of passing time, and transform that into gauging the rhythm of passing and accumulating seconds, groups of seconds, and fractions of seconds. In this way, we can tap into the perception of time increments used in recording production, and we can understand the passage of time as it relates to the timelines of critical- and analytical-listening processes.

We can remember tempos because they bring us to re-experience pieces of music. We "know" the right tempo for certain pieces of music when we know those pieces well. We "know" when the tempo is right, when the rhythms come together to provide the correct beat, pulse, groove—the core impression of the pulse of the music that is part of the music's form. Once we figure out the metronome marking of that tempo and remember it, we have a point of reference.

Exercise 5.3 will bring the reader to explore a piece of music they know well, to identify its metronome marking (MM) or the speed of its pulse, or metric grid. This is also found labeled as beats per minute (bpm). Once that tempo is known, it is then possible to use that knowledge to find MM:60 or MM:120. How closely related the tempo is to 60 bpm or 120 bpm will determine how complicated it might be to make this transition. Once this transposition happens, however, the reader will have an internal clock pulse that can accurately calculate and/or track clock time. The more tempos or songs the reader knows, the more easily this reference can be accessed and used to arrive at clock time.

Seconds are pulses at 60 bpm, and half seconds are pulses at 120 bpm. Another division that is very helpful is tenths of seconds. Tenths of seconds are 600 bpm, or five equal divisions of 120. This can be arrived at by thinking of the tenth of a second as a quintuplet of the 120 bpm pulse. This is not simple for some people. Alternatively, one can learn the tempo of 75 bpm or 150 bpm, just as MM:60 was learned above. With this, the reader could learn to hear the pulse of tenths of seconds as four equal divisions of the 150 bpm pulse; thinking of the 150 bpm pulse as quarter notes, the tenth of second pulse (600 bpm) would be sixteenth-note pulses. Finding a way to perceive and remember this sixteenth-note pulse at 150 bpm would provide the reader with an important point of reference for many listening activities.

Exercise 5.4 will assist us in bringing this information into practice.

## Beginning to Hear the Relationships and Qualities of the Mix

We often do not hear when we listen. Learning to hear takes listening with a purpose, it takes listening with some intention and some knowledge, and it requires bringing the energy and awareness of full attention to what is heard. Interestingly, we often hear and remember more than we realize, not trusting our observations or not bringing what we have heard into our conscious thoughts. Listening takes place on many levels, and uses our memory of what we know and what we have heard, as well as what we are experiencing at that moment, and what we are expecting to have follow. Attention, awareness, and an openness to having the unknown present itself will assist in keeping your skills developing.

With the following exercises, we will begin our work toward recognizing what we hear when we listen to a mix. Fully engaging the listening process will hasten learning materials and experiencing all aspects of sound.

### Song Structure Exercise

Refer back to Chapter 3 and its discussion of song structure. Exercise 5.5 will lead the reader through identifying the structure of a piece of music and creating a timeline. This act of recognizing the structure of a song is extremely important to the recordist. It allows them to know where they are in the song at any moment in time, and how the various sections of the song relate to each other.

Remember: Do not write while you are listening. Listen when it is time to listen. Write what you have recognized and remembered once you have stopped listening, when you have stopped collecting information from the listening experience. Follow the sequence: Listen, stop the sound, and write what you know.

### Musical Memory

People remember more than they believe they do. If you will trust your memory and use it, your memory will develop. Your confidence will grow as well. On the other hand, we often do not pick up on all detail when we listen. The reader must find the proper balance between tapping into what has already been experienced and remembered correctly, and what was not perceived and processed during the previous listening experience(s).

Exercise 5.6 in musical memory development will assist the reader in both of these areas. The reader is encouraged to revisit Chapter 4's explanation of musical memory development before proceeding into this exercise.

This exercise should be continually modified to incorporate any sound element you need to evaluate. For example, stereo location could replace structure. The purpose of this exercise is to improve your memory for the perceived parameters and the aesthetic and artistic elements of sound—any of them and all of them, and at any level of perspective.

### Perspective

When we listen at different levels of perspective, we hear different degrees of detail. We extract different information from the same material at each level of detail. For example, in listening to a mix, we can bring our attention to any particular instrument or voice, or any group of instruments, or to the overall balance of sounds, or to just some small aspect of a particular instrument—such as listening for the qualities of the attack of an acoustic guitar.

Switching between perspectives can be an important activity at many times. Bringing focus to a single level of perspective and remaining focused on that single level for a set period of time will be critical to the success of many recording and mixing tasks. Holding two or more sounds with equal attention is important in balancing sounds, and in many other decisions. These are all acquired skills of the recordist, and ones that the perspective exercise, Exercise 5.7, will allow the reader to begin to address.

### General Musical Balance versus Performance Intensity

Exercise 5.8 brings us to explore how dynamic levels of the performance appear in the mix, understanding there might be a subtle or significant difference. The dynamic levels of the performance are reflected in the timbre or sound qualities of the instrument or voice, where loudnesses carry characteristic timbres. These sound qualities can appear in the musical balance of the mix at substantially or subtly different loudness levels. A review of Chapter 2's explanation may be of assistance here.

The purpose of this exercise is to become sensitive to general differences between the timbre or sound qualities of instruments and voices resulting from the intensity levels of the performances that were recorded in tracking. This will bring an understanding of how these sounds appear at altered loudness levels within the mix. Most multitrack recordings have substantial differences in these areas, and extreme subtleties as well; these will be explored later in more detail.

## General Impressions of the Sound Stage

To perceive spatial information correctly, be certain you have established a reasonably accurate playback system and environment. See the Introduction section, "Establishing an Accurate Playback of Recordings," for guidance as needed. A stable listening environment is needed to hear the sound stage consistently, which is critical for learning the characteristics and relationships of the sound stage; it is best to avoid headphones and especially avoid earbuds for all spatial listening.

The empty sound stage diagram of Figure 5.5 should be copied for your use in noting the locations of sound sources.

In this exercise the reader should first seek to identify the general left–right location of sound sources. We will be ignoring image-size concerns at this time. Identify the center of the sources, place a dot on the diagram at that point, and label the dot with the name of the instrument/voice.

Then go back and seek to identify the general distance location of sound sources. We will be concerned about general placement, and recognizing which instrument is closest and which is furthest, at this time. Ask yourself: "Can I touch the sound?" "Is it in the same room with me?" "Is it in the same building?" "Is it well beyond where I am?" Place a dot at that distance, behind or in front of the dot you wrote for the L–R observation. Erase the first dot. Move the label of the source to the new location, which represents both dimensions. Remember: Distance is determined by perceived timbral detail; it is not any other perception. And remember: We confuse a host of things as being distance related. The empty sound-stage diagram of Figure 5.5 should be copied for your use in noting the locations of sound sources here in Exercise 5.9.

**Figure 5.5** Empty sound stage for general observations.

## Hearing Subtle Qualities

The following will bring us to engage subtle aspects of sound that we do not normally consider. In fact, some (such as the content of spectrum) we have been conditioned to ignore. Identifying, recognizing, and holding some of these items in our attention can be challenging. The reader should expect to not be successful with some of these exercises, or some aspects of them, in their first attempts. Focused attention and listening with clear intention of what information is being sought will be of great benefit. This section will seek to allow the reader a gentle introduction to these skills.

### Harmonic Series

This exercise will ultimately bring the reader to develop skill in perceiving the spectrum of a sound, and identifying its content—after a few other follow-up exercises in later chapters. This skill is important to recognizing, controlling, and shaping the sound qualities of the recording.

Exercise 5.10 should be practiced thoroughly. In doing this, consider the harmonic series as a chord with a singular, characteristic voicing; the interval between the partials is fixed. Next, consider the harmonic series as a single timbre or overall sound quality. It will prove valuable to practice alternating perspectives to recognize the intervals between the partials that make up the harmonic series, and to recognize the overall quality of the timbre. Finally, listen for subtle changes within the spectrum of the sound, as partials change in loudness levels and tuning.

### Time Judgment and the Timbre of Time

The time unit between iterations of the same sound can create the impression of a timbre, or sound quality. The pulses fuse to create the sensation of an overall quality. This overall quality makes it possible to judge and identify time segments by the timbre that is created by the time unit, allowing the perception of time increments through learning "timbre" or unique sound qualities of short time units.

Learning to recognize short time units—such as several milliseconds—will take some practice. It is, however, quite possible to develop skill in recognizing the sound quality of these time units when they are presented as delay times of repeated sounds. The different time delays will have distinctly different sound qualities, once you are able to recognize and remember the individual qualities. These time delays will be heard as "timbres of time." With this exercise time units will actually take on a characteristic sound, as they transform the sound quality of the reiterated drum sound. These unique sound qualities of sound reiterations will transform all other sounds in a similar way.

Exercise 5.11 concludes this chapter. The fundamental skills developed in this chapter will serve as a foundation for our work in Chapter 6 and beyond.

## Conclusion

A system for evaluating sound has been devised and is presented in the following chapter and detailed in Part Three. It will provide a means for evaluating sound in its many forms and uses, and will provide a vocabulary that can communicate meaningful information about sound.

The audio professional needs to evaluate sound for its aesthetic and artistic elements and its perceived parameters, as they exist in critical-listening and analytical-listening applications and at all levels of perspective. The system for evaluating sound addresses these concerns, and more. That system will be presented in practice during Part Three. The exercises that follow below will prepare the reader to most readily learn and make use of the system that unfolds in the following chapters.

# Exercises

The following exercises should be practiced until you are confident in recognizing the material covered.

### Exercise 5.1

*Pitch Reference Exercise*

Development of a personal pitch reference is possible with daily attention over several weeks. Daily work will involve a number of 5-, 10-, or 15-minute work sessions throughout the day. With focused effort this exercise will yield significant results quickly. The companion website has some potential reference pitches.

1. The first step is to become aware of your personal pitch reference. This may most likely be accomplished as follows:
   a. Consider your performance experiences for pitches that have become emphasized in your awareness. Exploring your experiences as a performer can turn up a significant memory. People who have played an instrument will often have certain pitches in memory that are related to that instrument (especially when they are in the act of playing it). Guitarists often know the sound

of an "E"; trumpet players a "B-flat"; violinists an "A"; and so forth. You can make use of this pitch reference if you have this experience. Think carefully about the tuning pitch of your instrument or some other pitch level you feel drawn to, and clearly identify the pitch. Now play the pitch, listen to it carefully, and work to internalize it as a reliable reference.

b. Paying close attention to your natural speaking voice, speak carefully, but deliberately, and try to remove any stress from your voice box. Pay close attention to your inflections and notice when you are causing the pitch of your voice to go up and down. Finally, bring your attention to identify the pitch of your voice where it is operating without resistance, and where it is not being forced up or down in pitch. In this monotone, if properly produced, is the natural pitch of your voice. This can often serve as a reliable pitch reference.

c. Vocalize or sing freely to find the pitch level where your voice causes your chest cavity to resonate. This will require you to relax your vocal chords and pay close attention to the sensations of your body for a fullness to develop in the chest cavity. The voice should then be swept through your comfortable singing range to find the pitch that creates the greatest fullness. That level of greatest fullness could become "your" pitch, which might serve as a reference.

d. Other pitch references are possible, such as drums or specific pieces of music. These are unusually unique to the individual. At times these can be easily identified or refined.

2. If your pitch reference is based on your voice, make a note of the pitch level you identified. Repeat the process of identifying this level four or five times over two or three days to validate the level. You will eventually identify a specific and consistent pitch level.
3. Now, listen to your reference pitch often throughout several days. A few times per day, stop your normal activity to sing, play, and listen to that pitch level. Use a piano, pitch pipe, tuning fork, or another instrument to play your pitch. Sing it frequently and become accustomed to the placement of that pitch in your voice. Work to bring yourself to hear the pitch in your mind before singing it. With this, you are bringing your sense of pitch into your consciousness.
4. Next, try to consciously carry your reference pitch with you throughout the day. Take five minutes at four or five set times throughout the day to sing your pitch level and check it for accuracy. Before singing, quiet your thoughts and bring your attention to your voice or to your memory of the pitch. Do not sing a pitch unless you are confident you have a memory of the pitch. If you do not have a memory of the pitch, return to Step 3 for more practice.

Sing the pitch in your memory and check it. Don't allow yourself to get frustrated by wrong pitches. Everyone will make mistakes at this stage. By evaluating your mistakes you can learn from them. Keep a record of the pitch you sang—high or low, size of interval off, etc. Look for patterns and try to identify what caused the errors.

5. Begin each day recalling your pitch reference. Check it, and adjust for accuracy as needed. Carry the pitch with you throughout the day. Remember, this exercise will greatly enhance a skill you will use throughout your career in audio.

---

**Exercise 5.2**

*Pitch-Level Estimation Exercise*

The companion website presents sound files with the boundaries of the pitch registers played on a piano. These may be of assistance in learning these boundaries.

1. Working with a keyboard instrument or some type of tone generator, practice listening to the boundaries of the registers. Seek to remember the sounds of those boundaries, and remember the pitch names and frequency levels of the several pitches that make up those boundaries.
2. It will be helpful to work with another person. While this other person performs pitches at the boundaries between registers (on a keyboard or other device or instrument), identify those boundaries. Maintain a record of your mistakes so that you can evaluate them and make adjustments.
3. Once confidence has been established in recognizing the general areas encompassed by the registers, begin playing individual pitches against the pitch registers. Identify the pitch register of the sound.
4. Finally, work to identify where pitches are sounding within the registers (i.e., the upper third of low, or the lower quarter of mid).

Throughout these steps, you should rely solely on your memory of the pitch/frequency registers in making these judgments. Use an instrument to check yourself, to validate success or to make adjustments; base your estimations solely on your growing ability to recognize pitch unaided.

### Exercise 5.3

*Identifying Tempos and Transposing Tempo into Clock Time*

1. Identify at least two pieces of music or songs that you know well, and that are meaningful to you. These should be pieces you can call up in your conscious thoughts clearly and be consistently accurate.
2. Hold the first piece in your thoughts and "hear" it internally. In doing this, notice its tempo, and try to focus on the tempo of the music. It may help to "feel" its pulse in your body.
3. When this tempo is clear to you, identify the tempo in terms of beats per minute or a metronome marking. Use a metronome or some other device or software to identify this exact tempo; you may need a number of attempts to find the correct tempo, keeping checking your findings. Make a note of this tempo marking.
4. Modify the tempo in your thoughts, and notice how the music just isn't right at a bit slower or faster. Settle into the knowledge that this piece of music only sounds right at this certain bpm, and remember that number and that tempo. You can check this tempo in your memory just by bringing up the memory of this song.
5. Repeat Steps 2 through 4 for at least one other song, so you have more than one tempo reference.
6. To transpose one of the known tempos you have identified into clock time, examine the tempos for their mathematical relationship to 60 bpm or 75 bpm. For example, if one of your songs is at 90 bpm, you can arrive at 60 bpm by transposing the pulse of the song to its dotted quarter note. You may have to "stretch" the pulse to be a bit faster or slower to make some numbers work; if this is the case, practice this, and work to be consistent; check your success by listening to the original pulse and the 60 bpm pulse.
7. Determine a way to transpose one or both of the songs you have identified to both 75 bpm (which will generate tenths of seconds) and 60 bpm (which will generate seconds and half seconds). If you cannot be accurate and consistent, that song will not work for this exercise. Identify another song and repeat this process.

You will now be able to carry with you a way to access clock time.

**Figure 5.6** Simple rhythm for graphing clock time.

### Exercise 5.4

*Exercise in Graphing Clock Time*

Select a very simple and slow rhythm such as that found in Figure 5.6. Set up a metronome or a click track on a DAW you can use to establish and maintain a steady pulse.

1. Use a marking of 60 bpm to establish a pulse clearly in your thoughts and set up a timeline divided into pulse increments.
2. Use that pulse to place the notes of the rhythm against the timeline.
3. Repeat Steps 1 and 2 with more complicated rhythms, and at MM:120 and MM:240.
4. Keep working with these steps until you begin to internalize or "remember" the tempos of 60 bpm, 120 bpm, and 240 bpm.

This process will help you learn the "sound" and "feel" of these tempos that are directly related to clock time. Knowing these tempos will be important for understanding time issues in critical listening, and will also allow you to be able to calculate time units for compressors, delays, reverbs, and other devices. It might be helpful to note that many band marches are at MM:120; if a person can clearly remember one of these pieces, they may well be able to quickly and accurately establish a tempo of MM:120, or a pulse every 0.5 seconds, which could be very helpful in many situations.

### Exercise 5.5

*Song Structure Exercise*

The following exercise should be practiced on a variety of pieces of music until you are comfortable with the material covered. The purpose of this exercise is to create a timeline of a song, divided into major structural divisions and phrases within those divisions.

1. Select a recording of a song you know reasonably well and prepare a timeline with measures numbered, up to perhaps 100.
2. Listen to the recording to identify where the major sections fall against the timeline. Try following the timeline while tapping the pulse of the song, or conducting. When a major section begins/ends, make a mark on the timeline.
3. After listening to the song, write down the names of those divisions. Then, try filling in additional information, such as other verse or chorus beginning/ending points and phrase lengths.

4. Repeat listening to the recording, followed by writing down the information recognized.
5. The graph is completed when it includes all of the major structural divisions, the mid-level structural divisions, and the smallest uniform phrase. Incorporating text information is also helpful.
6. Following the timeline while listening may prove helpful in initial studies in identifying structural divisions. Remember to wait until the music is stopped before writing observations. Clearly separating the listening and writing activities will assist you in improving listening skills and in learning to evaluate sound. This will become increasingly important as this book progresses.

**Exercise 5.6**

*Musical Memory Development Exercise*

1. Select a recording of a song you know reasonably well and prepare a timeline with measures numbered, up to perhaps 100.
2. Before listening to the recording, sit quietly and try to remember as much detail of the song as you can.
3. Now, write down the song's meter. In your mind, listen to the piece from your memory. Then write down where the major sections begin and end. If you cannot come up with those divisions easily, you might well be able to deduce that information by thinking about the patterns of phrases in the introduction, verses, choruses, etc. Write down as much information as you can.
4. Think carefully about what you wrote and identify aspects you are not certain about—things that need to be determined when you listen to the recording.
5. Now you can listen to the recording. Listen intently for the information you have determined you need. Do not follow your graph. Listen with your eyes closed. Listen to remember what you hear. Do not write while you are listening and do not correct your graph while you are listening.
6. When the song has stopped, write down what you heard in your one listening and correct what you previously wrote. Then repeat Steps 4 and 5 until you have created a timeline and structure of the song—in as few listening sessions as possible. Finally, check your information one last time while following your graph. All of the information you wrote should be checked for accuracy; make corrections to your graph.
7. Do not get discouraged. Keep trying. If overwhelmed, take a break but return to the exercise in short order.

8. Select another piece of music, one that you know very well, and perform this exercise again.

This exercise can be performed whenever a timeline needs to be created. If faced with a new song, listen intently to the song once, immediately after sketching a timeline.

### Exercise 5.7

*Perspective*

Find a recording with three to five sound sources (instruments and/or voices) during the first two sections of the song.

1. Bring your attention to a single instrument. While playing back the recording, follow that single instrument from the beginning of the song through its first two sections. Try to ignore all other instruments or voices; maintain your attention on only the musical parts being performed by that instrument.
2. Next, bring your attention to that same instrument again. While playing back the recording, follow that single instrument from the beginning of the song through its first two sections with your attention focused on only one aspect of that sound—for instance, performance quality, dynamics, stereo location. It is very important to practice bringing your full attention to only that instrument and only that aspect of sound, learning to completely ignore all other aspects of sound within the instrument, and all other aspects of sound.
3. Perform Steps 1 and 2 on another instrument in the song. Take care to notice how well you are able to maintain your complete attention on the one item you are seeking to follow.
4. In this step we will conceptualize the two instruments we examined above as being equal in our perception. Listen to the recording again; this time try to follow the loudness levels of two instruments simultaneously. Repeat the task, and see if you have the same result. In order for us to perceive two sounds equally we must pull our attention to a higher level of perspective, one with less detail than we heard when focusing on a single sound. Bring your attention to a place where you can perceive the two sounds as being equal—knowing if you bring one to the center of your attention it will dominate and make comparisons of the sounds inaccurate.
5. Perform Steps 1 and 2 on a third instrument in the song. Take care to notice how well you are able to maintain your complete attention on the one item you are seeking to follow.

6. Bring all three instruments into your perception equally in the next listening. Find yourself at a level of perspective where you can bring your attention to all three sounds in the same instant. This is not a matter of switching between the sounds; it is a matter of hearing the three simultaneously.

The purpose of this exercise is to introduce the reader to the skill of bringing focus to various levels of perspective, and the different types of information each will generate. This exercise brings beginning skills for perceiving and crafting the mix.

**Exercise 5.8**

*General Musical Balance and Performance Intensity Observations*

Find a multitrack production that you know well and that begins with three to five sound sources that have been transformed by the tracking and mixing processes. Understanding this transformation may be hard to identify at this stage of your development. As the exercise progresses you will recognize whether or not you have selected an appropriate or easily accessible recording; you may need to switch to another recording at some point.

1. Listen carefully to identify the instruments and/or voices. Notice how each will change in performance intensity through the course of the first two major sections of the song. Follow each part separately to observe the conflicting levels/cues between the musical balance of the mix and the performance intensities of the parts as they were tracked.
2. This will require separate hearings for each sound source, focusing only on that source. In these succeeding hearings, find instances where musical balance is at a different loudness than the performance-intensity information of the instrument's sound qualities.
3. Listen again, while focusing attention on a specific instrument or voice you know well; follow that sound source carefully throughout the entire song, to make some general observations of performance-intensity cues.
4. Listen again and note the actual loudness of that instrument/voice in relation to the other sound sources. Think in general "louder than" or "softer than" terms at this point in time.
5. Finally, listen again for how these relationships change between major sections of the song (i.e., between verse and chorus).

The purpose of this exercise is to become sensitive to the differences between the timbre or sound qualities of instruments and voices resulting from the intensity levels of the performances that were recorded in tracking, and how these sounds appear at altered loudness levels within the mix. Many recordings have substantial differences in these subtleties, which will be explored later.

**Exercise 5.9**

*General Impressions of the Sound Stage*

Use the same multitrack production recordings you selected for Exercise 5.8 here—ones that begin with three to five sound sources.

1. Listen to identify the same three to five primary instruments and/or voices that you followed in the musical balance/performance-intensity exercise. During the first three sections of the song, follow one of those parts separately, and note the left–right placement of the sound source. Locate the L–R placement of the sound on a sound-stage diagram; use two or three diagrams, one for each section.
2. In succeeding hearings, find the locations of the other two, three, or four sources.
3. Listen again, while focusing attention on a specific instrument or voice you know well; follow that sound source carefully throughout the entire song. Make some general observations of any changes in L–R location. Use additional diagrams for specific sections of the song, as needed.
4. Now listen again to the first instrument/voice and locate the sound at a distance from you, as audience member. Take care to bring your attention to distance only; remember distance is timbral detail. Place the sound on the graph by moving the L–R location to an appropriate distance. Listen for changing distances between sections, and note those changes on the separate graphs.
5. In succeeding hearings, find the distance locations of the other two, three, or four sources.
6. Finally, listen again for how these relationships change between major sections of the song (i.e., between verse and chorus). Making additional diagrams might be necessary to keep track of all of the placements you find in your song.

The purpose of this exercise is to become sensitive to recognizing the left-to-right and the front-to-back dimensions of the sound stage. The observations here are intended to be general and preliminary. There will be much substantial information not captured in these diagrams; these materials will be explored later in much detail.

## Exercise 5.10

*Learning the Sound Quality of the Harmonic Series*

This exercise should be practiced in short sessions over time, and should be supplemented with writing out the harmonic series at a variety of pitch/frequency levels.

Tracks 1 and 2 on the companion website provide examples of the harmonic series. Learning the "sound quality" of the harmonic series will be valuable in learning to identify the spectral content of sounds.

1. Listen carefully to the harmonic series being constructed a single pitch at a time. Notice the spacing between the tones and the sequence of intervals of the series. Work to commit the sequence to memory—both the names of the intervals and the sound quality of the interval sequence should be learned.
2. Through 10 partials, practice recalling the sequence of intervals by writing them.
3. Next, practice playing the sequence on a keyboard, remembering the higher intervals do not align with the equal-tempered tuning of the keyboard.
4. Continue to listen to the harmonic series provided, and shift your attention to the perspective of the quality of the "chord" that is played after the individual pitches of the harmonic series. Listen carefully to the overall quality of this chord; then seek to identify each individual pitch within it.
5. Repeat this process to obtain confidence in spelling the harmonic series and recognizing its pitch succession and overall sound quality.
6. Repeat Steps 2 through 5 with a series through 16 partials. Practice until you are comfortable quickly conceptualizing each of the pitches in the harmonic series when you hear the chord at the end of tracks 1 and 2.
7. When some comfort in recognizing partials has been accomplished, listen to a single piano note being sustained. Take your focus to the perspective of individual harmonics. You will notice that partials change in loudness level and in pitch during the duration of the sound. Hold one, single harmonic in the center of your focus and follow its changes over the duration of the sound. Do this with several other partials during successive listenings to gain practice at hearing and focusing on spectral components.

The goal of this exercise is to bring the reader to recognize the "sound quality" (or timbre) of the "chord" that is created by the harmonic series, in its specific voicing (or spacing) of intervals and pitches. The spacing is important to recognize, as well as the overall quality, and the individual partials. This knowledge will be used as a template against which the reader can identify the partials of a sound's spectrum.

---

**Exercise 5.11**

*Time Judgment Exercise*

Using a digital delay unit and a recording of a high-pitch drum (such as the snare drum or high tom tracks on the companion website):

1. Route the signal to one loudspeaker and the processed signal to the other loudspeaker.
2. Delay the signal and listen to many repetitions of the sound while changing time increments. Listen until each delay increment is easily recognized. Try to move by the same time increment, such as 100 ms, during any given listening session.
3. As confidence is obtained in being able to accurately judge certain time units, move to other time units—both smaller and larger—and repeat the sequence in Step 2.
4. When you are confident in this ability, test your accuracy by routing both the direct and delayed signals to both loudspeakers (or to a single loudspeaker).
5. Continue to work through many repetitions of time increments in a systematic manner, comparing the qualities of the time relationships of each listening to previous and successive material. Using a logical sequence (a suggested pattern/sequence: 150 ms, 125 ms, 100 ms, 75 ms, 125 ms, 150 ms) that can be remembered will produce results faster.
6. Continue moving to smaller and smaller time units, until consistency has been achieved at being able to accurately judge time increments of 3 to 5 ms.

Tracks 26–33 on the companion website present time delays with a snare-drum sound. The delay times range from 50 ms down to 2 ms. Listening to those tracks will provide valuable support to learning this material and working through the above exercise.

---

# 6 A System for Evaluating Sound

The many different positions of the audio industry and their unique needs for sound evaluation create a need for a sound-evaluation system that can be readily transferred to a variety of contexts. It must easily yield meaningful and significant information to people of diverse backgrounds and job functions. The method must transfer between musical contexts and abstract, critical-listening applications.

The aspects of sound evaluated and shaped by people in audio cannot be described using our current vocabulary. No way to accurately talk about sound is available.

A system for evaluating sound will be presented over the next chapters. The system can be adapted to be useful to all people in the audio industry, and also for analyzing the technical and artistic qualities of recordings. The system will establish guidelines for talking about sound, by describing the physical dimensions of sound, as they have been perceived. Through this, these descriptions can be objective and accurate.

Outside language, scientific measurement, and music notation systems, sound has no written form. Being able to write down the qualities of sound will greatly assist people in audio in evaluating sound, describing sound, studying recording techniques and recordings, discussing sound with others, and keeping records. The system incorporates ways of graphing and notating sound's perceived parameters and the artistic elements. This will greatly aid the listener in achieving these goals, and in understanding, recognizing, or evaluating sound.

## System Overview

The system for sound evaluations was created to supply objective information on the listening experience. It seeks to give the listener the tools to define what is being heard. This will lead to a better understanding of the unique qualities of recorded/reproduced sound, better communication between people discussing sound, and enhanced control of the artistic aspects of making music recordings.

The elements of sound are all evaluated independently, using a variety of techniques. These isolated evaluations may then be related to evaluations of other elements, to observe how they interact. The standard *X-Y* graph used in so many different scientific contexts has been adapted for many of these evaluations, especially those that take place against time. Other evaluations use unique diagrams such as the sound stage.

The system seeks to describe and define the activities of the five physical dimensions of sound, as they are used in recording production/reproduction. The system examines the changes of state and value of those dimensions of sound, as they appear in perception and in artistic expression. Table 6.1 outlines how the various evaluations of the system relate to the aesthetic and artistic elements or perceived parameters of sound.

**Table 6.1** Evaluation techniques for the elements of sound

| ***Element of Sound*** | ***Evaluation Graphs and Processes*** |
|---|---|
| Time | Timeline of song; with structure, phrase, and text indications<br>Sound sources against timeline |
| Pitch | Melodic contour<br>Pitch area<br>Pitch density |
| Dynamics | Dynamic contour<br>Musical balance |
| Sound quality | Performance intensity<br>Sound-quality evaluation<br>Timbral balance |
| Spatial properties | Distance location<br>Stereo location and surround location<br>Sound stage<br>Perceived performance environment<br>Environmental characteristics of sources |

The system began with foundational skills in the last chapter, and now builds on them. Skill in recognizing musical materials and building timelines leads to the development of skills in pitch-related perception and dynamics. Interspersed throughout the system are exercises to build skills preparing the reader to undertake sound-quality evaluations. Finally, skills in recognizing spatial properties and environmental characteristics are addressed.

A complete listing of exercises appears after the table of contents. The exercises are arranged in what is usually the most effective order for skill development, and the reader is encouraged to work through the exercises in the presented sequence. The exercises appear at the end of the chapters that contain explanations of the material. Some skills will take longer to acquire than others, and the reader should be careful in assessing their progress. The assistance of someone who is already a skilled listener or teacher will at times be valuable. Any one exercise should be learned well before progressing too far ahead, though mastery of skills is not necessary before moving ahead. Indeed, mastery of some of these skills might take years, and the reader is encouraged to return to those exercises throughout an extended period of time.

Notating or writing down the characteristics of a sound can greatly assist the listener in understanding the sound. These notations are written representations of the sound. They can also be used for communicating with others about the sound, for evaluating the sound, for remembering the characteristics of the sound, and even for recreating the sound. While the reader will certainly not seek to perform a written evaluation of all sounds during their career, performing a detailed evaluation of a sound will provide information that might otherwise go unobserved, especially early in one's development. This process also will greatly increase accuracy level in sound evaluations, and speed at recognizing materials. Notating sound material in graph form will be used for finely developing the reader's perception and sound-evaluation skills. It will also provide the reader with a useful resource to assist in evaluating sound.

Creating the graphs of the following chapters will guide readers to place their focus on a specific element. This will bring them to discover and recognize the characteristics of the element in greater and greater detail as their skill develops. The reader will develop listening skills much more quickly by taking the time and effort to create the graphs than would happen otherwise. Further, the graphs will aid the listener in comparing one sound or mix to another, and by such comparisons learn more about the artists, producers/engineers, and the recordings. The graphs also allow the reader to "record" their observations for further study in the future; it will be interesting for the reader to observe how these graphs change in accuracy over a short amount of time and focused attention.

The system for evaluating sound has much in common with traditional forms of music-related ear training. Some of the skills learned by musicians will transfer to this process. An ability to take traditional music dictation will be beneficial to learning the process of evaluating sound, but is not required. Traditional listening skills emphasize pitch relationships in musical contexts. This comprises a very small part of our concerns about sound in audio. The skills of making time judgments and an awareness of activities in pitch, dynamics, and timbre will need to be developed much further than traditional approaches allow.

Many musicians start their studies by mimicking or repeating music on recordings. They often learn music by listening to recordings and trying to play back what was heard. Many people have even learned to play musical instruments almost solely by listening to recordings. Repeated listening to the same recording is something commonly done, whether to learn something or for enjoyment. This experience will be important in the many exercises that will develop skill in evaluating sound.

It is important that all information extracted from sound evaluations be objective. Audio professionals need to communicate about the characteristics of sound. Communications about how the sound makes them feel or whether or not they like the sound may come from clients or non-industry people and need to be interpreted into the audio professional's work activities, but the information is not relevant or valuable in the evaluation of sound.

The reader must learn never to use subjective impressions or descriptions of the sound event in the evaluation process. Expect that this may be difficult if you have described sounds in this way in the past. Try to remember, such impressions are unique to each individual and cannot be accurately communicated between individuals (they mean something different to all people); they do not contribute to an understanding and recognition of the characteristics of the sound event. Subjective impressions or descriptions do not contribute pertinent, meaningful information about the sound, and will not contribute to understanding the characteristics of sound. They have no place in the sound-evaluation process.

## Sound-Evaluation Sequence

The sound-evaluation process will follow a sequence of activities:

1. Perceiving an element of sound or an activity of material to be analyzed at a defined perspective
2. Identifying the material
3. Defining the material
4. Observing the characteristics present in the material, and/or between the material and its musical context.

The evaluation of sound begins with perceiving the sound event or sound object that comprises the musical material.

A sound event/object can be any sound, aspect of sound, or sequence(s) of sounds that can be recognized as forming a single unit. The event/object may be at any hierarchical level of musical context or of sound-quality analysis—from a distant perspective (such as the shape of the overall piece of music) to a detailed perspective with a focus on some nuance (such as a small change in the spectral content of timbre). The sound event/object must, however, have a specific and defined perspective. Each sound evaluation will have a single focus on a specific perspective. This perspective must be well defined in the listener's mind.

Next, the listener must recognize and, in some way, identify the sound object/event. This act is necessary to differentiate it from the material that precedes it, follows it, or is occurring simultaneously. The sound event/object will have dimensions within which it exists, and through which it is defined. It will have points in time where it begins and ends. It will be perceived within the musical/communications context or in isolation. The sound event/object will be defined through an understanding of the unique states and activities of the components of sound (artistic elements or perceived parameters) that comprise the sound.

Whatever the content of the sound object/event, the listener must perceive it as a single unit. This will be accomplished through identifying and recognizing the boundaries within which the sound event exists.

Third, the listener defines the sound event/object. This definition process will seek to compile information on the activities or unique qualities of the materials of the sound event/object that make it separate and distinct from the materials that preceded it, followed it, and/or that occurred simultaneously with it. Defining this activity (calculating what is happening to or in the various artistic elements or perceived parameters) is often the most challenging task of sound evaluation.

Any number of repeated hearings of the sound event/object will be needed to define all of the information it contains. This skill will need to be developed over time and with practice. As the listener's evaluation and listening skills improve, the number of hearings required will reduce significantly.

The final step is to seek to make sense of the information that accumulated in defining the sound event/object. In comparing the information of its components, meaningful observations can be made. The listener can compare materials recently experienced and materials that are well known to the listener. These other sound events/objects are evaluated for their relationship to the defined sound event/object. The listener will be looking for same, similar, and dissimilar states of activity and other attributes in the other known sound sources as those that defined the sound source, to assist in making pertinent observations about the sound event/object. The process is complete when the listener has compiled the information necessary to make the needed observations of the event and its context.

The sound-evaluation system is a clear set of routines. It is directly related to the listening experience. The routines follow the order:

1. Identifying perspective, with suitable alteration of the listener's sense of focus
2. Defining the boundaries of the sound event or sound object
3. Gathering detailed information on the material and activity
4. Making observations from the compiled information.

The perception of the individual sound event/object occurs at a specific perspective. The listener must consciously decide the level of detail at which the sound event/object will be evaluated—the perspective. The sound event/object and its component parts can then be identified and isolated from all other aspects of sound. The perspective at which the listener has identified the sound event/object becomes the reference level of the hierarchy, or framework, for the individual *X-Y* graph (discussed below).

Second, the sound event/object will be defined by its boundaries. The boundaries are the extreme levels of its elements and speed of their activities. The sound event/object is most often defined by (1) when it exists (its timeline), (2) its most significant sound elements (providing the event with its unique characteristics—the levels of the elements of sound that comprise the sound event), (3) the highest and lowest levels (boundaries) within those sound elements (the extremes of levels of activity to be mapped against the timeline, or that do not change over time—levels), and (4) the relationships of how the sound event's characteristics change over time (amount of change and rate of change of levels mapped against the timeline).

To define the sound event/object, we need to determine or identify:

1. The timeline: (a) beginning and ending points in time of the sound event/object, and (b) the suitable time increment to allow the activity of the components of sound that characterize the event to be clearly presented.
2. The elements of sound that hold the significant information that characterize the sound event/object. These are the components that supply the information that define its unique characteristics. These are the components that must be thoroughly evaluated to understand the content of the sound event/object.

3. The boundaries of the components/elements of sound being evaluated. These will be maximum and minimum values or levels found in each of the components of sound.
4. The speed at which the fastest change takes place. This will assist in defining the most suitable smallest time increment of the timeline.

The third step compiles detailed information on the material and activity. This will add detail to the fourth step.

Make a listing of the sources or items to be analyzed. Creating this list will draw the listener into the evaluation process quickly and directly. This should become one of the very first steps in collecting detailed information for the evaluation of sound.

Determine the precise levels of the components next, and place them accurately on the timeline. This is plotting or notating the activity of the component parts of the sound. The components of the sound will be closely evaluated, with as much detail as possible, to determine their precise levels throughout.

Most often the component of sound being evaluated changes over time and must have its levels related to a timeline. This information will be plotted on a two-dimensional graph (discussed below); this allows the information to be written/notated. This process involves all of the skills of taking music dictation. In fact, this process is a type of music dictation for some new and some previously ignored aspects of sound.

A written form of the sound will be created through the process of following Steps 1 through 3, above. Graphing the sound event/object makes it much easier to compile the information that will allow the listener to recognize and understand the characteristics of sound. When sound has been notated, it is possible for the listener to check previous observations for accuracy, to focus on particular portions of the sound event/object, and for the listener to be able to continue examining information on the sound out of real time.

The fourth and final activity in evaluating sound is examining the compiled information to make observations about the sound event. The type of observations made will vary considerably depending on context —such as either a music mix or a microphone technique.

For example, if the observations are being made concerning the functioning of a particular piece of audio equipment, the evaluations will center on the aspects of sound that the particular piece of equipment acts upon. Observations might be focused around the effectiveness of the piece of equipment, the integrity of the audio signal, any differences between the input and the output signals, and how the device acts on the various dimensions of sound.

The listener/evaluator will formulate questions and will use the information compiled in the steps above to answer those questions. Which questions to ask will be determined by their appropriateness to the purpose of the evaluation. The answers acquired through this process will be ones of substance and will be directly related to the sound event/object. The answers produced by this process will not produce subjective impressions or opinions.

The observations made in this final evaluation process need not be profound to be significant. Often the simplest, most obvious observations offer the most significant and important information concerning a sound event.

## Graphing the States and Activity of Sound Components

The traditional two-dimensional line graph is quickly understood and easily designed and used by most people. Therefore, it has been selected as the basis for notating the various artistic elements and perceived parameters that create sound events and sound objects.

The line graph will nearly always be used with time as the horizontal *(X)* axis. In this way, values of states (levels) of the component parts of the sound can be plotted with respect to time. This allows the sound to be observed from beginning to end at a glance, out of real time.

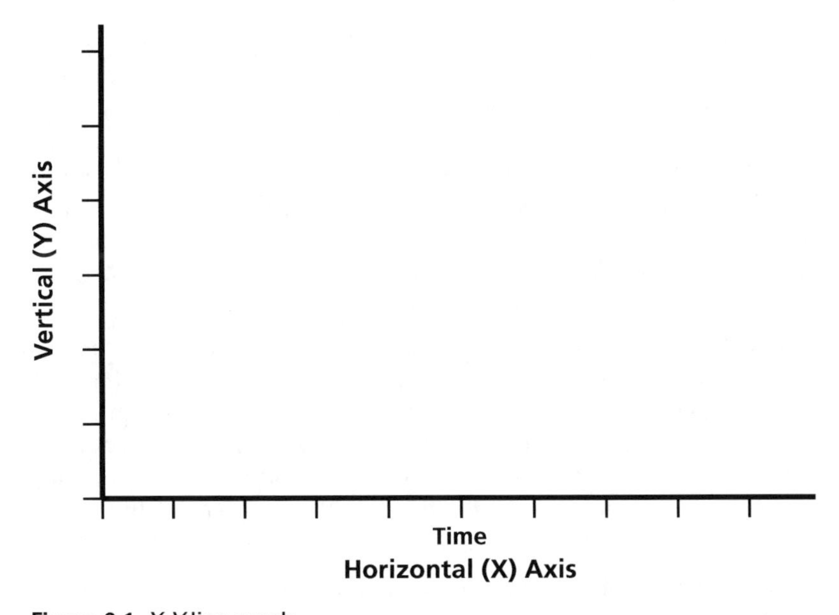


**Figure 6.1** *X-Y* line graph.

## Timeline

The length of the material that can be plotted on a single graph (occupying a single page) will be determined by the divisions of the time axis, or *timeline*. Events of great length (and little detail) may be plotted on a single graph, and events of short duration (and great detail) may be plotted on a single graph. A balance must be found in selecting the appropriate time increment for the timeline. For the graph to be of greatest benefit, the sound should be easily observed in its totality (from beginning to end), and the graph should have sufficient detail to be of use in observing the qualities of the material—this may require a graph spanning several pages.

Time increments will be selected for the *X*-axis that are appropriate for the sound. Time increments will take one of two forms: (1) units based on the second (millisecond, tenths of seconds, groups of seconds, etc.), and (2) units based on the metric grid (individual or subdivisions of pulses, measures, or groups of measures).

If the sound material is in a musical context, the metric grid will nearly always be the appropriate unit for the time axis. Remember, we judge time increments most accurately with the recurring pulse of the metric grid acting as a reference.

In general, when the sound evaluation utilizes the metric grid, a process of analytical listening is occurring. Critical-listening evaluations most often use real-time increments and not the metric grid. The difference is one of context and focus.

If the sound material being evaluated is not in a musical context, increments based on the second must be used. It will be common to use increments based on the second in the evaluation of timbre relationships (including sound quality and environmental characteristics). Envisioning a pulse of MM:60 (or an integer or a multiple thereof) will provide some reference to the listener in making time judgments without a metric grid, but this activity may not always be appropriate. It may distort the listener's perception of the material, and the reference may be unstable, as the listener's attention will rightly be focused elsewhere.

A stopwatch might assist in evaluating larger time units (to the tenths of seconds). The ability to judge time relationships can be developed. It is recommended the reader practice the time exercises found in Chapter 5 (Exercises 5.3, 5.4, and 5.11). They will allow the reader to refine their skills in accurately making time judgments, by learning to recognize the unique sound qualities (timbres) of various time units and tempo of clock time.

With practice, the listener will develop the ability to make accurate time judgments of a few milliseconds within the context of known, recognizable sound sources and materials. This skill will be invaluable in many of the advanced sound-evaluation tasks regularly performed by audio professionals.

The time unit used in any line graph will be that which is most appropriate for the sound event or sound object. The time increment selected must allow the graph to depict the example accurately. The smallest perceivable change in the components of sound being analyzed must be readily apparent, and yet as much material as possible should be contained on the single graph.

**Listen**

to tracks 26–33

for exercise in developing skills in judging (recognizing) short time units.

### Vertical *(Y)* Axis

The components of sound and the boundaries of levels and activities of those components are next determined. In the initial two stages of the sound-evaluation process, the listener determines those components of the sound event that provide it with its unique character. These components will be the ones most appropriately evaluated by plotting their activity on the line graph.

The component of the sound event to be evaluated will be placed on the vertical *(Y)* axis of the line graph. The second step of the sound-evaluation process (above) is now followed. The listener will now determine the maximum and minimum levels reached in the sound event, in each of the components of sound to be graphed. These maximum and minimum levels will be slightly exceeded when establishing the upper and lower boundaries of the *Y*-axis.

Exceeding these perceived boundaries allows for errors that may have been made during initial judgments of the boundaries and allows for greater visual clarity of the graph. Boundaries should be exceeded by 5 to 15 percent, depending on context of the material and the space available on the line graph.

Next, the minimum changes of activity and levels are determined. Through Step 3 of the sound-evaluation sequence described above, the listener will determine the smallest increment of level change for the components of the sound event or sound object.

This smallest increment of levels will serve as the reference in determining the correct division of the *Y*-axis. It is necessary for the *Y*-axis to be divided to allow the smallest value of the component of sound to be clearly represented, just as the *X* (time)-axis of the graph was divided previously so the fastest change of level would be clear.

The division of the vertical axis must allow the graph to depict the material accurately. The smallest significant change in the components of sound being evaluated must be immediately visible to the reader of the graph, and yet the vertical axis must not occupy so much space as to distort the material. The reader of the graph must be able to identify the overall shape of the activity, as well as the small details of the activity of the component the graph represents. A balance between limitations of space and clarity of presentation of the materials will always be sought.

### Multitiered Graphs

It is not always desirable for each component of the sound event/object to have a separate line graph. At times several components will be included on the same graph and plotted against the same timeline. *Multitier graphs* allow several components to be represented against the same timeline, and provide the advantage that these characteristics can be more easily related to one another. This will lead to an easier understanding of the sound's qualities.

In multitier graphs the vertical *(Y)* axis of the line graph is divided into segments. Each segment is dedicated to a different component of sound. Each segment will have its own boundaries and increments.

Plotting a number of components of sound against the same timeline not only makes efficient use of space on the graph, it also allows a number of the characteristics of the sound (perhaps the entire sound) to be viewed simultaneously.

The person reading the graph will be able to extract information more quickly from a multitier graph than from a series of individual graphs. In addition, when several components of sound appear against the same timeline, the states and activities of the various components of the sound event/object can be compared in ways that would be difficult (if not impossible) were these components separated.

Specific multitier graphs will be used for certain evaluations in later chapters. In those cases, the graphs will always appear in a predetermined format, and greatly assist evaluations of components such as sound quality, musical balance versus performance intensity, and environmental characteristics.

### Graphing Multiple Sound Sources

Multiple sound sources within the same tier will also need to be graphed. It is quite common for more than one aspect of a component to be taking place at any one time (such as the sound of harmonics and

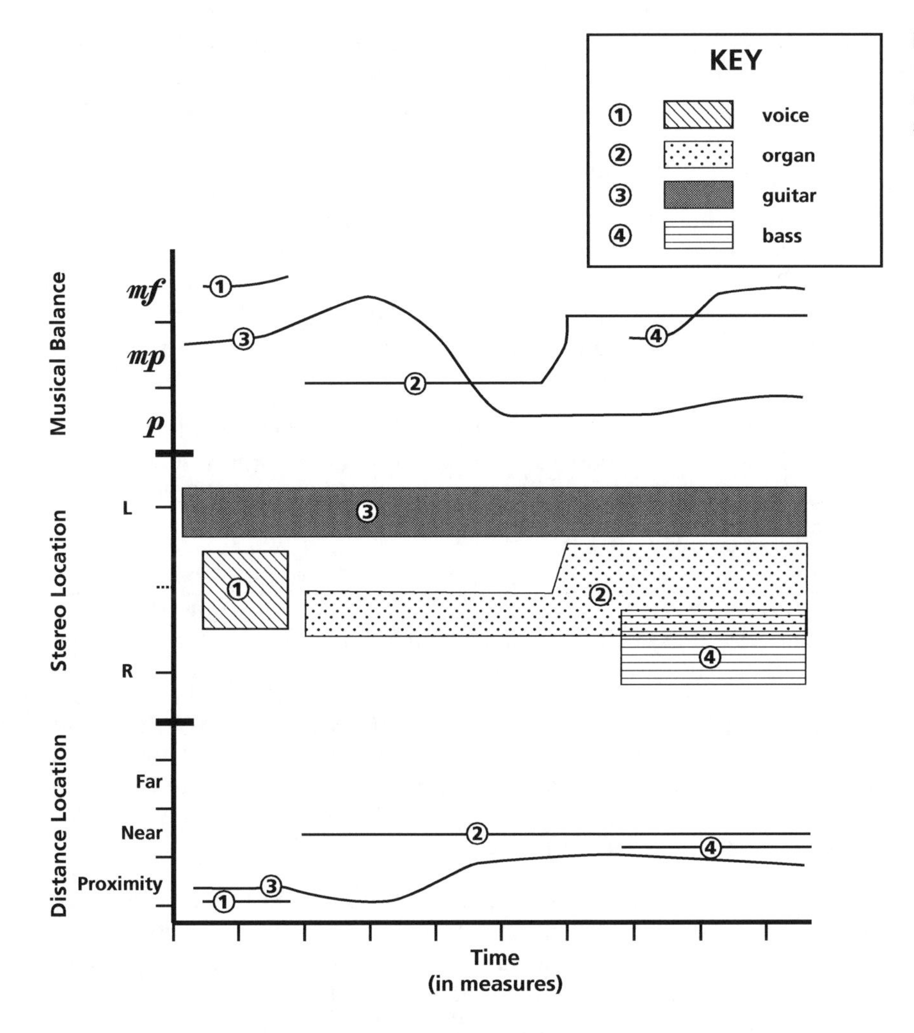


**Figure 6.2**
Multitier graph with multiple sound sources.

overtones within the spectrum of a sound). Without combining sources on the same tier, this activity would require a separate tier of a multitier graph for each sound source in each component of the sound event/object being evaluated. The line graph would quickly become large and unclear.

As long as the segment of the graph can remain clear, it is possible for any number of sound sources to appear on any graph. When more than one sound source appears against the same two axes, the activities of each sound source must be clearly differentiated from the others. Sound sources may be differentiated in a number of ways. Each of these ways may be useful, depending on the situation—what is available to the reader, the nature of the sound, or the context of the material.

The lines that denote each sound source may be labeled. Placing a number or the name of each sound source in or near the appropriate line on the graph can be clear in allowing easy identification (as in Figure 6.2). This type of differentiation is useful for graphs that contain relatively few sound sources.

Providing a different line configuration for each sound source is rarely a suitable way of differentiating a number of sources on the same tier of an *X-Y* graph. Combinations of dots and dashes, or the insertion of geometric shapes into the sourcelines, may be useful for differentiating sound sources on the same graph, but this usually distorts the material being plotted.

When sound sources are assigned lines of different colors, the graph can clearly display the largest number of sources. Only the number of easily recognized colors available then limits the number of sources that can be placed on the same graph.

The use of different colors has the further advantage of being able to define groups of sound sources by assigning a color to the group and assigning a different line configuration (combination of dots and dashes) to the individual sound source.

Using lines of various thicknesses to differentiate sound sources is not an option. This approach will obscure the information on the graph. Varying line thickness will cause the sound to visually appear to occupy an area of the vertical axis. This is a state that is only accurate for a few select components of sound.

The use of color is not always feasible, but it is the preferred method of placing a number of sound sources on the same graph. Using numbered lines for each sound source is the next most flexible and clear method of differentiating sound sources. Individual sound sources must always be easily distinguished on line graphs. Readily identifiable lines that have been precisely defined (by using a key, as described below) will ensure the clarity and usefulness of the graph.

The same sound sources may be depicted on a number of tiers of a multitier graph. In this case, care must be taken to define each sound source and to depict the sound sources in the same way on each tier (either by the same number or color). This will allow someone reading the graph to quickly and accurately determine the states and activities of all of the sound sources (or aspects of the sound sources) over time. A *key* of the sound sources plotted should be created to ensure this clarity.

A key lists all sound sources of the example and presents how they are represented on the line graph (see Figure 6.2). This listing of sound sources with their designations must be included in each line graph that contains more than one sound source.

The listing of sound sources is one of the first activities undertaken in the entire evaluation process. A listing of the sources (elements to be analyzed) will draw the listener into the evaluation process quickly and directly, and it should become one of the very first steps in evaluating sound.

## Plotting Sources against a Timeline

Plotting the individual sources against the timeline, without concern for levels and rates of activity of the component parts of a sound, will allow the listener to compile preliminary information on the material without getting overwhelmed by detail. This process is also an excellent first step in getting acquainted with the activity of writing down material that is being heard (the taking of dictation). It may become a common initial activity each time the listener undertakes a detailed evaluation of a sound event. A reliable ability to place sources against a timeline (and, of course, correctly identifying the timeline) will be assumed throughout the remainder of the book. This process will be repeated, at least conceptually, before almost all future exercises. This is also an excellent exercise for learning to identify all the sound sources (instruments and vocals) present in a mix—something that sounds simple but often proves otherwise.

Listing sound sources is an important first step in many evaluations and will need to be undertaken as a first step toward plotting sources against a timeline. It is important that all individual sources be identified and listed separately. These individual sources often act independently and were usually recorded with some degree of separation, giving the recordist an independent control of the sound that will be evaluated in many ways. Lists of sound sources should identify all independent vocal parts separately. Groups of background vocals presenting one musical idea should be listed as a single sound source; similarly, groups of stringed instruments playing one line or musical idea would also be labeled as "strings." Instruments should be listed by names. When more than one instrument of the same type is used, the instruments should be numbered either by order of appearance or by range, with the highest instrument usually the lowest number. Sounds should not be listed by descriptive terms (lush guitar, happy flute, etc.). If the listener is at a loss as to what to call a sound, using terms such as "unknown 1" would be appropriate until the sound is recognized. Performing a sound-quality evaluation (even a general one) would allow the listener to further define the sound as "unknown 1, with long final decay." When the listener does not know the names of instruments, or the sound sources are very unique, listing sound sources must be undertaken with care, and will take effort. For example, it would be a great undertaking to list all sound sources from "Tomorrow Never Knows" by The Beatles. Many of the song's sound sources would need to be described in terms that addressed the sound source or the sound quality.

In working through many of the exercises and graphs of the following chapters, the reader will frequently (1) create a listing of the sound sources in the example, and (2) create a timeline of the event. By adding a third step of plotting the listed sound sources against the timeline, they will be able to focus more intently on the material being graphed. The reader should practice Exercise 6.1 (at the end of the chapter) and become comfortable with the process. The recordist must be able to quickly recognize sound sources and focus on the activity of each.

Figure 6.3 provides an example of sound sources plotted against a timeline. The listener will be able to follow the figure while listening to The Beatles' "She Said She Said." Whenever a sound source (an instrument or voice) is sounding, a line is drawn across that half measure against the timeline. As an additional activity, complete the graph by plotting the presence of the high hat part against the timeline. Listen carefully to the recording, following each sound source in a separate listening to the entire section graphed. Then, scan the texture for the possibility of additional sound sources and note any you discover; notice how you are now listening at a different level of perspective.

## Notating Sounds in Snapshots of Time

Some important components of sound might not change over time. While they are static, their status may well be a very important

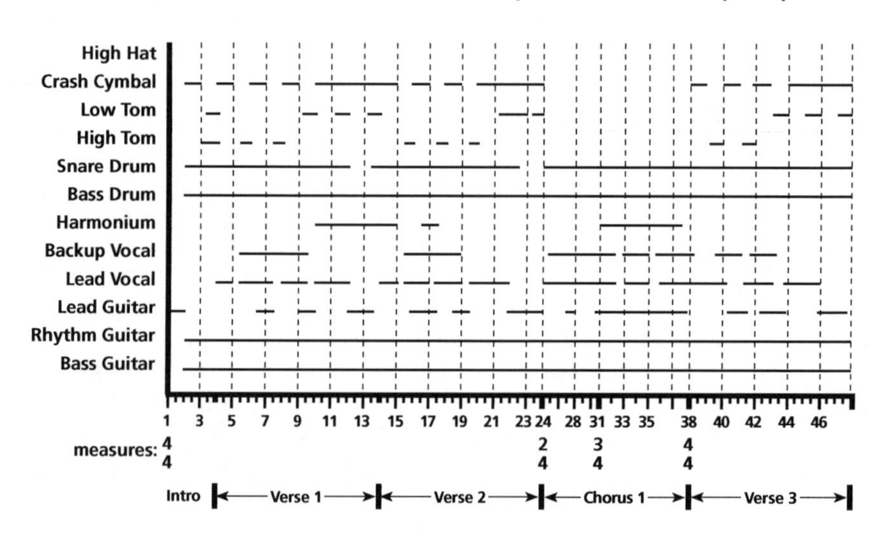


**Figure 6.3** Sound sources placed against timeline, The Beatles' "She Said She Said."

characteristic of the sound event/object. These sound components can be evaluated without a timeline and understood as elements that remain unchanged throughout a defined time period. Sounds are examined for qualities of their component parts that remain unchanged over the time period, as if examining a snapshot of the sound's existence over the time period.

The location of instruments on a sound stage is one such possibility, as explored in Exercise 5.9. While it is very possible for sound sources to change lateral location or distance location, sound sources often remain in the same location for extended periods of time (entire verses or choruses, for example). It is also quite common for some sound sources to remain in the same location throughout an entire song. It is important to note: Changes in the mix can and do cause changes to the perceived sound stage even when locations are not directly altered; these are often subtle and very important to the quality of the recording.

The graphs used for plotting the components can also be used to show the static, unchanging states. The *X*-axis can be used to provide a place for each sound source that is present, instead of a timeline. Many sources can then be placed against the same vertical axis. In addition, the diagrams of the perceived performance environment can also be used to show placement of sound sources on the sound stage, when appropriate (as will be discussed in Chapters 10 and 11).

## Summary

Graphs will often need to be supplemented by verbal descriptions to complete evaluations of sound events or sound objects. Simply graphing a sound does not complete the evaluation.

The contents of the graphs need to be reviewed and then described through observations of the content and activity of the materials. In all instances, the language and concepts used to define and describe the sound must be completely objective in nature. All descriptions refer to the actual levels, and any changes of levels, of the components of the sound. The descriptions lead to overall observations about the qualities of the sound itself and its relationships to the other events/objects and the entire recording.

The listener must never use subjective impressions or descriptions of the sound in the evaluation. Such impressions are unique to each individual, cannot be accurately communicated between individuals (they mean something different to all people), and do not contribute to an understanding and recognition of the characteristics of the sound. Subjective impressions or descriptions do not contribute pertinent, meaningful information about the sound for the audio professional.

Part Three will take the reader through the individual artistic elements and perceived parameters of recordings, working through this system for evaluating sound. The sequence of activities outlined above, and the process of notating (graphing) sound, will be core activities of those chapters.

In order to benefit the most from those chapters, the reader should take the time to work carefully through the examples and exercises. The many nuances of sound contained in recordings will gradually become more apparent, and the reader will gradually acquire an ability to talk about sound in ways that have clear meaning to others. Further, the reader will find they are better able to understand their own recordings, and better able to craft recording projects creatively.

Learning the materials of the exercises and going through example recordings carefully will guide the reader through the process of discovering and learning about the sound qualities of recording. Very often in beginning studies, learning the techniques of seeking information is more important than finding the "right" answer. Learning the process and sharpening the skills of observation will carry you throughout your career, and far outweigh the goal of finding "the answer."

## Exercise

The following exercise should be practiced until you are comfortable with this material covered. This exercise will save the reader much time if created before doing the exercises in Part Three.

### Exercise 6.1

*Exercise for Plotting the Presence of Sound Sources against the Timeline*

Graph the first few major sections (verse, chorus, etc.) of a piece of popular music, using the following steps:

1. Compile a list of all the sound sources of the song. Individual percussion sounds and vocal parts should be listed separately.
2. Create a suitable timeline by:

    a. determining the pulse (metric grid) of the song
    b. grouping the pulses into measures (weak and strong beats)
    c. plotting those measures on the horizontal *(X)* axis in increments that clearly show the material being graphed.

3. Plot the individual sound sources against the timeline. Each sound will have its own location on the vertical axis, making it unnecessary to make distinctions between the lines of each sound source on the graph. When an instrument is playing, place a line for that instrument in the appropriate location against the timeline. If the instrument is playing in the measure, extend the line through the entire measure. Alternatively, you can change this resolution to make note of instruments appearing every half measure. If still more detail is sought, a smaller time increment could be used.
4. After several days return to the graph. Check the timeline for accuracy and listen several times again for sound sources. It is not unusual for sound sources to appear in the music that were not heard in previous hearings.
5. Listen several more times to check the entrances and exits of the instruments against the timeline.

The reader will use this skill at the beginning of many exercises in the following chapters. Learning to do this well now will prove very helpful. Further, identifying sound sources and what they are doing is a very important part of tracking and mixing. This exercise will also improve a skill the reader will use in production work.

# Part Three

---

# Understanding the Mix:

## Developing Listening and Sound-Evaluation Skills

# 7 Evaluating Pitch in Audio and Music Recordings

Pitch relationships shape musical materials more than the other artistic elements. This is obvious when we consider melody as a succession of pitch relationships, and chords/harmony as simultaneously sounding pitch relationships; the main melody of a song can make a lasting impression on the listener.

With the exception of percussion instruments, nearly all musical instruments were specifically designed to produce many precise variations in pitch, far fewer variations in timbre, and most have a continuously variable range of loudness. Most Western music places great emphasis on pitch information for the communication of the musical message.

Pitch is the central artistic element of most music, with the other artistic elements most often supporting the activities occurring in pitch relationships. Like all elements, pitch levels and relationships are present at all levels of perspective, and have the potential to be the most significant element, or play a central supportive role. At various levels of perspective it will present itself differently. A lower level of perspective might bring a simple bass line pattern repeating; a higher level might contain harmonies and melodic ideas; higher still we perceive phrases containing the melodies of verses and choruses, and their unfolding relationships. All of these are commonly explored in traditional music theory studies.

In evaluating pitch in audio and music recordings, the recordist will work well beyond traditional concepts of pitch as melody and harmony. An acute sense of pitch will bring the recording professional to recognize pitch levels, and to identify pitch areas and frequency bands. We have begun developing this skill with our initial work in pitch reference and recognizing pitch/frequency levels.

Pitch evaluation is used in both analytical and critical listening. In critical listening, perceived pitch is often transferred into frequency calculations. The analytical-listening process most often relates pitch relationships to the musical qualities, ideas, and message.

The reader will be developing ways of analyzing musical sounds that will also develop critical-listening skills. The same sensations of sound are perceived in each process. How the sound is evaluated (critically versus analytically) will be the difference.

The pitch-analysis concepts presented in this chapter are of particular importance to sound recordings and for developing the skills of the recording professional. The information gathered will help one to understand the piece of music (analytical listening), and to evaluate sound quality (critical listening), depending on how the information is applied. Concepts of pitch that the reader has had much experience with are built upon in this chapter to introduce new experiences. We will begin by transferring traditional melodic dictation onto pitch registers and a metric grid timeline, to use a skill the reader may have already developed (or used slightly differently by learning songs from listening to recordings) to assist their first attempts at placing sounds against a timeline and against pitch registers.

Next we will transform recognizing harmonies into identifying pitch areas, and making general observations of the placement and content of pitch areas, and then relationships of pitch areas. Pitch-area evaluation is an important first step in understanding and evaluating sound quality, and forming objective descriptions of what is happening within an isolated sound.

This chapter also presents the processes for evaluating sounds, writing down observations during listening sessions, and making a timeline. These are important for many activities in upcoming chapters. The reader is encouraged to work through exercises and examples with attention to detail of the information presented and to establish a thorough method of working.

### Relevant Fundamental Skills

The reader is encouraged to return to Chapter 5's internal pitch reference exercise (Exercise 5.1) to ensure the development of that skill is understood, and progress is being made. This material should be refreshed now, to make the materials of this chapter more accessible and relevant. The reader's internal pitch reference will be greatly enhanced if they engage this skill briefly, immediately before attempting the pitch-related exercises that follow.

Next revisit the concepts for recognizing pitch levels. Work through those materials and exercises of Chapter 5. Using the boundaries between pitch registers as reference levels will greatly assist in refining the reader's skills in identifying the placements of pitch areas.

These fundamental skills will allow the reader to benefit most deeply from the following discussions and evaluations. With an internalized pitch reference and an ability to even generally recognize pitch levels, the reader will be able to explore and learn the concepts and skills of this chapter most easily and meaningfully.

## Analytical Systems

Numerous analytical systems have been devised to explain pitch relationships in music. These systems are made up of evaluation criteria that vary considerably between systems and are more or less specific to certain types of music. Generally, then, these systems can only be useful for examining certain styles or types of music. Any single analytical system may or may not yield information pertinent to the music being evaluated.

The recordist will need to recognize and apply the appropriate analytical system to study the pitch relationships of a particular piece of music, if such an evaluation is expected of them. This is rarely required. Many recordists do, however, innately sense the relationships that are explained through musical analysis. Understanding pitch relationships, and especially harmonic progressions, can often greatly assist the recording professional in crafting a recording.

Information about the artistic element of pitch levels and relationships will be related to (1) the relative dominance of certain pitches (which relate to tonal/key centers and chord progressions), (2) the relative register placement of pitch levels and patterns (related to arranging and orchestration; these will be later discussed as pitch density and timbral balance), or (3) pitch relationships: patterns of successive intervals (motives and melodies), relationships of those patterns, and patterns and relationships of simultaneous intervals (chords and progressions).

Study of the many systems used to analyze pitch relationships is well beyond the scope of this book. Theories about music attempt to explain the analytical-listening experience, and to extract basic information about the materials, structure, and form of the music. The recordist can use such insights to control and craft the sound qualities of the recording so they will support and directly enhance how the recording process can best deliver musical ideas to the listener.

## Melodic Contour

Graphing melodic contour will assist the recordist in a number of ways. First, at the early stages of the listening-skill development, graphing melodic contour will help develop skills in placing pitch/frequency changes and levels against time. These same skills will later be used in the much more detailed (and initially more difficult) task of evaluating pitch-related information in timbre/sound quality and environmental characteristics evaluations.

Learning to plot melodic contours against a timeline will be productive in developing the skills of recognizing pitch levels, of perceiving metric units and rhythm of time, and of mapping pitch contours. These skills directly transfer into many of the listening functions of the audio professional.

Second, recognizing the contour (or shape) of the melodic line is important to understanding certain pieces of music or musical ideas. In certain pieces of music, the contour of a melodic line is perceived instead of the individual intervals. When the melodic line is performed very rapidly (as is easily accomplished with technology), the perception of the line fuses into an outline or shape. The series of intervals that comprise the melodic line are not perceived. The contour of the melodic line is instead perceived.

The melodic contour graph allows the contour of the musical idea to be recognized and evaluated for its unique qualities.

### Graphing Material against a Timeline

Nearly all exercises in the book graph material against a timeline. The sequence for establishing a timeline will now be examined. This should be studied carefully, as the recording professional is continually engaged in listening to material to notice changes in sound over time, and how the changes relate to time. This process will define the activities of any artistic element against a timeline. This sequence of activities will be applied to both musical (analytical) and critical-listening contexts. It will be only slightly modified for each exercise in the following chapters, and the reader should become familiar with the order of activity:

1. During the first hearing(s) of the material, focus listening activity to establish the length of the timeline. In analytical listening we will be organizing the timeline according to the metric grid: measures and beats per measure or portions of beats per measure, depending on the resolution needed to accurately present the material.
2. Check the timeline for accuracy and make any alterations. Establish a complete list of sound sources (instruments and voices), and then place those sound sources against the completed timeline.

3. Next, observe the activity of the material being graphed for boundaries of levels (here, the highest and lowest pitches of the example, and the smallest changes between pitch levels) and speed of activity (noting the fastest changes of levels). The boundary of speed will establish the smallest time unit required to clearly show the smallest significant change of the material (melodic contour). The boundaries of levels of activity will establish the smallest increment of the *Y*-axis required to plot the smallest change of the material. This step will establish the perspective of the graph, which is the graph's level of detail. The *Y*-axis should allow some space above and below the highest or lowest level for the material to be clearly observed, and for the possibility of adding new material or corrections in that area of the graph.
4. Begin plotting the activity of the material (melodic contour) on the graph. First, identify prominent points of reference within the activity; these reference points might be the highest or lowest levels, the beginning and ending levels, points immediately after or before silences, sustained pitch levels, repeated sounds, or any other points that stand out from the remainder of the activity; place these levels correctly against the timeline. Use the points of reference to calculate the activity of the preceding and following material, filling in the graph around these points of reference.
5. To complete the plotting of the activity of the artistic element, alternate focus on the contour, speed, and amounts of level changes.
6. The evaluation is complete when the smallest significant detail has been heard, understood, and added to the graph.

## Listening and Writing

Many hearings of the example will be involved for each of the steps above. Each listening should seek specific new information and should confirm what has already been noticed about the material. Before listening to the material, the listener must be prepared to extract certain information. Listeners must have a clear idea of what they will be listening to and/or for, and work to keep from being distracted. Attention should be focused at a specific level of perspective and on a specific task. Beginning listeners should focus on either confirming their previous observations or checking what they have already accomplished; or they can be receptive to new discoveries about the example, listening to identify other aspects of the example or sound. As skill develops, the listener may check previous observations while seeking new information, all the time keeping a clear intention of what they are seeking to accomplish.

Listening to only small, specific portions of the example may assist certain observations at certain points in the evaluation process. In these situations, the listener should intersperse listenings to the entire example with the rehearings of small sections, to be certain consistency is being maintained throughout the evaluation and to maintain proper perspective.

The reader should very rarely write observations while listening. Instead, the reader should concentrate on the material and attempt to memorize their observations. This activity will develop auditory memory and will ultimately greatly reduce the number of hearings required to evaluate sounds, and improve skill. When the recording/sound has stopped, the reader should recall what was heard and only then write. If the material is not clearly remembered, listen again, perhaps to a shorter segment.

One must first recognize what has been heard before it can be written down. This process is about recognizing what is heard and making a written record of the experience. The reader should try to make the sequence "hear, recognize, remember, write" automatic.

## Melodic Contour Graph

The shape of the melodic line is plotted on a *melodic contour graph*. The graph is composed of:

1. Pitch-area register designations for the *Y*-axis, using the registers needed to clearly show the activity of the musical example
2. The *X*-axis of the graph is dedicated to a timeline divided into appropriate time units of a metric grid or real time (depending on the material and context)
3. Each sound source is plotted as a single line against the two axes (the melodic contour is the actual shape of this line)

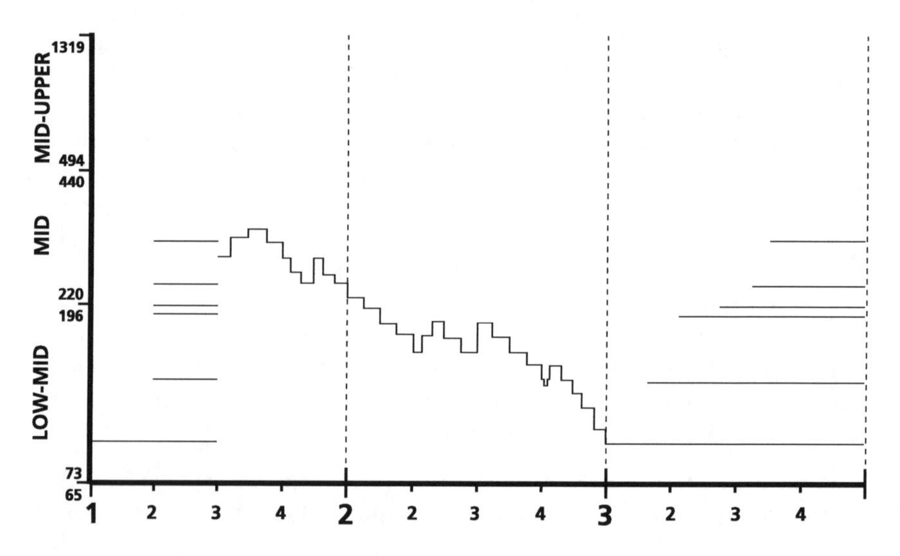


**Figure 7.1** Melodic contour graph of the guitar sound in The Beatles' "Wild Honey Pie," beginning at 0:53.

4. If more than one sound source is plotted on the same graph, a key should be devised and included with the graph to identify the lines with the sound sources.

Consider the melodic contour of the classical guitar line from *The Beatles* (the "White Album"), that concludes "Wild Honey Pie" and provides transitional material to "The Continuing Story of Bungalow Bill" (Figure 7.1). The intervals of the melodic line pass too quickly to be heard individually. Instead, this type of fast melodic gesture fuses into a shape or a contour. As an exercise, check Figure 7.1 for accuracy of shape, detail, and register placement. Determine the tempo of the passage and attempt to identify some of the graphed pitch/frequency levels. Note that sustained pitches remain at the same level; silences bring an absence of a line; chords begin and end the example, and are clearly shown as several lines occurring simultaneously against the timeline.

The Melodic Contour Analysis Exercise (Exercise 7.1) will refine this skill. The reader is encouraged to work through several additional recorded examples to become comfortable with this activity. This graph is an important bridge between the graphs and the processes of this system and traditional music dictation and traditional music notation.

Melodic gestures of this type are very common in music. They commonly appear, from heavy metal guitar solos, to eighteenth- and nineteenth-century keyboard music (especially the works of Chopin, Liszt, and works in the Rococo style). Transcribing fast melodic passages into melodic contour graphs will provide the reader with practice in developing the following skills required of recordists:

- Pitch recognition (estimation)
- Placing pitch/frequency levels correctly into pitch registers
- Recognizing and calculating pitch changes
- Placing pitch changes against a timeline.

These skills will be used and further developed in evaluating sound quality and environmental characteristics in later chapters.

## Pitch-Area and Frequency-Band Recognition

Recordists must often bring their attention to a specific range of frequencies, or a frequency band, to identify some aspect of sound quality or equipment performance. This section will present a rough equivalent of this activity, but within musical contexts, that can be used to develop this skill and more.

Many percussion-related sounds occupy a *pitch area*, or a range of frequencies, but not a specific pitch. These sounds have noise-like qualities and waveforms that are nonperiodic in many respects, but still contain some pitch quality. This allows us to recognize some drums as being "higher" or "lower" in pitch than others.

These sounds are perceived as existing in an area of pitch, existing between two pitch/frequency boundaries. The boundaries may, at times, be unstable and changing in pitch and/or dynamic level. One area of pitches will dominate a sound, and secondary pitch area(s) will typically also be present. Sounds often exhibit some harmonic relationships (often inexact) between the primary and secondary pitch areas; the primary pitch area will provide a sense of a dominant frequency/pitch "area" and not a precise pitch level.

These sounds can be evaluated and defined by:

1. The width and register of the dominant, or primary pitch area (the distance between the two boundary pitches)
2. The density (or the amount and closeness of spacing) of pitch information within the pitch area
3. The presence of secondary pitch areas (and their boundaries and densities)
4. The dynamic relationships of the primary and any secondary pitch areas.

Most sounds will have several separated pitch areas. The different pitch areas of the sound will be at different dynamic levels and have different densities (the amount and closeness of spacing of the pitch information within the pitch areas). One pitch area will dominate the sound and be the primary pitch area (similar to a fundamental frequency). The other pitch areas will be secondary pitch areas. These are components of the sound's spectrum. Some areas dominate, and others will be softer, or less dense, or otherwise less prominent.

The bandwidth of pitch areas (the distance between the lowest and highest frequencies that create the boundaries of the pitch area) is a region of frequency/pitch information and activity that we recognize as being a single, unified quality with density (amount of information) and loudness (has a reasonably uniform level). The reader should focus on the perspective of the spectrum of the sound source, listening for areas of density, the boundaries of those areas, and the gaps between them. In these initial studies, we will calculate density in general terms on a rough scale between "very dense" (like white noise) to "very sparse" (as a sine wave). See Exercise 7.2, Pitch-Area Bandwidth and Density Exercise, now to engage this material.

Evaluating the pitch areas of several percussion sounds will develop a number of important listening skills. Frequency and pitch estimation (recognition) will be refined, and the listener will take those skills to a more detailed level of perspective. The focus will now be on identifying pitch/frequency levels within the spectrum of the source—using sound sources that allow us to approach this information in a more noticeable and higher level of perspective.

This is a first step toward identifying subtle aspects of sound quality and spectrum (that will be greatly refined in Chapter 9). Further, this study will be accomplished without the added tasks of a timeline. This is significant in simplifying this initial work in recognizing spectrum. The graph will sum spectral information throughout the sound's duration. This spectral information will be recognized in general observations of the density of the pitch areas, and their registers and boundaries. The reader will also make some general observations on dynamic levels. This skill will also be greatly refined later.

**Listen**

to tracks 19–25

for isolated drum and cymbal sounds that can be used as source material for pitch-area evaluations.

It is likely that pitch-area analysis will be unlike anything the reader has done in the past. Few reference points will be available for the listener to draw upon, other than the pitch-estimation skills acquired above. Difficulties and frustrations are to be expected, along with great satisfaction with acquiring a very useful skill that will be used and improved throughout one's audio career.

To perform an analysis of pitch areas, sounds are plotted on a *pitch area analysis graph*. The graph incorporates:

1. The register designations for the *Y*-axis
2. A space on the *X*-axis of the graph that is dedicated to each sound instead of the passage of time (since this evaluation sums all information during the sound's duration)
3. The pitch areas, which are boxed off by upper and lower boundaries of their bandwidths, in relation to these two axes
4. The density of each pitch area designated by a number within each box that relates to a relative scale from very dense to very sparse
5. Assigning a number to the relative dynamic levels of pitch areas, especially important for identifying the predominant (primary) pitch area (dynamics will receive detailed coverage in the next chapter).

Figure 7.2 presents the pitch areas of the prominent bass drum sound found in The Beatles' "Come Together" (*1* version) at 0:34.

The percussion sound is composed of four pitch areas. The primary pitch area is the second from the bottom. It is moderately dense and is the loudest and dominant area. The density of the lowest pitch area is a bit more dense, but considerably softer. The area between 167 and 265 Hz is moderately dense and a bit louder than the lowest area,

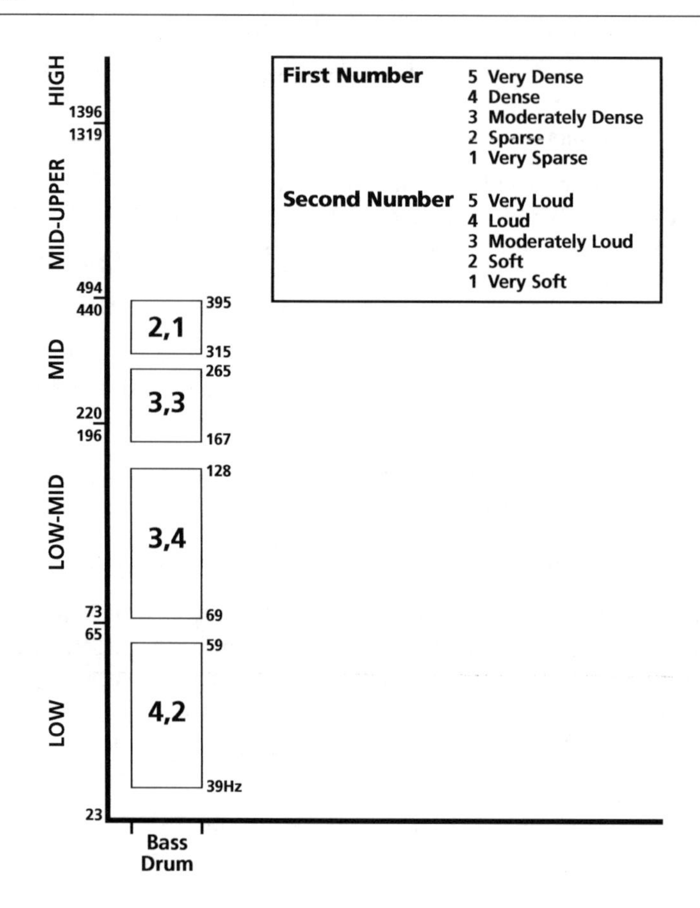


**Figure 7.2** Pitch-area analysis of the bass drum sound from The Beatles' "Come Together" *(1).*

and not as loud as the primary pitch area. The highest pitch area (approximately 315–395 Hz) has a rather sparse density and is at the lowest loudness level of all four pitch areas. It is interesting to note that the lower boundaries of the four pitch areas are nearly whole number multiples and harmonically related, but far enough away to create strong noise elements. These types of relationships are common for drum resonances and head vibrations.

The pitch-area graph and the objective descriptions of density and dynamic relationships of the pitch areas provide much useful and universally perceived information about the sound. Meaningful communication about this sound is possible with this information.

Next, the Pitch-Area Analysis Exercise (Exercise 7.3 at the end of the chapter) should be performed until the material is learned. The reader is encouraged to evaluate drums and cymbals of a variety of sizes directly from music recordings, after working through some isolated percussion sounds (such as the ones found on the companion website).

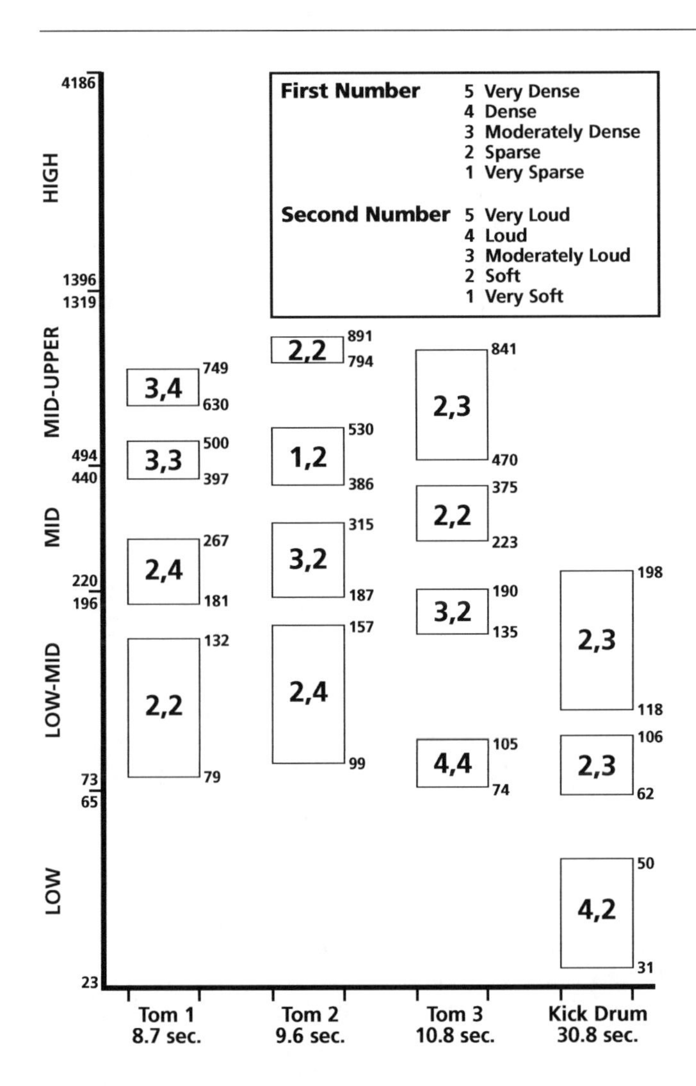


**Figure 7.3**
Drum sounds from The Beatles' "The End."

A number of percussion sounds from the drum solo of The Beatles' "The End" appear in Figure 7.3. The reader should examine the dynamic relationships of the pitch areas and observe their densities. Study the example carefully, seeking to identify the boundaries of the pitch areas, and confirm the density information presented. Once the pitch areas can be identified, the reader can observe the dynamic relationships of the pitch areas.

As an exercise, determine the approximate pitches of the frequencies of the pitch areas' boundaries. Notice any harmonic relationships, even if approximate.

The reader will be able to apply the skills and concepts of pitch area to the recognition of *frequency bands* in critical-listening applications.

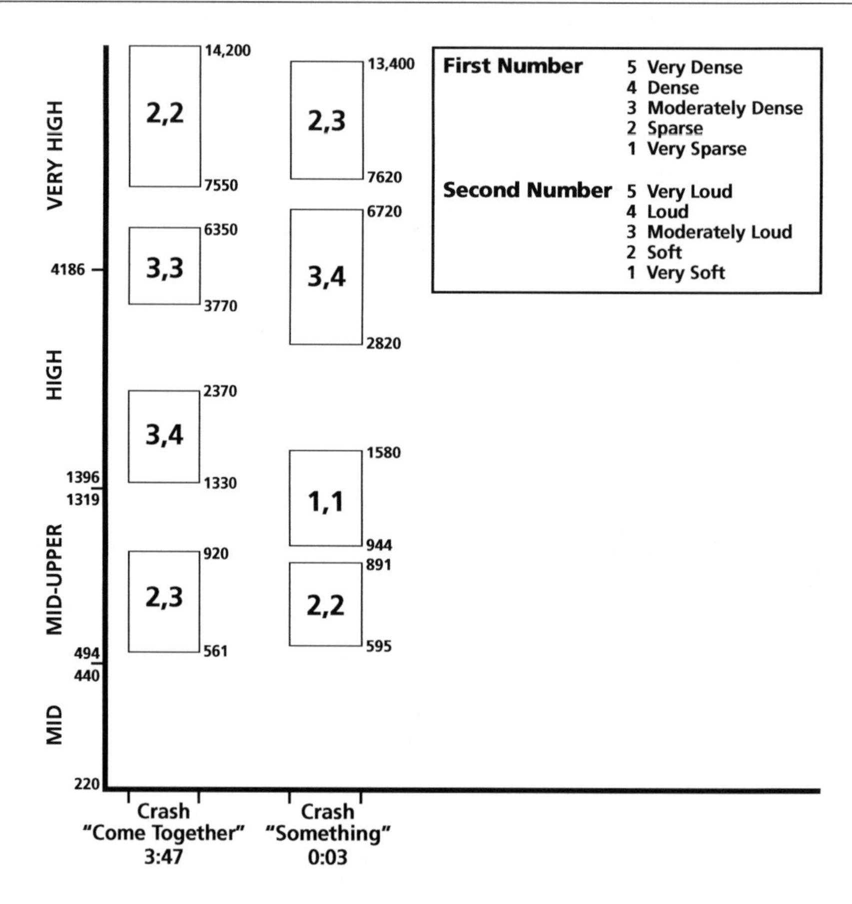


**Figure 7.4**
Crash cymbal sounds from The Beatles' "Come Together" *(1)* and "Something" *(1)*.

The information for evaluating the states and changes of frequency levels is readily deduced from perceived pitch information. The skills gained through the recognition of pitch areas and frequency bands will later be directly applied to the evaluation of timbre and sound quality. In fact, these pitch area analyses are rudimentary timbre (sound quality) analyses.

A common critical-listening situation is to compare two sounds. Figure 7.4 compares the crash cymbal from two different songs. Notice the characteristics of the sounds through careful listening to identify pitch areas and their densities. From the graph we can see the areas are similar in location and nearly identical in their densities. Some areas have markedly different relative dynamic levels.

As skill increases, the listener will gradually come to understand and recognize slight differences present in the spectra of similar sounds. Eventually the reader will arrive at the point where evaluations of the high hat and crash cymbal sounds from four different releases of the song "Let It Be" can be compared (listed in the Discography). The reader is encouraged to listen carefully to these sounds. The sounds change within each song and are slightly (to markedly) different in each recording. The reader will notice many striking differences and perhaps a few of the many subtleties. These differences will gradually seem more and more significant. Ultimately, the reader will recognize the sounds as very different and will perceive the details of their sound qualities.

### Pitch Area and Pitch Density

Pitch-area evaluation of a single percussion sound leads us to consider the similar concept of pitch density for all instruments. Pitch density is the range of pitches spanned by a musical idea plus the spectrum of the sound source playing it. In effect, it is the pitch area of the musical idea. It is understood as the contour of pitches of the musical idea plus the characteristic and dominant spectral information of the instrument/voice presenting the idea. This connection between pitch area as the spectrum of a sound and the musical material being presented is important to many activities in mixing and to understanding the frequency content of recordings.

Figure 7.5 presents four percussion sounds from *Abbey Road*'s "Here Comes the Sun." Listen to this recording and identify the pitch areas of these instruments. In your next hearings, notice how the instruments compare to one another, and to others, in terms of their pitch areas—or spectral content. You will be listening at the level of perspective of the individual sound sources when identifying the pitch areas of the instruments, then at one level higher where you can compare sounds and perceive each source as being equal.

The final step is hearing all of the pitch densities of all of the sounds present, and being able to notice their content and relationships. This is at the highest level of perspective, and the overall texture of the recording. This highest level where all pitch densities come together is timbral balance.

Listen again to the recording, and notice:

- the ways the percussion sounds blend with others that occupy similar pitch areas
- the way sounds are more easily distinguished when they occupy pitch areas that are different from others
- that certain pitch registers of the overall texture have more information than other registers
- that the amount of activity in certain pitch registers, within the overall texture, changes from one moment to the next
- that different "left" to "right" locations of the stereo field have different pitch areas present and absent.

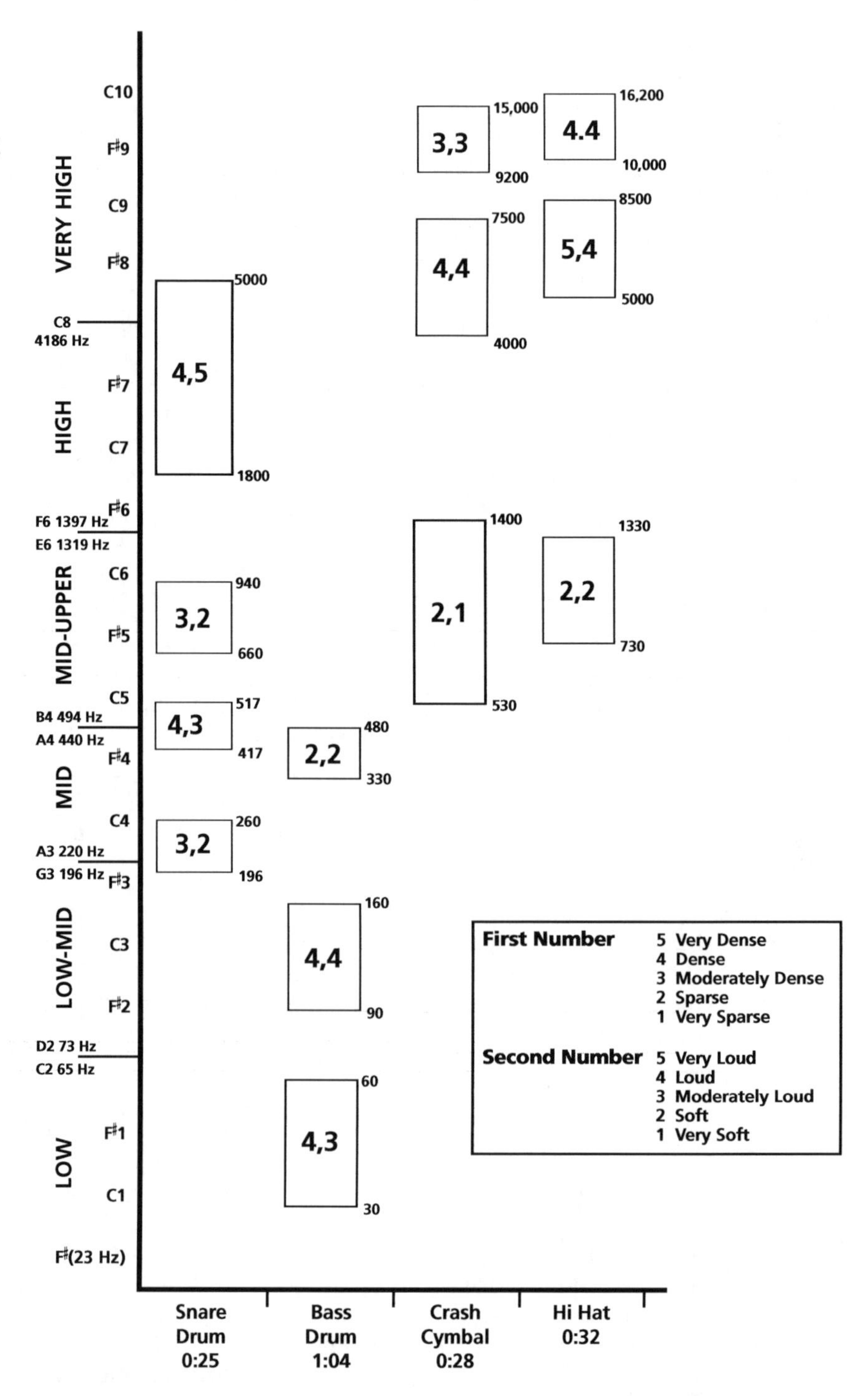


**Figure 7.5**
Four percussion sounds from The Beatles' "Here Comes the Sun," (*Abbey Road,* 1969).

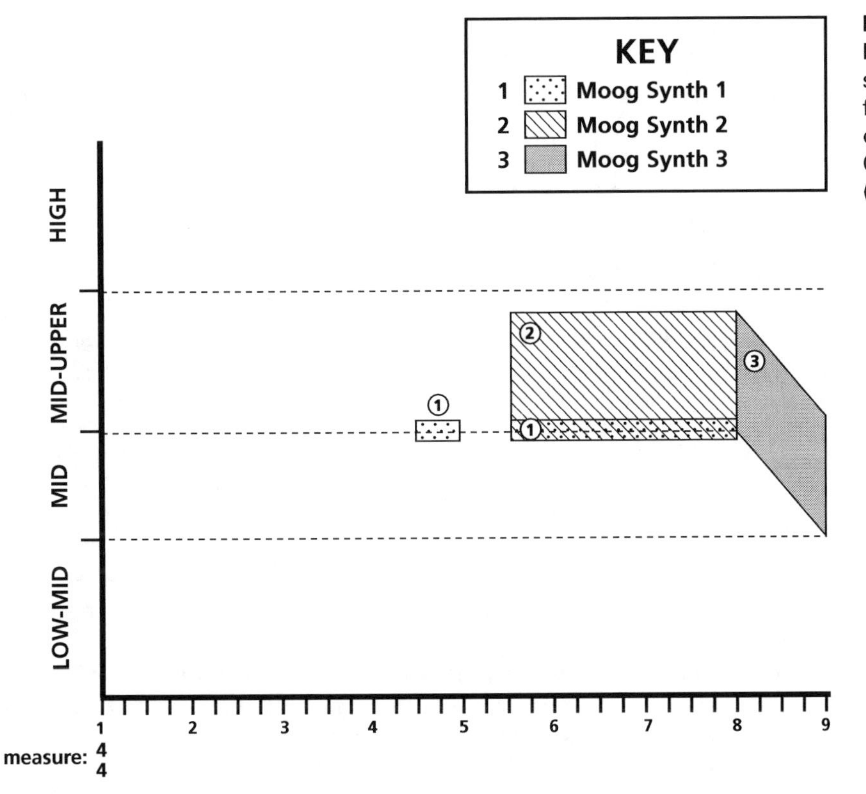


**Figure 7.6**
Pitch density of Moog synthesizer sounds from the introduction of The Beatles' "Here Comes the Sun" (*Abbey Road*, 1969).

Figure 7.6 is the pitch density of the three Moog synthesizer sounds from the introduction to "Here Comes the Sun." Notice how the first sound has a narrow bandwidth, and is barely audible. Sounds 2 and 3 have a similar bandwidth, but rather different spectral content. Notice that Sound 2 presents a melodic line within the lowest portion of its pitch density area, and the remainder of the area is its spectral content. Sound 3 is the pitch density of the glissando sound; this sound will appear later as a sound-quality evaluation in Figure 9.4.

To understand the timbral balance of the introduction, we would need to also graph the guitar lines and taps. All of these on the same graph would represent the introduction's timbral balance.

Pitch density and timbral balance will be explored in much greater depth in Chapter 12, in later readings and exercises. Listen one last time to the four drum sounds as the music progresses; recognize their dominant pitch areas and their boundaries, and compare those pitch areas to those of the other sounds happening simultaneously. This is your beginning work in recognizing pitch density and timbral balance.

## Exercises

The following exercises should be practiced until you are comfortable with the material covered.

### Exercise 7.1

*Melodic Contour Analysis Exercise*

Find a recording with a melodic line that is performed too quickly to be heard as individual pitches. It is best for the melodic line to be at least two measures, or five seconds, in duration.

1. Determine the timeline of the example, including the appropriate time units (clock time or meter) and the length of the timeline. Then make note of beginning and ending pitch/frequency levels.
2. Begin plotting the melodic contour against the timeline by identifying highest, lowest, and other prominent pitch levels, and placing them at precise locations on the timeline. Check the timeline for accuracy of length.
3. Work to establish as many reference pitches as possible. The highest and lowest pitches of the line will establish the upper and lower limits of the *Y*-axis. Identify the fastest change of pitch level; this will become the smallest time unit the graph needs to clearly present, and will determine the appropriate division of the *X*-axis.
4. Draw the melodic contour graph using the *X*- and *Y*-axes determined in Step 3.
5. Locate the reference pitches on the graph at the appropriate locations against the timeline.
6. Fill in the remaining pitch information, making certain to check observations regularly. The graph is completed when the last noticeable pitch change is incorporated into the graph.

### Exercise 7.2

*Pitch-Area Bandwidth and Density Exercise*

Work at a piano keyboard, with or without a partner.

1 Play the succession of intervals in number 1, in Figure 7.7. Allow the sound to ring out completely, with a pause between them: major second, major third, minor third, perfect fourth.
2. Listen intently to the sound qualities of the intervals, and seek to understand the two notes as the outer boundaries of a pitch area.

**Figure 7.7**
Intervals, chords, and clusters for exploring the sound qualities of bandwidth and density.

Get accustomed to the notion of the "width" of the pitch areas. Notice the sound qualities of the intervals without notes between the pitches.

3. Now, play the chords in number 2 of Figure 7.7. Allow the sound to ring out completely, with a pause between them. Strive to play all pitches with equal loudness.
4. Listen intently to the sound qualities of the chords, noting the different densities of pitch information within the chords, and how the density is moving from least dense to most dense.
5. Finally, play the chords in number 3 of Figure 7.7, in the same manner, with pauses between the chords. Strive to play all the pitches of all the half steps between the outer pitches with equal loudness. Note, this is the greatest density possible in equal temperament.
6. Listen intently to the chords, and note the stable densities while the bandwidth changes. Transpose the chords to different octaves and repeat Steps 1 through 5.
7. Seek to understand the sound qualities of the different densities. Use these observations within the pitch-area analysis exercise that follows.

---

**Exercise 7.3**

*Pitch-Area Analysis Exercise*

Identify an instrument you want to evaluate—most people find drums are easiest in beginning attempts. The companion website has a number of drum and cymbal sounds that will serve this purpose very well. You might wish to establish a way to repeat the track quickly and with little distraction.

1. Determine the most prominent pitch area by defining the lower boundary first, then the upper boundary of the area (a steep filter can be helpful in determining these boundaries during beginning studies).

2. Determine any secondary areas of concentrated activity (these will be identified by either width, density, or dynamic prominence of the pitch area), by identifying the lowest and then the highest boundary.
3. Repeat the process for any other pitch areas present. The specific frequencies/pitches of the boundaries are often audible, despite the sound not having an audible fundamental frequency. These pitches/frequencies should be identified and noted on the graph.
4. Evaluate the densities of the pitch areas and incorporate that information into the graph (this is the general amount of pitch/frequency activity within the pitch area and is noted on a numbering scale from very dense to very sparse).
5. Finally, identify the dynamic relationships between the pitch areas within the sounds. Describe this information as part of the analysis (these are the general dynamic relationships of the area and are noted on a number scale identifying the relative loudness of the pitch areas).

# 8 Evaluating Loudness in Audio and Music Recordings

Loudness has traditionally been used in musical contexts to assist in the expressive qualities of musical ideas. This function of dynamics helps shape the direction of a musical idea, helps delineate separate musical ideas (usually in relation to their importance to the musical message), assists in creating nuance in the expressive qualities of the performance, and/or it may add drama to the musical moment. In all of these cases, dynamics have functioned in supportive roles in the communication of the musical message.

The recording process has given the recordist more precise control over dynamics than exist in live acoustic performances. This control has brought audio recordings to have additional relationships of dynamics and the potential of placing more musical importance on dynamic relationships. An example of a new relationship of dynamics is the occurrence of contradictory cues between the loudness level at which a sound was performed during the initial recording (tracking) and the dynamic level at which the sound is heard in the final musical texture (the mix). The recordist must be aware of all relationships of dynamic levels, both those that are naturally occurring and those caused by the recording/reproduction process, and of the possibility that dynamic levels and relationships may function on any hierarchical level of the musical structure.

When most people think of mixing, they think of setting loudness levels of instruments and voices (sound sources). While this is significant, there is much more to loudness in recording. The dynamic alterations in mixing are only part of what is covered in this chapter. Program dynamic contour and overall levels are important aspects of the recordist's work and are covered here. In Chapter 9 loudness will be considered again as a part of sound-quality/timbre evaluation, as we look at dynamic envelope and spectral envelope after shifting our level of perspective. No matter the level of perspective, a reference dynamic level (RDL) is required for judging loudness levels.

The recording process places unique critical- and analytical-listening requirements on the recordist in the area of the evaluation of loudness. The recordist must be able to focus on changes in dynamics at all levels of perspective and to quickly switch focus between perspective levels. The recordist must also be able to use the skills of identifying loudness relationships, switching between listening to the sound itself out of context (using critical listening) and listening to the sound within its musical context (analytical listening).

Much confusion often accompanies beginning attempts to perceive, follow, and graph dynamic changes. The listener must remain focused on the act of perceiving and defining changes in loudness levels. In the mix these are the dynamic levels of instruments/voices and their relationships to one another. Listeners must not allow themselves to be distracted or misled.

The listener must be conscious, and ever mindful, of not confusing other, easily misleading information as changes in dynamic levels. Some aspects of sound that are often confused for dynamics are distance cues, timbral complexity, performance intensity, sound-source pitch register, any information that draws the attention (focus) of the listener (such as a text, sudden entrance of an instrument), environmental cues, speed of musical information, and more.

We readily confuse prominence with loudness. Prominence is not loudness. Prominence is simply that which is dominating our awareness, or is at the center of our attention. It is common for sounds most prominent in the listener's focus to not be the loudest sounds in the musical texture. Loudness itself does not create or ensure prominence of the material. It comes as a surprise to many people that the most prominent sound (or aspect of sound) in the listener's attention is often *not* the sound with the highest dynamic level.

This chapter seeks to define actual loudness levels in musical contexts (dynamics), with the exception of the final section. In that section, the actual loudness of the musical parts as musical balance will be compared to performance intensity (the loudness of the sound sources when they were performed in the recording process).

## Reference Levels and the Hierarchy of Dynamics

Dynamics are traditionally described by imprecise terms such as very loud (fortissimo), soft (piano), medium loud (mezzo forte), etc. These terms do not provide adequate information to define the loudness level of the sound source. They merely provide a vocabulary to communicate relative values.

The artistic element of dynamics in a piece of music is judged in relation to context. Dynamic levels are gauged in relation to (1) the overall, conceptual dynamic/intensity level of the piece of music, (2) the sounds occurring simultaneously with a sound source, and (3) the sounds that immediately follow and precede a particular sound source. In this way, loudness is perceived as relationships between sound sources and in relation to a reference level. Evaluation is more precise, and it is only possible to communicate meaningful information about dynamic levels when a reference level is defined.

### Reference Dynamic Level

The impression of an overall or global intensity level of a piece of music will be the primary reference level for making judgments concerning dynamics. This level is arrived at through our impression or recognition of the intensity level of the performance of the work as a single idea. It is the *perceived performance intensity* of the work as a whole, conceptualizing the entire work as a single entity out of time. The work's form and essence has a dimension that is the energy, drama, urgency, and expression of its singular perceived performance intensity. This is the work's *reference dynamic level.*

Every work can be conceived as having a single, overall reference dynamic level (RDL). This is the dynamic level that characterizes a piece of music when the form of the work is envisioned. The RDL is a single, specific dynamic level that represents the intensity and expression of the piece. This specific dynamic level is an understanding of the intensity level of the performance of a piece of music, as a whole and as realized in an instant (its form). It is the inherent spirit of the music/recording as a level of exertion, expression, mood, and sometimes message combined into a single concept. When some people talk about a song's "groove," they may actually be talking about its RDL.

Performers establish this reference dynamic level in their minds before beginning a performance of any piece of music. Often this occurs intuitively. Composers also retain this level in their thoughts (at least subconsciously) throughout the process of writing a piece of music. Recordists will go through a similar process in production work. Recordists establish this level as a reference from which they are able to calculate all other dynamic levels and relationships. In recordings, this level often needs to be consistent for hours, days, weeks, or longer, as sessions for a piece of music progress at their own pace.

Performance-intensity cues are related to timbral changes of the sound sources. Sound sources will exhibit different timbral characteristics when performed at different dynamic levels and with different amounts of physical exertion. The impressions the listener receives, related to the intensity level of the performance (of all of the musical parts individually and collectively), will be related to actual dynamic relationships of the musical parts.

The perceived performance intensity in the recording and the perceived dynamic relationships of the musical parts will directly shape the listener's impression of the existing RDL of the recording. Dynamic and performance-intensity cues (including expressive qualities of the performance) play significant roles in determining RDL, as does tempo. The sound sources that present the primary musical materials (or that are at the center of the listener's focus) will often have proportionally more influence than sources presenting less significant material.

Through the perception of these cues and the influence of tempo, a single, conceptual dynamic level will be determined. This is the RDL of the performance (recording/piece of music)—the level at which it is envisioned as existing. This is the reference level that will be used to calculate the dynamic levels of the individual musical parts in relation to the whole, as well as the dynamic contour of the overall program.

The RDL does not change within a piece of music. It is an unchanging reference. Conceptually this is similar to the decibel. The decibel is a ratio, identifying the variable's relationship to a reference level. In dynamics, levels can change at any moment and by any amount, and just like the decibel, this changing level does not change the reference level. The reference dynamic level serves to unify all of the dynamic levels and relationships of a piece of music.

Every work will have a specific RDL. When a work is divided into separate major sections, even separate movements (such as a symphony), the work will have an RDL that allows all sections to be related to the overall concept. Even a 90-minute symphony will have only a single RDL. The RDL becomes one of the factors that allow the listener to perceive the work as a single entity, composed of many related parts.

The reference dynamic level can be perceived as existing anywhere from very loud *(fff)* to very soft *(ppp).* The RDL will be established as a precisely defined dynamic level that will serve as a reference level throughout the work. This will be discussed further.

The RDL will be used for evaluating/defining:

1. Dynamic relationships of the overall dynamic contour of the program (piece of music)
2. Dynamic levels of the musical ideas (sound sources) of the work
3. Dynamic relationships of the individual dynamic contours of the musical balance of the work.

## Performance Intensity and Dynamic Markings

Timbre changes of performance intensity are important cues in our perception of dynamic levels of acoustic performances. We apply these same cues to recorded sounds to imagine a live performance, despite the medium. In recordings we gauge performance intensity solely by sound quality, as the visual cues of a live performance are not present—though they are still imagined. Our reference for performance intensity is our knowledge of the instrument's timbre, as it is played at various levels of physical exertion, with different types of expression, and with various performance techniques.

The listener recognizes the amount of physical exertion required to produce a certain sound quality on an instrument. This understanding becomes the perceived performance intensity. Performed sounds that require expending energy are perceived as moderately loud (mezzo forte, ***mf***), or above. Performed sounds that appear to be withholding energy are perceived as moderately soft (mezzo piano, ***mp***), or below. When the listener imagines a considerable amount of energy (or perhaps an excessive amount) was required to produce a perceived sound quality, the performance intensity will be forte ***(f)***, or perhaps more. This becomes simpler when remembering to relate dynamic markings and performance-intensity cues to energy expended by the performer.

The threshold between mezzo piano ***(mp)*** and mezzo forte ***(mf)*** is critical to this understanding. This is the energy level the performer can theoretically sustain indefinitely. Conceptually, at this level the performer is neither putting forth energy, nor holding back—no energy is being exerted. Above this threshold, energy is being consumed by pushing forward, becoming more assertive—even if only a very small amount. Below the threshold, energy is being held back, or being withdrawn, if only in a small amount. Moving further above or below the threshold, the perception becomes a matter of degree, or magnitude, of how much energy is being expended or withheld. The difference then between ***ff*** and ***fff*** is the level of intensity and the expectation of the length of time that level of energy can be sustained. Likewise, the level of restraint and the likelihood of the length of time that restraint might be sustained distinguish ***pp*** and ***ppp***.

The traditional terms for dynamic levels can continue to be of use with a well-defined RDL based on performance-intensity information. With a defined RDL, the comparative terms can have more significant meaning. The terms will remain imprecise, but they will be more meaningful. Dynamic levels are more precisely defined when sound sources are compared to one another and placed on the appropriate graph, making the use of the traditional terms a mere starting point for evaluation of loudness levels.

The terms retain their meanings from musical contexts, whereas the dynamic terms (such as mezzo forte) describe a quality of performance and an amount of physical exertion and drama on the part of the performer, as well as being a description of the loudness level. When placed on a graph (see Figure 8.1), these general terms are transformed into areas where sounds can be precisely defined against the RDL and in relation to performance intensity.

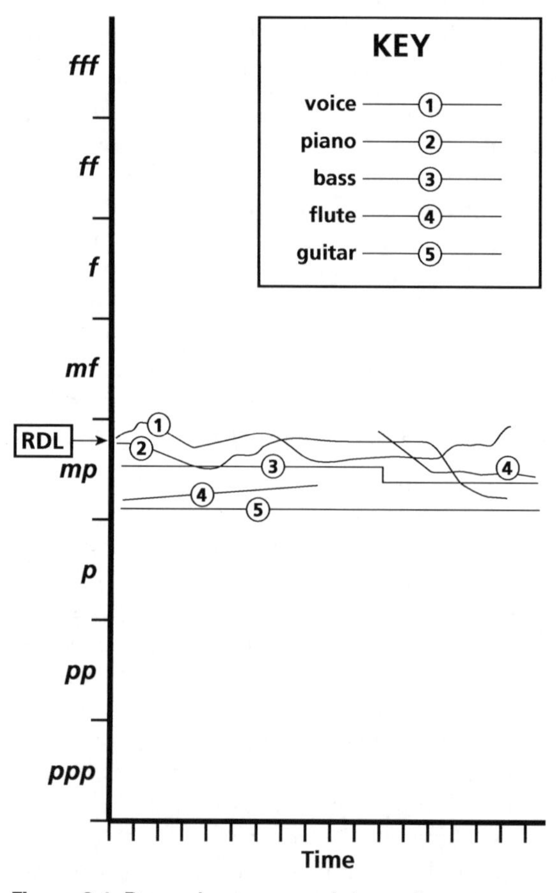


**Figure 8.1** Dynamic ranges and dynamic contour.

## Dynamic Levels as Ranges

Dynamic levels do not exist as a specific level of loudness in musical contexts. Dynamics are understood as ranges or areas. Dynamic markings refer to a range of actual loudness levels.

The dynamic marking "mezzo forte" does not refer to a precise level. It refers to a range of dynamic levels between "mezzo piano" and "forte." Many gradations of "mezzo forte" may exist in a certain piece of music. Many instruments can be performing at different levels of loudness, yet be accurately described as being in the "mezzo forte" dynamic range. Entire musical works or performances *may* take place within the range of a certain dynamic marking, yet exhibit striking contrasts of loudness levels.

Figure 8.1 presents the vertical axis that will be used for all graphs plotting dynamics. The dynamic markings are centered within the ranges. A number of sound sources are plotted on the graph, and the reference dynamic level is designated on the vertical axis. Each sound source can be compared to the dynamic levels and contours of the other sources, and to the RDL. The unique characteristics of each source are readily apparent from the graph.

If limited to describing the sound sources by traditional dynamic-level designations, some sounds would be a "loud mezzo piano," some sounds a "moderate mezzo piano," and others a "soft mezzo piano." The graph, in this instance, circumvents the need for these vague and cumbersome descriptions.

The most extreme boundaries of dynamic ranges, for nearly all musical contexts, will extend from ***ppp*** to ***fff***. Musical examples that contain material beyond these boundaries are rare. The individual graph should incorporate only those ranges that are needed to accurately present the material, and to leave some vertical area of the graph unused above and below the plotted sounds, for clarity of presentation.

## Determining Reference Dynamic Levels

The perceived performance-intensity level that serves as our reference for evaluating the dynamic relationships of the work is, again, the reference dynamic level. This single RDL will be used (1) at the highest hierarchical level to calculate the dynamic contour of the entire program (the complete musical texture) and (2) at the mid levels of the structural hierarchy to calculate the dynamic contours of the individual sound sources in musical balance.

At the lower levels of the structural hierarchy, the reference level switches to the performance-intensity level of the individual sound. The intensity level of a specific appearance of the sound source is used as the reference level to determine the dynamic contours of the individual sound and its component parts. At these levels of perspective, dynamic contours are plotted of (1) a typical appearance of the overall sound source (the dynamic envelope) and (2) the individual components of the spectrum (spectral envelope). This dynamic contour information seeks to define the sound quality or timbre of the sound source, and will be explored in the next chapter, in that context.

The RDL of the piece of music is the reference for determining the dynamic levels of the sound sources and the composite musical texture. In order to make these evaluations, the RDL must first be defined.

The RDL is a precise level that can be clearly defined. It is not subjective. All listeners putting forth the effort to perceive it will arrive at the same level. The listener will recognize when they have identified the correct level. The level will cause all other dynamic relationships to be understood, to make sense. The listener will perceive the piece of music as existing at the precisely identified dynamic level. The dynamic level is envisioned as a dimension of the core essence of the piece of music.

A listener's first attempts to define RDL are typically difficult. The concept itself eludes many people at first. The reader must remain conscious of trying to understand this important concept. It will require many hearings of the recording/piece of music to learn it well enough to try to define something requiring this depth of understanding. After achieving this level of understanding, the listener can become more comfortable with formulating the impression of a single dynamic level that *is* the dynamic/intensity level of the piece. The listener will recognize the RDL of the piece once it has been experienced and understood. It is likely they will then not forget it. Listeners often experience the RDL even in passive listening for entertainment and do not realize it; indeed it is often the song's overall dynamic/intensity (energy, intensity, motion, mood, expression—"groove"—combined) that draws a listener to identify strongly with a song.

It is common for some information to get in the way of formulating this impression of the RDL. Musical materials, lyrics, tempo, and instrument timbres all give cues that the listener will be tempted to factor prominently into this observation. Skilled musicians are even prone to drawing conclusions based on what they would *want* to be present in the music, rather than listening to what *is* present. Some instruments may well be performing at intensities that send conflicting cues when considered against the listener's ideas of the potential RDL; this conflicting information enriches art, but makes defining it more difficult. A magic formula does not exist for determining the RDL. This is one of the significant artistic dimensions of a piece of music that defies

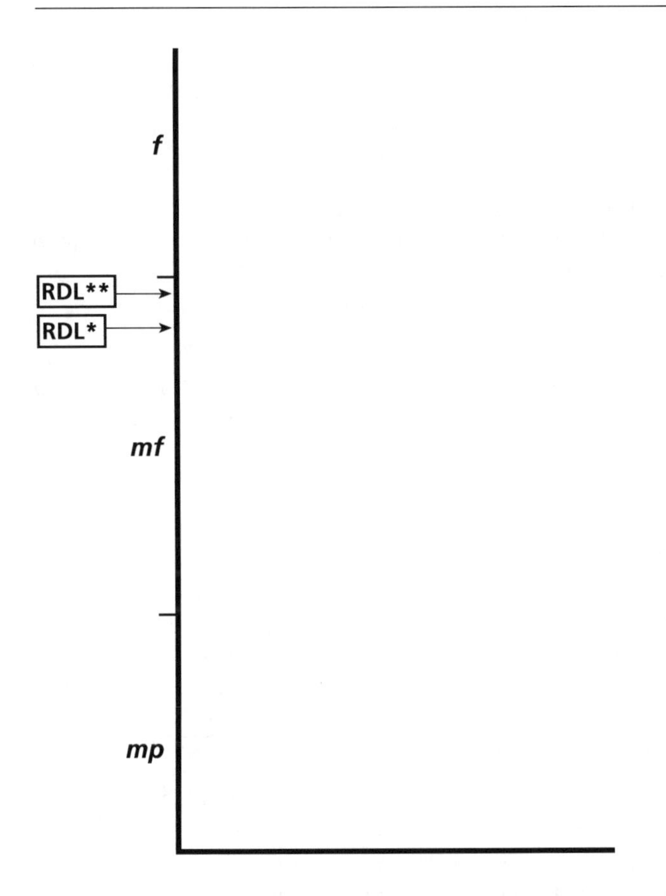


**Figure 8.2**
Reference dynamic levels of two versions of "Lucy in the Sky with Diamonds" (*original version from *Sgt. Pepper's Lonely Hearts Club Band*, and **1999 *Yellow Submarine* version).

theoretical analysis and instead uses the sensibilities of the listener. This does not make it subjective, only that it cannot be predicted by measured calculation, but is determined through the experience of the music and arrived at through an understanding of the piece. Again, the RDL will be a precise level that all listeners can agree upon (± 2 percent) given enough attention to the task.

Figure 8.2 identifies the RDL of "Lucy in the Sky with Diamonds." Two levels are shown: one for the original version from *Sgt. Pepper's Lonely Hearts Club Band* and the other for the 1999 *Yellow Submarine* version. The two versions have slightly different reference dynamic levels. The different levels are caused by the different mixes and presentations of materials, among other factors such as the sound qualities imparted during mastering. The essence of the piece has been slightly altered by the slightly different sound qualities of the sound sources and the recordings. The reference dynamic levels are perceived as clearly within mezzo forte's moderate expending of energy. They are both beyond the level midway between the beginning of ***mf*** and the threshold of ***f*** (where exertion moves beyond moderate). In fact both exceed the three-quarters level of the area that comprises **mf**. After much listening and contemplation, the levels can be understood as being in the upper 15 percent of the **mf** area, with not more than 10 percent of the area separating the two levels.

RDL is calculated after getting to know the composition, recording, and/or performance well. Two different performances of the same piece by the same performer may each have a different RDL. Each of two different interpretations will almost certainly have a different RDL, if only slightly.

An exercise in determining the reference dynamic level of a recording/piece of music appears at the end of this chapter, Exercise 8.1.

## Program Dynamic Contour

Changes of dynamic level, over time, comprise dynamic contour. As noted, the dynamic levels and relationships occur at all hierarchical levels. The broadest level of perspective will allow the dynamic contour of the overall program to be perceived and plotted. This is the single dynamic level/contour of the composite sound, the result of combining all sounds. This *program dynamic contour* can be envisioned as a monometer, following the dynamics of the entire program.

Skill in recognizing the dynamic level of the overall program is developed through plotting program dynamic contour. Recordists use this skill in many listening evaluations. This high-level graph, or the associated listening skill alone, will be applied to many analytical- and critical-listening applications.

Dynamic contour must not be confused with performance intensity, with distance cues, or with spectral complexity. These aspects of recorded sound often present cues that contradict actual dynamic (loudness) level or that alter the perception of the actual dynamic level.

Program dynamic contour information is plotted on a program dynamic contour graph. The graph incorporates:

1. Dynamic area designations for the *Y*-axis, distributed to complement the characteristics of the musical example
2. The reference dynamic level designated as a precise level on the *Y*-axis
3. *X*-axis of the graph dedicated to a timeline, divided into appropriate increments of the metric grid or of real time (depending on the material and context)
4. A single line plotted against the two axes (the dynamic contour of the composite program is the shape of this line).

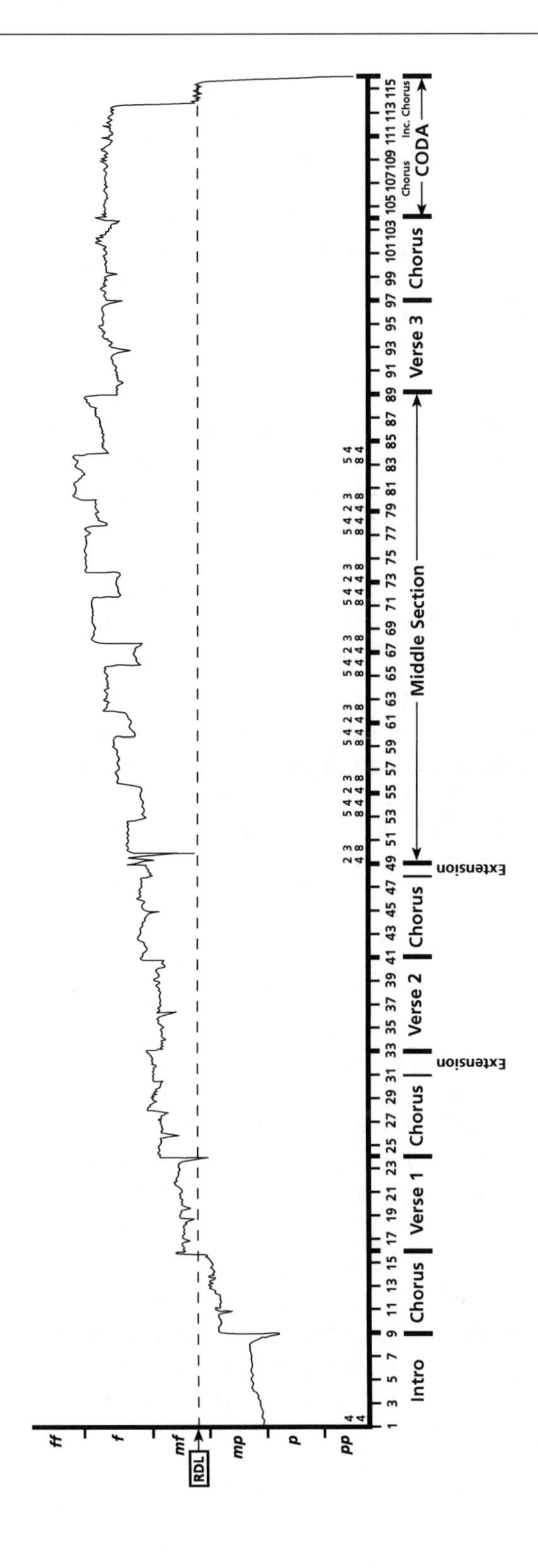

**Figure 8.3**
Program dynamic contour—The Beatles' "Here Comes the Sun," (*Abbey Road*, 1969).

The program dynamic contour of The Beatles' "Here Comes the Sun" is plotted in Figure 8.3. The program dynamic contour graph shows the changes in overall dynamic level throughout the work. In listening to the work, the striking changes in the overall dynamic level will be evident. The wide dynamic range goes through many large and many subtle changes of loudness level. Patterns of shapes within the dynamic contour will likely also emerge. Pertinent to reference dynamic level, a sense of arrival happens at the end of the piece. This occurs when the song settles at the RDL for a moment and then the song is over.

It is common for a piece of music to arrive at its RDL as an important occurrence of the work. It might appear in the introduction, the dramatic climax, the final chorus, or the final verse. These are all common places an RDL might be reached, but any location is possible. It is also possible that the song's RDL is never sounded—purposefully leaving the listener unfulfilled in this regard. It is possible the song's RDL will be prevalent in a piece or heard only once, as in "Here Comes the Sun." Many possibilities exist, including silence, as some songs only fully make sense when the silence at the end arrives and brings introspection.

Listening to the loudness contour and levels of program dynamic contour is an important skill for the recordist. It is used extensively for live sound, film sound, mastering, and broadcast at times, when one is concerned about the level of the overall program and how it changes over time. During mixdown and tracking, it is also necessary to be aware of the overall program level and dynamic contour in order to control the quality and level of the recording.

## Musical Balance

The plotting of all the individual sound sources in a musical texture by their dynamic contours provides the *musical balance graph*. The graph will show the actual dynamic level of the sound sources, in relation to the RDL. Each sound source will have a separate line on the graph that will allow the dynamic contours of the sources to be mapped against a common timeline. This graph will clearly show the loudness levels of all sounds and will represent the mix of the work.

The musical balance graph will not include information on performance intensity, sound quality, or distance. These cues are often confused with perceived dynamic level, and are purposefully avoided here. This graph is solely dedicated to evaluating and understanding the dynamic levels, dynamic contours, and dynamic relationships of the sound sources. These will be the sole focus of the listener here.

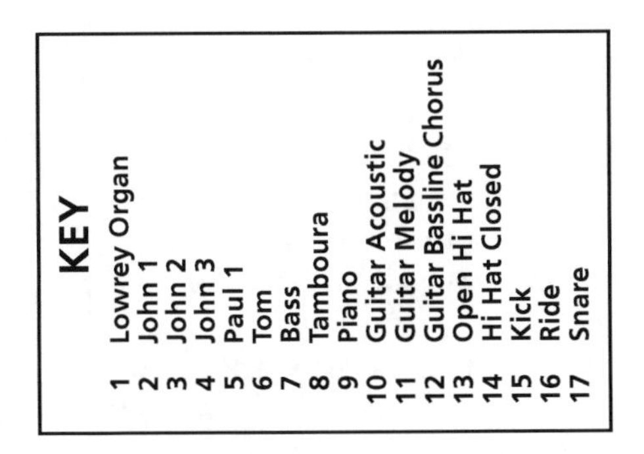


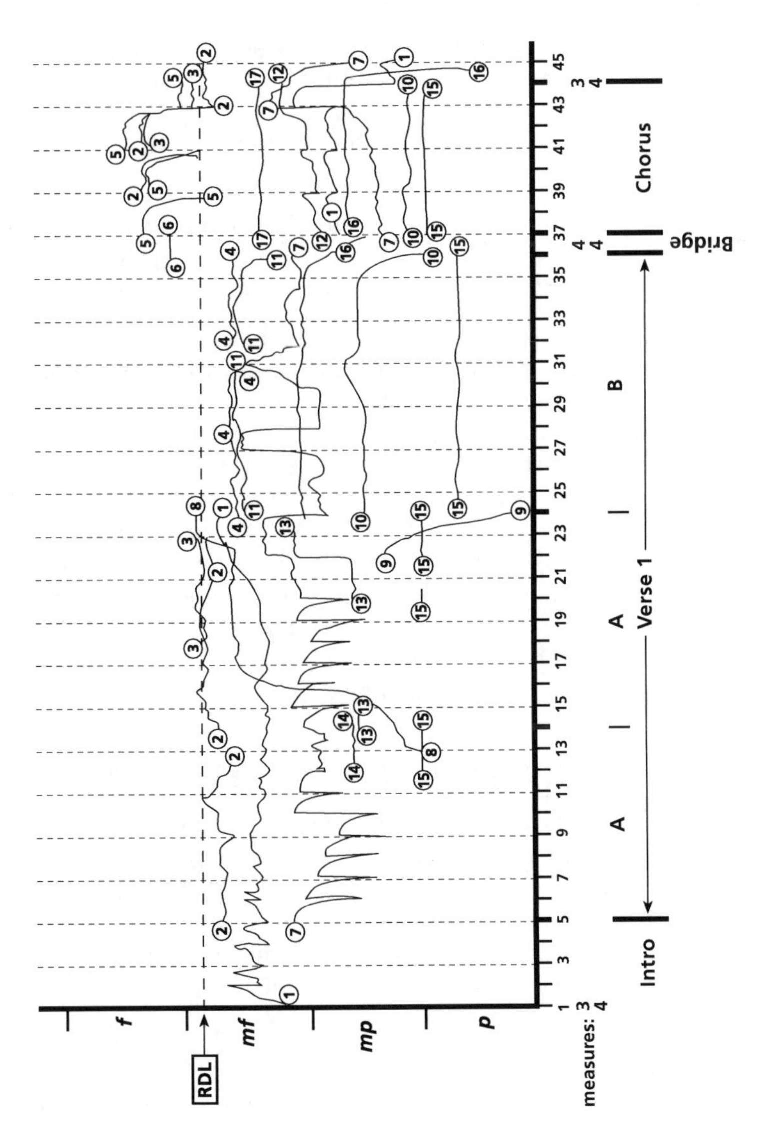


**Figure 8.4** Musical balance graph—The Beatles' "Lucy in the Sky with Diamonds" *(Sgt. Pepper's Lonely Hearts Club Band)*.

The musical balance graph incorporates:

1. Dynamic area designations for the *Y*-axis, distributed to complement the characteristics of the musical example
2. Reference dynamic level designated as a precise level on the *Y*-axis
3. *X*-axis of the graph dedicated to a timeline, divided into appropriate increments of the metric grid (measures, beats within measures, etc.)
4. A single line plotted against the two axes for each sound source
5. A key relating the names of all sound sources to a unique number or color to identify all source lines on the graph.

Figure 8.4 presents a musical balance graph of some of the sound sources in The Beatles' "Lucy in the Sky with Diamonds," *Sgt. Pepper's Lonely Hearts Club Band* version. These are the actual loudness levels of the instruments. The listener will notice that the loudest instrument/voice is not always the most prominent.

Exercise 8.3 at the end of the chapter will lead the listener through the process of creating a musical balance graph. It is suggested the reader become well acquainted with this exercise, as it is one of the most important production and basic listening skills required of the recordist.

It is necessary to clearly envision the correct perspective to accurately hear musical balance relationships. To compare the levels of two or more instruments (or sound sources), one must focus on the perspective one level higher than that of the individual instruments; at that level the sound sources can be perceived as being equal, and compared without bias. When listening at the level of the individual sound source, a single instrument will become the center of one's attention and it will be emphasized in one's mind, thereby causing loudness judgment to be skewed. One does, however, listen at this perspective of the individual sound source to identify the level of a sound source in relation to the work's RDL, and also to recognize the sound source's dynamic contour.

## Performance Intensity versus Musical Balance

Performance intensity is the dynamic level at which the sound source was performing when it was recorded. In many music productions, this dynamic level will be altered in the mixing process of the recording. The performance intensity of the sound source and the actual dynamic level of the sound source in the recording will most often not be identical and will send conflicting information to the listener.

The dynamic levels of the various sound sources of a recording will often be at relationships that contradict reality. Sounds of low performance intensity often appear at higher dynamic levels in recordings than sounds that were originally recorded at high performance-intensity levels. This is especially found in vocal lines. This conflicting information may or may not be desirable, and the recordist should be aware of these relationships.

**Listen**

to tracks 37 and 38

to identify and recognize one mix that closely aligns performance intensity and musical balance, and a different mix of the same performance that radically alters the musical balance of the original performance.

Important information can be determined by plotting performance intensity against musical balance. This will often provide significant information on the relationships of sound sources and the overall dynamic and intensity levels of the work, as well as the mixing techniques of the recording.

Performance intensity is plotted as the dynamic levels of the original performance. The listener will judge the intensity of the original performance through timbre cues. The listener will make judgments based on their knowledge of the sound qualities of instruments and voices when performed at various dynamic levels.

The reference for performance intensity is the listener's knowledge of the particular instrument's timbre, as that instrument is played at various levels of physical exertion and with various performance techniques. A reference dynamic level is not applicable to this element, therefore the song's RDL is not used here.

Musical balance is plotted as the dynamic levels of the sound sources, as the listener perceives their actual loudness levels in the recording itself. This was discussed immediately above.

The *performance-intensity/musical-balance graph* incorporates:

1. Dynamic-area designations in two tiers for the *Y*-axis, distributed to complement the characteristics of the musical example (one tier is dedicated to musical balance, and one tier is dedicated to performance intensity)
2. Reference dynamic level on the musical-balance tier, designated as a precise level on the *Y*-axis (an RDL is not relevant to the performance-intensity tier)
3. *X*-axis of the graph dedicated to a timeline, divided into appropriate increments of the metric grid
4. A single line plotted against the two axes for each sound source, on each tier of the graph (each sound source will appear on both tiers; the same number or color line is used for the source on each tier of the graph)
5. A key used to clearly relate the sound sources to their respective source line (the same key applies to both tiers of the graph).

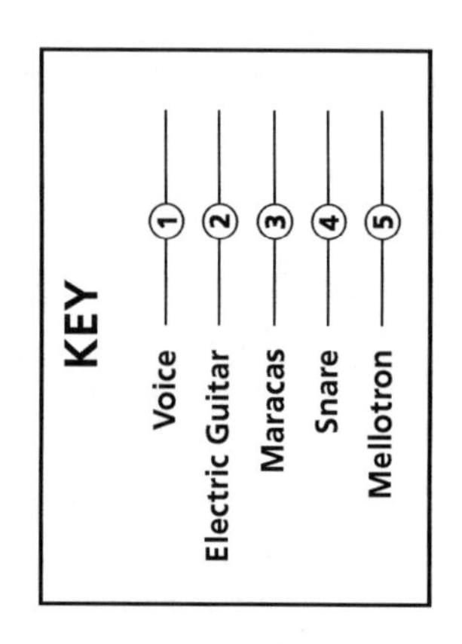


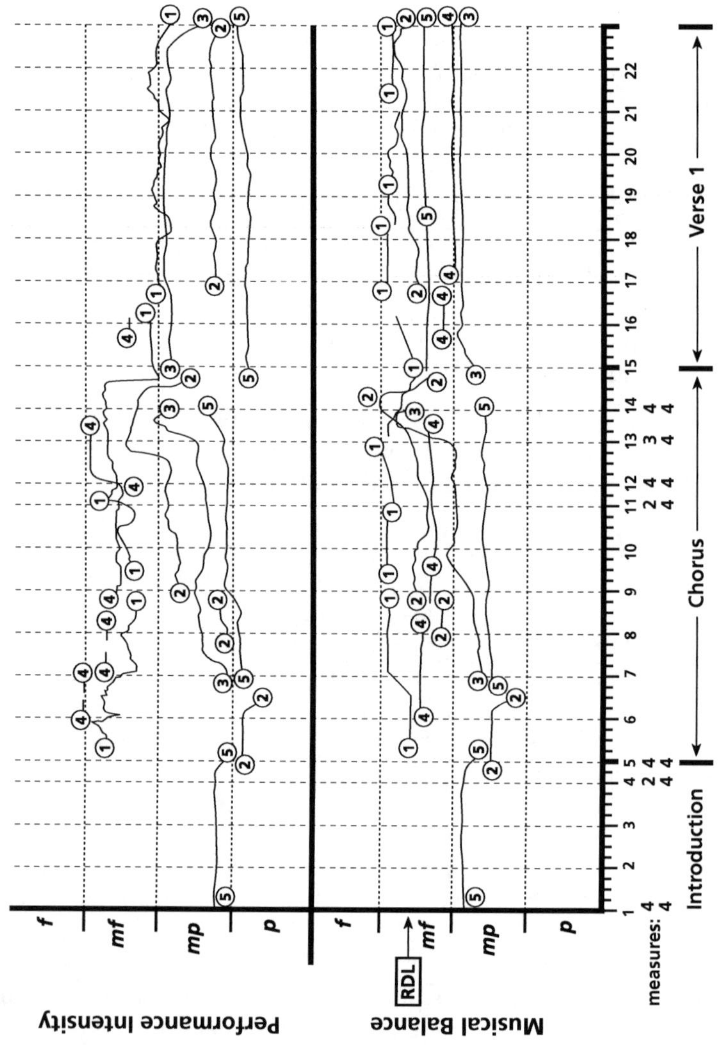


**Figure 8.5**
Performance intensity versus musical balance—The Beatles' "Strawberry Fields Forever" *(Magical Mystery Tour).*

Figure 8.5 will allow the listener to observe some of the differences between the recording's actual loudness levels and the performance intensities (loudness levels) of the sound sources when they were recorded. A few primary sound sources are graphed from The Beatles' "Strawberry Fields Forever." Some sound sources are at very different levels in each tier, and others show no significant change between tiers (little change between how the sounds were performed during tracking versus how their levels exist in the mix). Some sources contain subtle changes of dynamic levels and/or many nuances of performance-intensity information, and the Mellotron exhibits few gradations of dynamics and intensity.

An exercise to develop skills in recognizing and evaluating performance-intensity versus musical-balance information appears as Exercise 8.4 at the end of the chapter. As an additional exercise, the reader might determine the musical balance and performance intensities of the other sound sources in "Strawberry Fields Forever" during the measures of Figure 8.5.

The musical balance and the performance-intensity graphs will function at the same level of perspective as pitch density (see Chapters 7 and 12). When evaluated jointly, these three artistic elements will allow the listener to extract much pertinent information about the mixing and recording processes, and the creative concepts of the music.

## Exercises

The following exercises should be performed with care. You should work methodically to become comfortable with the material covered.

When performing loudness/dynamic evaluations, and evaluations of many of the other elements that follow, work to identify clearly what you know for certain. Ask yourself if the sound is at extremes of the dynamic range, and work toward focusing in on the correct level. Continue to feel comfortable that you have knowledge of what the sound level *is not*, and work toward finding what the level *is*.

### Exercise 8.1

*Reference Dynamic Level Exercise*

Select a recording you know well for your initial attempts at determining the reference dynamic level of a piece of music. It would be best for the work to be less than four minutes' duration.

1. Before listening to the piece, spend some time thinking about the overall character of the piece; consider the overall energy level, performance intensity, concept or message, and other important aspects of the song.
2. Listen to the song several times to confirm that the observations in your memory are reflected in the actual music and recording.
3. Reconsider your observations with each new hearing of the recording.
4. Attempt to determine a precise dynamic level for the RDL. Begin this process by working from the extreme levels—***ppp*** and ***fff***—asking if the level exists in those areas. Eliminate dynamic areas where the RDL is obviously not present. This will focus your efforts.
5. Once the dynamic area has been determined, work to define a precise level by asking if the RDL is below 50 percent in the level, or above. Continue to work toward a specific level by narrowing the area further.
6. Leave the example and your answer for a period of time (several hours or several days). Listen to the song again. Reconsider the RDL previously defined.

If you do not know a piece of music, many hearings will be required before initial observations can be made.

### Exercise 8.2

*Program Dynamic Contour Exercise*

Select a short song for your initial attempts at creating a program dynamic contour graph. The entire song should be graphed for overall dynamic contour. The dynamic level of the entire recording (the composite dynamic level of all sounds) will be the focus of this exercise. To aid in developing this skill, initial attempts at this exercise should use a song with large, sudden changes of dynamic level. Repeat the exercise using a song with changes that are smaller, or subtler. Determine a general shape of the dynamic contour before attempting to grasp all of the subtle details.

1. During the first hearing(s), listen to the example to establish the length of the timeline. At the same time, become acquainted with the character of the song to begin formulating an idea of its RDL.
2. Check the timeline for accuracy and make any alterations. Establish the RDL of the work by working through the previous exercise.
3. Notice the activity of the program dynamic contour for boundaries of levels of activity and speed of activity. The boundary of speed will establish the smallest time unit required to accurately plot the

smallest significant change of the element. The boundary of levels of activity will establish the upper and lower thresholds of the graph. Next, determine the smallest increment of the *Y*-axis required to plot the smallest change of the dynamic contour.

4. Begin plotting the dynamic contour on the graph, continually relating the perceived dynamic level to the RDL. First, establish prominent points within the contour. These reference points will be the highest or lowest levels, the beginning and ending levels, points immediately after silences, and other points that stand out from the remainder of the activity. Use the points of reference to judge the activity of the preceding and following material. Focus on the contour, speed, and amounts of level changes to complete the plotting of the contour. Notice any patterns or recurring shapes within the contour, and changes of levels between sections of the song.
5. The evaluation is complete when the smallest significant detail has been perceived, understood, and added to the graph.

**Exercise 8.3**

*Musical Balance Exercise*

Select a popular song with at least three instruments and voices. To make this exercise more meaningful and take less time, you may wish to use one of the songs from Exercise 7.1 or 7.2. Evaluate the first 32 bars for musical balance. This exercise will graph the dynamic contours—actual loudness levels—of all sound sources against the song's RDL. Musical balance is the relationships of sound sources to one another. Initial attempts should use pieces of music with only a few sound sources.

The exercise will follow the sequence:

1. During the first hearing(s), establish the length and structural divisions of the timeline. At the same time, notice prominent instrumentation and activity of their general dynamic levels against the timeline.
2. Check the timeline for accuracy and make any alterations. Establish a complete list of sound sources (instruments and voices), and sketch the presence of the sound sources against the completed timeline. A key should be created, assigning each sound source with its own number, or color.
3. Determine the reference dynamic level of the sound, using the process previously presented.
4. Notice the activity of the dynamic levels of the sound sources (instruments and voices) for boundaries of levels of activity and speed of activity. The boundary of speed will establish the smallest time

unit required to accurately plot the smallest significant change of dynamic level. The boundary of levels of activity will establish the smallest increment of the Y-axis required to plot the smallest change of dynamics.

5. Begin plotting the dynamic contours of each instrument or voice on the graph. Keeping the RDL clearly in mind, establish the beginning dynamic levels of each sound source. Next, determine other prominent points of reference. Use the points of reference to judge the activity of the preceding and following material. Focus on the contour, speed, and amounts of level changes to complete the plotting of the dynamic contours.
6. You should periodically shift your focus to compare the dynamic levels of the sound sources to one another. This will aid in developing the dynamic contours and will keep you focused on the relationships of dynamic levels of the various sources. The evaluation is complete when the smallest significant detail has been incorporated into the graph.

It is important to remain focused on the actual loudness of instruments, making certain your attention is not drawn to other aspects of sound.

As you gain experience and confidence in making these evaluations, songs with more instruments should be examined and longer sections of the works should be evaluated. You will begin to notice striking subtle changes as your skill level improves.

### Exercise 8.4

*Performance Intensity versus Musical Balance Exercise*

Select a multitrack recording of a popular song with at least five sound sources. Evaluate the first 16 bars for performance intensity and musical balance. Select five sound sources to graph for this exercise. The graph will have two tiers: one will graph musical balance (the actual loudness levels in the recording), and the other performance intensity (the loudness levels of the instruments when they were recorded).

1. The musical balance exercise should first be completed as in the previous section. This will generate the graph's timeline as well.
2. Performance intensity will now be determined for each sound source, for the performance-intensity tier. Sound sources will have the same number, or the same color, as on the musical balance tier.
3. Begin plotting the performance intensity of each sound source on the graph. Start by establishing the beginning performance-intensity levels of each sound source. Next, determine other prominent points

of reference. Use the points of reference to judge the activity of the preceding and following material. Focus on the contour, speed, and amounts of level changes to complete the plotting of these performance-intensity contours. The evaluation is complete when the smallest significant detail has been incorporated into the graph.

You can now compare the two tiers, and learn significant information on how the voices and instruments were tracked, and how the recording and mixing processes altered the sound sources. These provide insights into the musical and production decisions that went into "crafting the mix" of that recording.

As you gain experience in making these evaluations, incorporate all of the sound sources into your evaluation and work on larger portion (at least 32 measures). The song should have a significant change in the mix, and the section should move between verse(s) and a chorus. Notice how the sound sources in the mix change (or do not change) when moving from one section of the song to another (i.e., verses to choruses).

# 9 Evaluating Sound Quality

This chapter will present a process that can be used to evaluate sound quality and timbre. This process can bring anyone in the audio industry to communicate meaningful information about sound.

By directly describing the physical dimensions of sound, the process can be easily adapted to evaluate any sound. The reader will learn to describe sound quality (accounting for the various contexts within which sounds exist), and to evaluate timbre (sound out of context as an abstract sound object).

The critical-listening process and the technical areas of the audio industry are often juxtaposed with creative applications and analytical-listening processes. These differences are most apparent in examining the characteristics of sound quality and timbre, and will be articulated here. In addition, the perception of sound quality at various hierarchical levels will be explored.

Sound quality as an overall shape or character exists at a number of levels in perspective and types of listening. These concepts are central to the evaluation process, as they allow us to understand sound as an object (available for evaluation out of time) at all levels of perspective.

Communicating information about sound quality is important to all facets of music production. Nearly all positions in the entire audio industry need to communicate about the content or quality of sound. Yet a vocabulary for describing sound quality and a process for objectively evaluating the components of sound quality do not exist. Meaningful communication about sound quality can be accomplished through describing the physical components of timbre and sound quality. Sounds will be described by the characteristics that make them unique. These characteristics are the activities and states that occur in the component parts of the sound source's timbre. The component parts have been reduced to the definition of fundamental frequency, dynamic envelope, spectrum (spectral content), and spectral envelope. Meaningful information about sound quality can be communicated through describing these characteristics. This can be done in great detail or in a general way. Information can be gathered and communicated in a more detailed and precise manner through graphing sound quality. The following sections will lead the reader to develop skill and language to objectively describe sound.

## Sound Quality in Critical-Listening Contexts

Describing sound quality in and out of musical contexts are skills that are important for recordists. In both contexts, sound-quality evaluations occur at all levels of our perceptual hierarchies—at all perspectives.

Outside musical contexts, we are concerned with understanding the characteristics of sound quality for its own sake and as the integrity of the audio signal. These are perhaps the most prominent approaches to sound quality for the recordist.

This first approach to examining sound quality looks at the unique character of a particular sound. This character of the sound may be what separates one microphone's imprint from another, one person's snare drum sound from another's, or even one monitor speaker from others. Through critical-listening skills, sound quality of a particular sound is evaluated for its unique qualities. The evaluation examines the activities of the components of timbre that occur within that sound only. This type of sound-quality evaluation seeks to define the individual sound so it can be understood. This information can then be put to use in various ways. One might evaluate a sound to examine how a microphone is altering the characteristics of an instrument, to determine if the microphone produces the desired sound. Perhaps two or more sounds are evaluated and compared—the crash cymbal sounds from Chapter 7's Figure 7.4 are an example of this type of comparison.

Critical-listening evaluation is also often concerned with the technical quality, or integrity, of the recording. The technical quality or the integrity of the signal is of paramount importance, and is out of the musical contexts. It is usually the goal of recordings to be of the highest technical quality, and to be void of all degradations of signal quality/integrity and all unwanted sound (although Internet compression is modifying this goal by accepting degradations of signal in exchange for speed of transmission). The technical quality of recordings is degraded by unwanted sounds and can also be impacted by the numerous ways the dimensions of frequency/pitch, amplitude/loudness, and time/phase are undesirably altered by the recording and reproduction equipment and processes (often malfunctions, mismatches, or miscalibrations). These must all be heard and understood by the recordist, and are only perceived through sound-quality evaluations.

The focus of the listener may need to be at any level of perspective, with the listener needing to quickly and accurately shift perspectives. The listener may be evaluating the technical quality of the sound for information related to frequency response (or spectral content, or spectral envelope), or the listener may be listening for transient response (or dynamic envelope, or dynamic contour of a specific frequency area). As examples of extremes of perspective, the listener may focus on the sound quality of the overall program or may be listening at the close perspective of focusing on the sound quality of a particular characteristic of a single, isolated sound source.

Critical listening involves the evaluation of sound quality to define what is physically present, to identify the characteristic qualities of the sound being evaluated, or to identify any undesirable sounds or characteristics that influence the integrity of the audio signal. This process and conceptualization is performed without consideration of the function and/or meaning of the sound and without taking into account the context of the sound.

## Sound Quality in Analytical-Listening Contexts

Analytical listening involves the evaluation of sound quality to identify its characteristic qualities in relation to the context of the sound. Analytical listening will seek to define the sound quality in terms of what is physically present, but will then relate that information to the musical context in which the sound material is presented and perceived. It involves defining sounds and then comparing the sound to others.

Pitch density is an evaluation of pitch-related characteristics of sound quality, at the perspective of the individual sound source. This evaluation will use sound-quality information and relate it to a musical context. Pitch-area analysis (such as the ones performed on percussion sounds in Chapter 7) is another evaluation of pitch-related characteristics of sound quality, only at a lower perspective within the characteristics of the individual sound. This may not necessarily take place within the musical context, and if so it is a critical-listening process. When the information is used to compare how a pitch area compares to other sounds in comprising the overall texture (timbral balance) of the recording, this is in musical context and is analytical listening.

The pitch density and timbral-balance evaluations in Chapter 12 are in musical contexts. The pitch-area analyses of the percussion sounds in Chapter 7 are out of musical context.

Both of these studies are simple (or more general) approaches to timbre and sound-quality evaluation. They both define the pitch-component information that leads to defining timbre, or evaluating sound quality. This process is related to evaluating the spectral content of a sound. In the same way, dynamic contour analyses are related to sound-quality analysis, at various structural levels.

The sound quality of the entire program, or of any individual sound source, may be supplying the most significant musical information in certain pieces of music. This concept of music composition (which can be explained through equivalence) is quite prominent in many very different styles of music. Throughout the twentieth century, a type of writing, *sound mass composition*, evolved through the work of composers Edgar Varèse, George Antheil, Krzysztof Penderecki, Karlheinz Stockhausen (whose photo appears on the album cover of *Sgt. Pepper's Lonely Hearts Club Band*), and Luciano Berio (to name only a few). Their music of this type places an emphasis on the dimensions of the overall musical texture (or sound mass) and/or on the sound-quality relationships within the overall musical texture.

The concept of giving musical significance to the sound quality of the entire program, the relationships of sound qualities, and to pitch density can be found in a wide variety of popular works from the 1960s on. Many examples of these ideas exist, although this concept is used in isolated areas in most works.

The Beatles' "A Day in the Life" is one such work. The concept of sound quality and pitch density is what shapes the music and the dramatic motion of the song's transition section and its conclusion. The sound-mass concept of pitch density is the primary musical element, causing timbre/sound quality to be the dominant artistic element during those sections.

## Sound Quality and Perspective

We will remember that sound quality is our recognition of a sound as a single, unique entity. The listener conceives sound in this way by its overall character, which is composed of the component parts: dynamic envelope, spectrum, and spectral envelope. Further, we recognize sound quality in this way at any number of levels of perspective (perceived detail).

We are able to recognize sound quality most often at the following perspectives for analytical listening:

1. As timbral balance of the overall program (the entire musical texture)
2. As groupings of similar or similarly acting sounds within the overall program (such as a brass or a string section within an orchestra, or the rhythm section of a jazz ensemble)
3. As the interrelationships of the pitch densities of individual instruments
4. As the characteristic overall impression of an individual sound source (instrument, voice, synthesizer patch, special effect)
5. As specific sounds generated by individual sound sources (individual expressive vocal sounds, a specified voicing of a guitar chord, a single sound's timbre characteristics)
6. As environmental characteristics—the sound quality of an environment (explored in the next chapter).

Critical listening also engages all levels of perspective in a similar manner. The recordist is concerned about the integrity of the signal at the overall texture, the individual sound source, any groupings of sound sources (such as a drum mix), an isolated sound, the subtle qualities of spectrum or spectral envelope within a sound, and more.

At these very different levels of perspective, we recognize sound quality as the concept that makes a sound a single, unique entity composed of recognizable characteristics. This global quality is evaluated to determine specific information, to identify the aspects that make each sound unique.

## Evaluating the Characteristics of Sound Quality and Timbre

Sound-quality evaluations will examine spectral content, spectral envelope, and dynamic contour, in both musical contexts and in critical-listening evaluations. Sound-quality evaluation will be approached in this way at all levels of perspective.

In analytical listening at the highest levels, the individual sound sources that make up the overall texture can be conceived as individual spectral components. At the lowest level, individual spectral components are evaluated, and individual evaluations may be performed for each occurrence of a sound source. Individual sound sources may be analyzed for their contributions to the sound quality of the overall program. In musical contexts, this evaluation will compare the sound sources to the overall sound quality through their individual dynamic contours (creating musical balance), their pitch area (creating pitch density evaluations), and their spatial characteristics (Chapter 10).

In critical listening, the contributions of the individual sounds to the overall program can be approached in relation to the same dimensions, but without relation to musical time or context.

Often, the audio professional is concerned with evaluating the individual sound source. Individual sound sources are evaluated for their unique sound quality as sound objects. The sounds are evaluated to define their characteristics through an evaluation of the states and activities of their component parts, out of the musical context.

This is the most widely applied use of the evaluation of sound quality. It is used for many activities from setting signal processors to evaluating the performance of audio devices, and from defining the general characteristics of a sound source (such as the sound quality of a guitar part in a recording) to a detailed evaluation of a particular guitar sound. Many other similar examples can readily be found.

Often, a sound-quality evaluation is performed on a single, isolated presentation of the sound source. An isolated presentation will have its unique pitch level, performed dynamic level, method and intensity of articulation, etc. Among many uses, examining a particular isolated presentation of a sound source allows for meaningful comparison between different performances of the same source, or of a different sound source performing similar material. In practice, several or even many isolated sounds are often compared to determine the most desirable sound or to try to identify the source of a distortion or noise.

Evaluations of sound quality will seek to define and describe the states and activities of the sound source's (1) dynamic envelope, (2) spectral content, and (3) spectral envelope. It will also make use of the listener's carefully evaluated perception of (4) pitch definition.

We will focus our study on timbre or sound-quality evaluations of single sounds. This will allow us to explore the most common application of sound-quality evaluation, and to learn this skill in perhaps the most straightforward manner.

## Defining a Timeline

Most often sound quality and timbre evaluation will take place out of musical contexts. In these critical-listening applications, clock time is used to evaluate the characteristic changes that occur over time. While we need to conceive sound quality as the shape of the sound in an instant, or out of time, sound only exists in time. Sound can only be accurately evaluated as changes in states or values of the component parts, which occur over time. Sound quality in critical-listening applications approaches the sound as an isolated, abstract object. The sound is understood as an object that has a characteristic shape that unfolded over time.

The evaluation may also take place within musical contexts. In these instances, the time of the metric grid will be used, when it is present. Evaluation of sound quality will be focused on the musical relationships of the material, and usually takes place at a higher perspective than critical listening's timbre evaluations.

**Listen**

to tracks 26–33 again

for the exercise in learning the sound quality of small time units. This will assist you in defining timelines for timbre and sound-quality evaluations—and much more.

Sound-quality evaluations of single sounds will nearly always use clock time in tenths or perhaps hundredths of seconds in the timeline. The timeline of the sound is determined first in any sound-quality evaluation. Next, increments within the timeline and suitable reference points within the timeline will be identified. Skill needs to be acquired in performing these tasks. The time exercises (5.3, 5.4, and 5.11) of Chapter 5 will greatly assist this development, and the reader is encouraged to revisit them now. While determining a timeline will require a number of hearings, this number will be reduced with experience and increased ability.

### Defining the Four Components of Sound-Quality Evaluations

The physical dimensions of the sound source (dynamic envelope, spectral content, and spectral envelope), in their unique states and levels, are used to define sound quality. The definition of fundamental frequency (pitch definition) aids in defining information of those values, especially the loudness level of the fundamental frequency in relation to the remainder of the sound's spectrum and the dominance of harmonic partials. These four components of sound-quality evaluations are examined throughout the duration of the sound material and are plotted against a single timeline. It is important to note that pitch definition and dynamic envelope exist at the perspective of the overall sound, and that spectrum and spectral envelope are internal components of the sound and are recognized only at a lower level of perspective.

#### *Pitch Definition*

Pitch definition will become the focus immediately after the timeline has been drafted. This definition of the fundamental frequency is useful in making preliminary and general observations of a sound. Definition of fundamental frequency is often somewhat stable during the sustain portion of a given sound. Changes in pitch quality are most commonly found between the onset and the body of the sound. Pitch quality is placed on a continuum between the two boundaries of well defined in pitch or precisely pitched (as a sine wave), through completely void of pitch or nonpitched (as white noise). A dominance of harmonics will bring the sound to have a more defined pitch quality. With increased presence of overtones (in either number or loudness level) comes a more nonpitched character to the sound.

Often the definition of fundamental frequency is verbally described as having a certain pitch quality for a certain portion of its duration, then another certain quality for the remainder of its duration. In effect, a contour of pitch definition is present. The pitch-definition tier of the sound-quality characteristics graph is not always required, but this aspect of the sound should always be addressed, if only during the early stages of the evaluation.

Examining pitch definition supplies many clues to the content of the spectrum and the spectral envelope. The less pitched the sound, the more prominent the overtone content; the recordist can then be drawn to determining what overtones are present and when they occur. With focus, one can trace how pitch definition changes over the sound's duration; then one can seek information on spectral content and spectral envelope, using this information as a point of departure. Pitch definition and dynamic envelope are also often linked during the onset of a sound or when the spectrum becomes active or dense.

### *Dynamic Contour*

The dynamic contour of the sound, or the sound's overall dynamic level as it changes throughout its duration, is reasonably apparent at first hearings. Difficulties may arise with confusing loudness changes and spectral complexity changes. The listener must remain focused on actual loudness and not be pulled to other perceived parameters of sound.

A reference dynamic level (RDL) will be required for mapping the dynamic contour. The RDL will be determined by the intensity level at which the source was performed. The intensity level itself is the RDL; it will be transferred to a precise dynamic level (a specific point in a dynamic area such as "mezzo piano" or "forte"). The same reference dynamic level will be used in the spectral-envelope tier, explained below. In this way, the same reference level functions on two levels of perspective, just as one reference dynamic level functions for both program dynamic contour and for musical balance (in Chapter 8).

The dynamic contour is readily described. This is accomplished by discussing the shape and speed of the dynamic envelope, and its dynamic levels at defined points in time. By discussing how loudness changes and by defining the levels and speed of those changes, the listener is describing this physical element as others are experiencing it. This is an important component of the unique objective character of the sound.

***Spectral Content***

Readers may hear few or no spectral components at the beginning of their studies. Harmonics and overtones fuse to the fundamental frequency, and we have been conditioned to perceive all of this information as part of a whole (the global sound quality). To a great extent, to evaluate sound quality we must work against many learned listening techniques and our previous listening experiences. Much patience and focused attention will be required. Practice and repetitive listening must be undertaken to acquire the skills of accurately recognizing spectral components and of accurately tracking dynamic contours of the components that make up spectral content.

The harmonic series can be an important tool to assist in identifying spectral components. The listener can learn to envision the harmonic series as a chord above the fundamental frequency—a chord with specific voicing, or sequence of intervals. Once the listener has learned the sound of this chord, it will be possible to focus attention on the individual pitches of the harmonic series while listening to a sound. The listener will then be in a better position to identify any pitches/frequencies of the harmonic series that might be present. Frequencies/pitches other than harmonics will also be noticed; the listener will ultimately be able to quickly calculate where the overtones fall in relation to the envisioned harmonic series. Frequencies/pitches that lie between harmonics will be able to be identified as being "between the fourth and fifth harmonic," for example, and a bit more attention will bring the listener to recognize the partial's frequency/pitch more precisely. In this way, the harmonic series is used as a template, to which the frequencies/pitches present can be compared and identified. An internal sense of the sound of the harmonic series can be a point of reference that makes evaluating spectral content much more approachable.

**Listen**

to tracks 1 and 2

for harmonic series played in individual frequencies and pitches, and as a chord. Work to recognize this pattern and spacing of intervals, and learn the "sound quality" of the chord that comprises the harmonic series.

The reader is encouraged to return to the discussion of the harmonic series in Chapter 1, to study its content and to spend time learning the sound of the series in Exercise 5.10 in Chapter 5. The reader is also encouraged to review the conversion of pitch levels into frequency levels and the reverse.

The reader should feel free to use whatever devices are available to assist in identifying partials. This will especially prove helpful in initial studies. A tone generator or keyboard may be used to assist in identifying frequency levels or pitch levels of prominent harmonics and overtones by matching pitches. A filter can be used to eliminate all but the fundamental frequency, and partials can be added one at a time; it can also be used to eliminate the fundamental frequency so the harmonics and overtones can be heard more directly. An equalizer can emphasize certain frequency bands to assist the reader in determining where the spectral components exist. It is important to remember to perform activities that engage the mind and the ear in searching for the information, learning to listen at the perspective of the individual partials, learning the sound of the spectrum, using one's knowledge of the content and sound quality of the harmonic series. DAWs and spectral-analysis software can sometimes assist one in evaluating sounds; this could be useful in helping one *acquire* this listening skill, but one must be wary of not letting the software *replace* this skill. It is important for the reader to be actively seeking the information in a thoughtful and methodical manner, and to engage the ears and mind in the process.

The reader will be able to describe much about a sound by addressing the spectrum. Defining which harmonics are present and those that are prominent, as well as indicating overtone content, provides a significant amount of objective information about a sound.

### *Spectral Envelope*

Describing the entrances and exits of the partials (harmonics and overtones) as well as the individual dynamic contours of these partials (the spectral envelope) will provide additional important information on sound quality. The spectral envelope will be calculated against the reference dynamic level that was identified for the overall dynamic contour and against a common timeline. All of the spectral components identified, as spectral content, will be present in the spectral-envelope tier, including the fundamental frequency.

The spectral envelope maps the dynamic levels and contours of all partials. Much significant information on pitch definition and the character of the sound is contained in the spectral envelope. By describing this activity, the listener is communicating very pertinent and meaningful information about the sound, which is completely objective, and can be experienced and understood by others.

If individual spectral components were very difficult to separate from a fused spectrum, hearing how those components change in terms of loudness over time can be extraordinarily difficult. When one becomes adept at hearing individual partials (especially the more prominent lower partials), one can begin to make observations on how those partials change in amplitude over time and how their level relates to the RDL. The information of how the spectrum changes over time is very important, and even general observations that stem from the pitch-definition tier can lead one to understand a bit of this information.

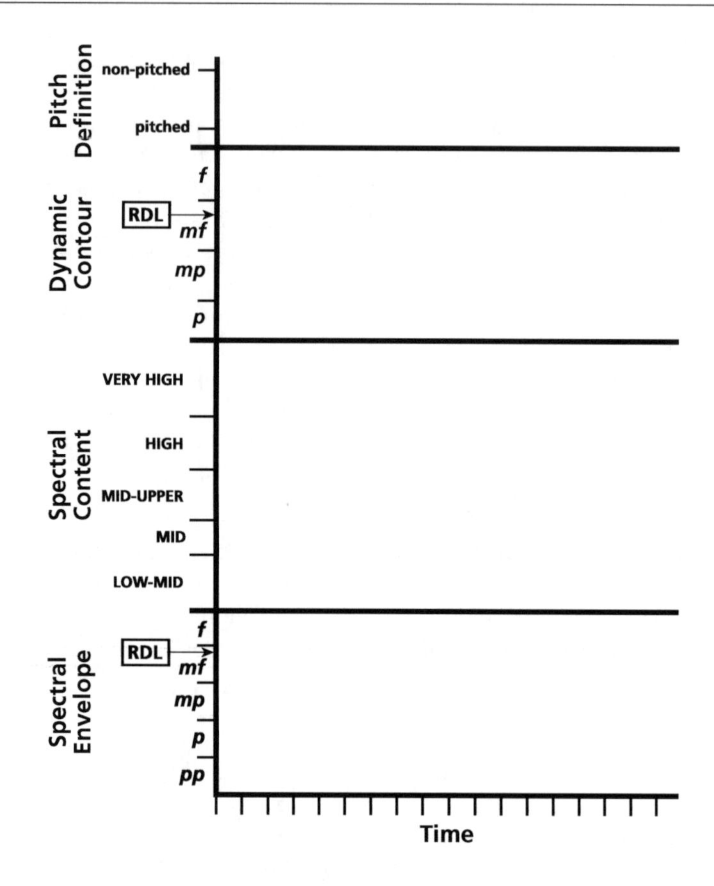


**Figure 9.1** Sound-quality characteristics graph.

## Process of Evaluating Sound Quality

The *sound-quality characteristics graph* will be used to help make sound-quality evaluations. The graph greatly assists in a detailed evaluation of a sound and allows the reader to record observations of any sound, whether or not a detailed evaluation is undertaken. Pertinent information can be written down that can provide a resource for objective descriptions formulated at a later time.

The process of evaluating sound quality should follow this sequence of events:

1. During the first hearing(s), listen to the example to establish the length of the timeline. During this time you may become aware of prominent states and activity of the components (especially dynamic contour and spectral content); if so, make some notes against the timeline.
2. Check the timeline for accuracy and make any alterations.
3. Notice the activity of the component parts of sound quality for their boundaries of levels of activity and speed of activity. The speed

boundaries will establish the smallest time units required in the graph to accurately present the smallest significant change of the element. The boundaries of levels will establish the smallest increment of the *Y*-axis required to plot the smallest change of each component (dynamic contour, spectral content, spectral envelope).

4. Pitch definition observations are next. Begin by making general observations about the pitch quality of the sound. Identify precise points in time when the sound is at its most pitched and least pitched states, and assign a relative value. Use these levels as references to complete this contour of pitch definition. The prominence of the fundamental frequency and the number and relative loudness levels of harmonics greatly influence the pitch quality of the sound source. Knowing when a sound is more pitched than other times provides information that can be used in determining the spectral content and spectral envelope of the sound.
5. Begin plotting the activity of the dynamic envelope on the graph. First, determine the RDL of the sound by identifying its performance intensity. Second, establish the beginning and ending dynamic levels of the sound. The highest or lowest dynamic levels are the next to be determined; then place them against the timeline. Use these levels as points of reference to judge the activity of the preceding and following material. Alternate your focus on the contour, speed, and amounts of level changes to complete the plotting of the dynamic contour. The evaluation is complete when the smallest significant detail has been perceived, understood, and added to the graph. The smallest time increment of the timeline may need to be altered at this stage to allow the dynamic contour to be clearly presented on the graph.
6. Plot the spectral content on the graph. First, identify the frequency/pitch levels of the prominent spectral components and the fundamental frequency. Knowledge of the sound of the harmonic series will prove valuable here. Map the presence of these frequencies against the timeline, clearly showing their entrances and exits from the spectrum. Finally, map any changes in the pitch/frequency levels of these partials against the timeline. Certain spectral components may not be present throughout the duration of the sound. It is not unusual for harmonics and overtones to enter and exit the spectrum. This evaluation is complete when all of the spectral components that can be perceived by the listener are added to the graph. Accuracy and detail will increase markedly with experience and practice on the part of the listener. With time and acquired skill, this process will yield much significant information on the sound. Initial attempts may not yield enough information to accurately define the sound source, but will improve substantially over time.

7. Plot the dynamic activity of the partials on the spectral-envelope tier of the graph. Use the same RDL as the dynamic-envelope tier. First, establish the beginning and ending levels of each of the spectral components that were identified in Step 6. For each of the spectral components, determine the highest or lowest dynamic levels and any other prominent points of reference within the dynamic contours. Use these points of reference to evaluate the preceding and following material. To complete plotting the activity, alternate focus on the contour, speed, and amounts of level changes. The dynamic envelopes of all of the spectral components are plotted on this tier. The evaluation is complete when the smallest significant dynamic-level change has been incorporated into the graph.

Many hearings will be involved for each of the steps above. Each listening should seek specific new information and should confirm what has already been noticed about the material. Remember to work from what is known, to determine what is unknown.

**Listen**

to track 3

for the harmonics and overtones of the sustained piano notes. You will notice changes in the spectrum of the pitches over their long durations.

Before listening to the material, the listener must be prepared to extract certain information, to confirm their previous observations, and to be receptive to new discoveries about the sound quality. A clear sense of the correct perspective and the specific information that is being sought will make the listening session more successful. The listener should check any previous observations often, although their listening attention may be seeking new information.

The sound-quality characteristics graph incorporates:

1. A multitiered *Y*-axis, distributed to complement the characteristics of the musical example: one tier with dynamic areas for dynamic envelope (with notated RDL), one tier with pitch register designations for spectral content, one tier with dynamic areas for spectral envelope (with notated RDL), and a fourth tier designating a pitched-to-nonpitched scale
2. The *X*-axis of the graph dedicated to a timeline that is divided into an appropriate increment of clock time (the increment will vary depending on the material)
3. Each spectral component plotted as a single line, against the two axes; its pitch characteristics on the spectral-content tier, and its dynamic contour on the spectral-envelope tier;
4. Each spectral component's line should have a different color or numbered line, to make each component the same on the two tiers.

## Sample Evaluations

Several sound-quality evaluations follow. Two are synthesized sounds and the other is a highly modified (feedback, etc.) electric guitar sound. These invented timbres are evaluated to determine their unique characteristics for critical-listening use and to better understand their relationships to other sounds in the music.

Each evaluation is of one identified sounding of the single sound source. A specific appearance of each sound source was selected and evaluated. The appearance was selected because the sounds' characteristics are not being masked by other sound sources and because most of the sounds' characteristics were audible in the specific example. This type of evaluation is common in recording production; although this type of detail and graphing is not present, recordists listen for this information and make observations, though they may not formulate the material in this way.

Figure 9.2 shows the great complexity of the pitch definition of the opening guitar sound from "It's All Too Much" (*Yellow Submarine*, 1999). The pitch definition changes are reflected in changes to the spectral envelope. They are closely associated. In observing the spectral-content tier, one can determine the dominance of harmonics and note the overtones present. The dynamic envelope exhibits many subtle changes in loudness level throughout the nearly 15-second duration of the sound.

Figures 9.3 and 9.4 are sound-quality evaluations of Moog synthesizer sounds from *Abbey Road*. The different waveforms used for the two sounds make for differences in spectral content. The simplicity of the early instrument is reflected in the basic contours of the dynamic envelope and spectral envelope, and the inclusion of only harmonics in the spectrum of each sound.

As an exercise, bring your attention to focus on each tier during four separate hearings. Try to identify all of the information present in these three sound-quality evaluations. Search the sound quality of the instrument to find the graphed information.

A "describing sound" exercise appears at the end of this chapter. All people in the audio industry talk about sound, and in doing so describe its qualities. This exercise will get the reader to think about sound for what is present, not how it makes them feel or how it reminds them of other senses or experiences. When the four components of sound quality are described, the content and character of the sound can be more precisely and clearly communicated.

The Moog sound from "Maxwell's Silver Hammer" that appears in Figure 9.3 could be described in this way. Pitch definition begins at nonpitched at the attack of the sound, and moves to pitched by 0.1 seconds, where

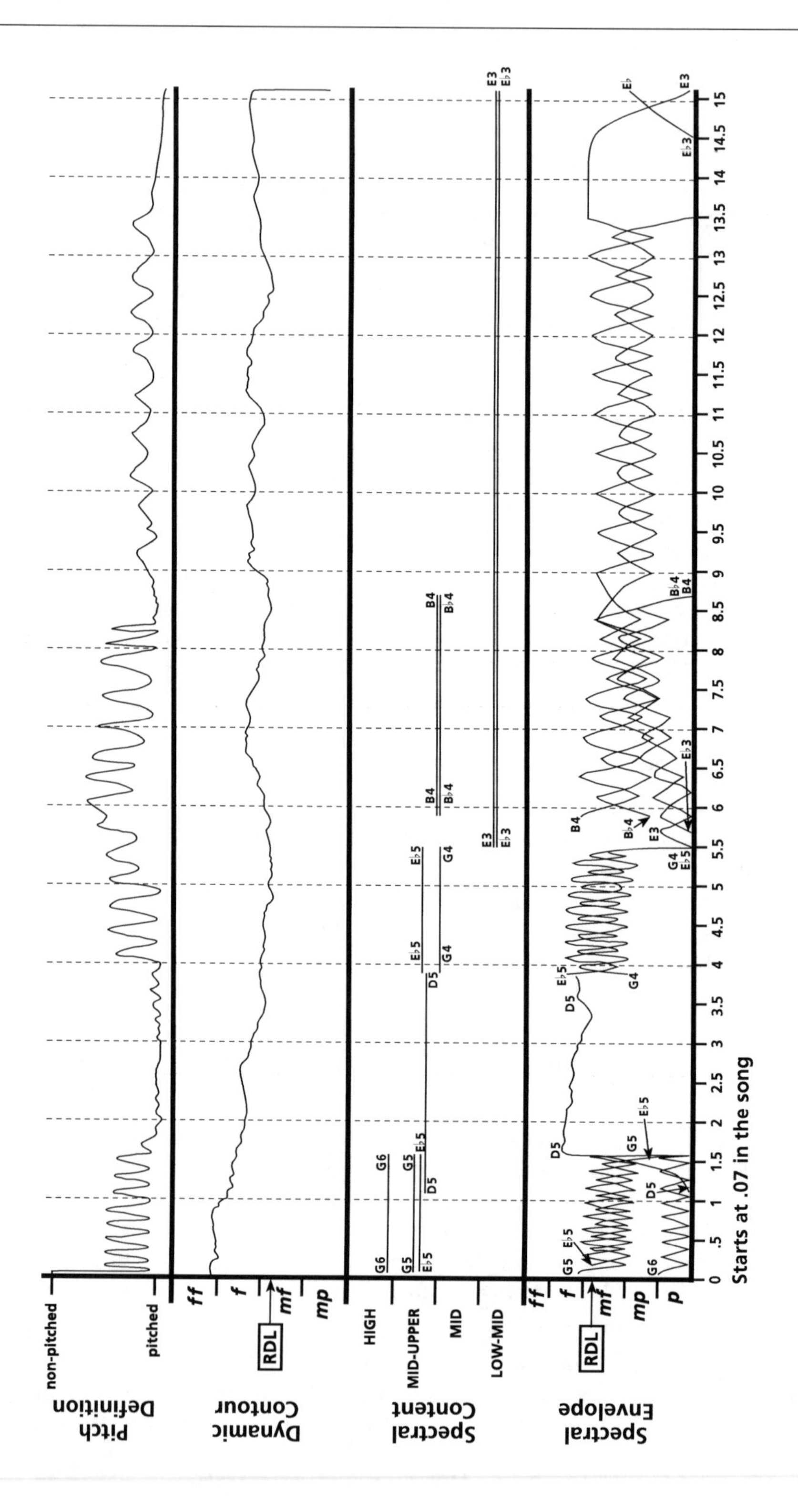


**Figure 9.2** Sound-quality evaluation of the opening guitar sound from The Beatles' "It's All Too Much" (*Yellow Submarine*, 1999).

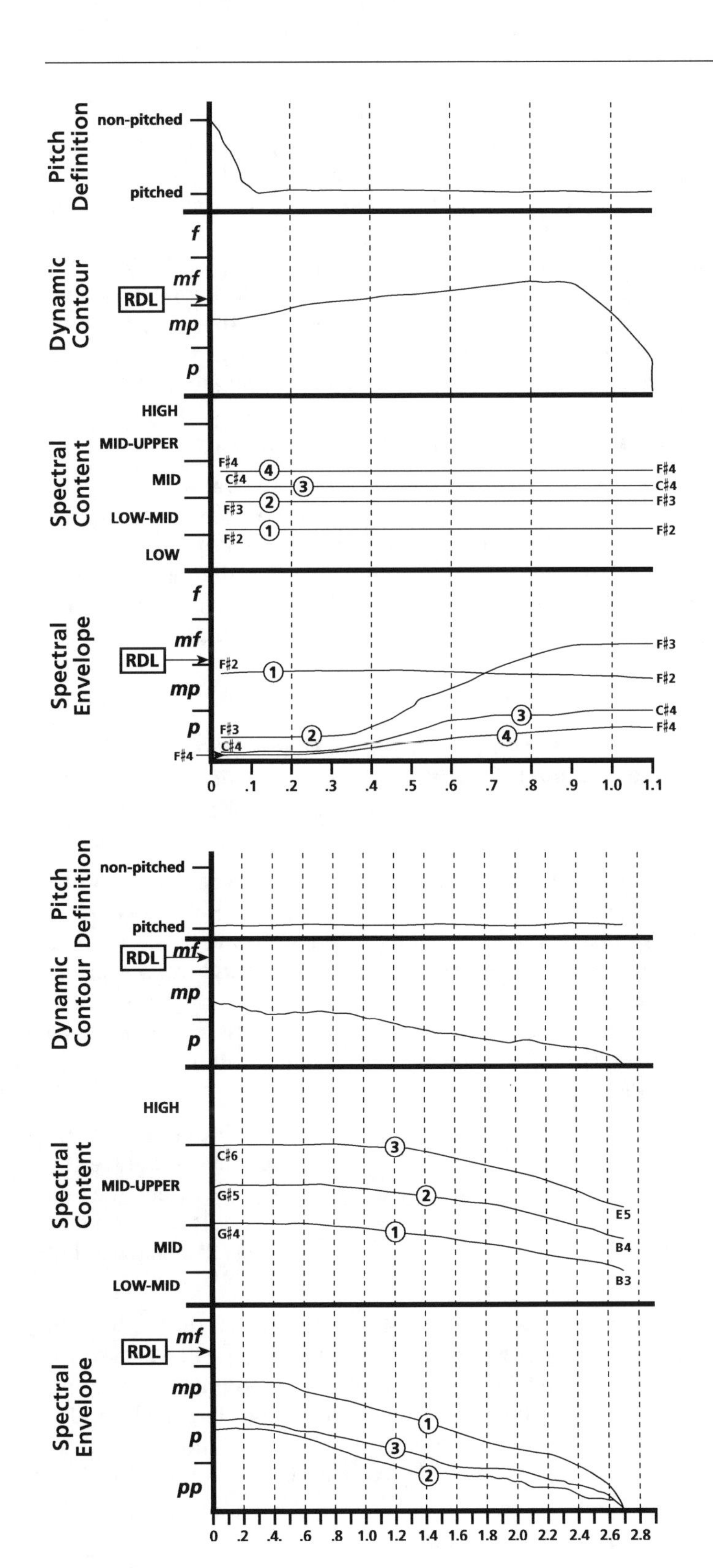


**Figure 9.3** Sound-quality evaluation of the Moog synthesizer sound from The Beatles' "Maxwell's Silver Hammer," at 51.1 seconds *(Abbey Road).*

**Figure 9.4** Sound-quality evaluation of the Moog synthesizer sound from The Beatles' "Here Comes the Sun," at 0:12 *(Abbey Road).*

it remains pitched for the remaining second of the sound's duration. Its dynamic envelope begins in ***mp*** and gradually rises to mid-***mf*** by 0.8 seconds, where it remains before beginning a gradual decay at 0.9 seconds; all dynamics are relative to the sound's RDL of about 10 percent into ***mf***. The spectrum of the sound is composed of the lower four harmonics based on a fundamental frequency of F#2. The spectral envelope demonstrates dramatic changes in level of the second harmonic (F#3), as it moves from much softer to louder than the fundamental, and the third and fourth harmonics increase loudness by lesser amounts but at the same time and speed as the second harmonic; the fundamental frequency is at the top 5 percent of ***mp*** at the beginning of the sound, and changes loudness in a slight arc over the sound's duration.

The reader should next perform the sound-quality evaluation exercise at the end of this chapter. A very important skill will be gradually acquired with some focused effort. As skill develops over a period of time, you will gain considerable awareness of the content of sound qualities. You will begin to hear things you previously could not imagine. A new world of sound will present itself, as will a new way of understanding that world.

## Summary

The ability to evaluate and communicate about sound quality and timbre is extremely important for the synthesist, sound designer, recording engineer, and producer. It is also required of nearly all positions in the audio industry.

Sound-quality evaluation is used in many activities in recording production. We evaluate sound when we create an equalizer setting or set a compressor; we compare sound qualities when we move microphones around to determine the most appropriate placement, and evaluate sound when selecting microphones. We evaluate sound to identify the quality of the signal we are recording, and to verify that quality (on all levels of perspective and in all of the parameters of sound) throughout the process of crafting a recording.

The sound-quality evaluations performed by following the method presented in this chapter will allow the reader to readily recognize the unique characters of sounds. Learning to recognize and describe the activities of the component parts of sound quality will allow the reader to talk about sound articulately and to share information that is pertinent and readily understood by others.

The skills gained through the previous chapters are brought together in the many steps of sound-quality (and timbre) evaluation. Pitch and pitch-area estimation, melodic and dynamic-contour mapping, and judging time increments are all now used for a more demanding task, and a very important one. These skills should become highly refined, and be continually developed through carefully considering how the recording process is altering, capturing, or creating sound quality. The audio professional will be continually engaged in evaluating sound. Recognizing and understanding the characteristics of sound quality are the first steps toward communicating accurate and relevant information about sound. Learning to hear and recognize the components of sound quality/timbre will allow the recordist to control their craft in shaping recordings.

## Exercises

### Exercise 9.1

*Describing Sound Exercise*

The purpose of this exercise is to develop an approach to talking about sound that discusses the sound's physical components.

1. Select a sound and record it, playing only one pitch. A sound from the companion website may be used.
2. Write down your observations of the timeline: How long does the sound last?
3. Make general observations of the pitch definition of the sound. Does it start with a burst of noise (like a piano)? Are there areas where the sound changes in pitch quality? Is the sound mostly pitched or mostly noise-like?
4. Next describe the dynamic envelope. How does the dynamic envelope change during the sound's duration? What is the speed of the attack and initial decay? What is the level of the attack? What is the sustain level in relation to the attack? How does the sound end?
5. Describe the spectrum of the sound. Is it dominated by harmonics? Where are overtones present in relation to the fundamental? Is there a different spectrum during the onset than in the body of the sound?
6. Describe the spectral envelope. Are some partials prominent? Is the fundamental louder than the remainder of the spectrum? How does the spectrum change over time?

Practice talking about sound in this way whenever you are working with an audio device. Ask yourself: What am I hearing related to the actual sound wave? What are its current qualities, or how are those qualities changing?

This will bring you to be able to quickly evaluate sounds in a meaningful way, and to be able to explain to others what needs to be done to obtain desired results, or what the wonderful qualities of your drum sounds are—specifically and understandably.

Practice describing sounds this way without first creating a sound-quality evaluation graph, and then after creating the graph.

## Exercise 9.2

*Sound-Quality Evaluation Exercise*

Find a sound that is a complex waveform (containing overtones as well as harmonics) with a duration of at least five seconds. A single occurrence of the sound should be identified and evaluated. A sound that is isolated from other sounds (does not have other sounds occurring simultaneously) will be easiest to evaluate. Make a recording of the sound to more easily repeat hearings of the sound.

Perform a sound-quality evaluation on the sound by using the following sequence of activities. See this chapter for more detail, as needed.

1. During the first hearing(s), listen to the example to establish the length of the timeline.
2. Check the timeline for accuracy and make any alterations.
3. Notice the activity of the component parts of sound quality for their boundaries of levels of activity and speed of activity, and establish the smallest increment of the *Y*-axis required to plot the smallest change of each component (dynamic contour, spectral content, spectral envelope).
4. Place pitch-definition observations on the graph.
5. Plot the activity of the dynamic envelope on the graph.
6. Identify the spectral content of the sound and place the partials on the graph.
7. Plot the dynamic activity of the partials on the spectral-envelope tier of the graph.

When completed, review your sound-quality evaluation graph and compare the activities of all tiers. Summarize and describe how the physical dimensions of the sound appear and change throughout the duration of the sound, as in Exercise 9.1, but in greater detail.

# 10 Evaluating the Spatial Elements of Two-Channel Sound

The spatial characteristics and relationships of sound sources are an integral part of music and audio productions. Spatial elements are precisely controllable in audio recording, and sophisticated ways of using these elements have developed in audio and music productions. Spatial relationships and characteristics often present significant qualities in current music recordings, and are also important considerations in critical-listening applications.

The evaluation of the spatial characteristics of stereo recordings covers three primary areas: (1) localization on a single horizontal plane in front of the listener, (2) localization in distance from the listener, and (3) the qualities of environmental characteristics. Surround-sound recording replaces localization in front with localization 360° around the listener, and will be considered separately in Chapter 11. The elements of environmental characteristics and distance illusions further interact and create other sound characteristics that must be evaluated in both formats.

The recordist must be able to evaluate these characteristics to properly evaluate recorded/reproduced sound. Many of the skills required to evaluate the spatial characteristics of a recording have been gradually developed throughout the previous four chapters. These skills of sound-quality evaluation, time judgments, pitch estimation, and dynamic contour mapping will be used again (from a new perspective) to recognize and evaluate the spatial elements of reproduced sound. The further development of these skills will again require patience and practice.

An accurate evaluation of the spatial elements is only possible under certain conditions. The listener must be located correctly with respect to the loudspeakers of the playback system. This is critical to accurately hear directional cues. The sound system must interact correctly with the listening environment to complement the reproduced sound. Reproduced sound can be radically altered by the characteristics of the playback room and the placement of loudspeakers within the room. Further, the sound system itself must be capable of reproducing frequency, amplitude, and spatial cues accurately.

Specific spatial aspects of two-channel, stereo recordings will be covered in this chapter. The spatial qualities specific to surround-sound recordings will be covered in the following chapter. The general aspects of this chapter will apply to both stereo and surround formats.

Many of the concepts of the spatial elements have not previously been well defined. The length of this chapter is the result of the number of important spatial elements of sounds in recordings, the methods one must use to perform meaningful evaluations of these elements, and the explanations required of new concepts.

## Understanding Space as an Artistic Element

Spatial characteristics and relationships are used as artistic elements in music productions. They are used as primary and secondary elements that help to shape the unique character of musical ideas. Space has the potential of being the most important artistic element in a musical idea, but most often serves to support other elements. It may support other elements by delineating musical materials, by adding new dimensions to the unique character of the sound source or musical idea, and/or by adding to the motion or direction of a musical idea. It is also among the primary qualities of the overall texture of the recording.

### Perceived Performance Environment

The listener acquires an impression of the spatial characteristics of the recording through the sound stage and imaging. They will imagine a performance space wherein the reproduced sound can exist during the re-performance of listening to the recording. The listener will perceive individual sound sources to be at specific locations within a *perceived performance environment.*

The recording represents an illusion of a live performance. The listener will conceive the performance as existing in a real, physical space, because the human mind will interpret any human activity in relationship to the known physical experiences of the individual. The recording will appear to be contained within a single, perceived physical space (the perceived performance environment), because in human experience we can only be in one place at one time.

The perceived performance environment will have an audible, characteristic sound quality that is established in one of two ways. The qualities of the perceived performance environment may be established by applying a set of environmental characteristics to the overall program (i.e., processing the final mix). Most often, the perceived performance environment is a composite of many perceived environments and environmental cues. In these instances, the listener formulates an impression of a perceived performance environment through interrelationships of many environmental characteristics cues. These cues may be (1) common or complementary between the environments of the individual sound sources, (2) prominent characteristics of the environments of prominent sound sources (sources that present the most important musical materials, or the loudest, the nearest, or the furthest sound sources, as examples), and/or (3) the result of environmental characteristics found in both of these areas.

Further, the listener will obtain a sense they are at a specific location within the perceived performance environment. The listener might be aware of their relationship to the sidewalls and any objects (balconies, seating, etc.) in the performance environment, to the wall behind and the ceiling above their location, and of their relationship to the front wall of the performance environment.

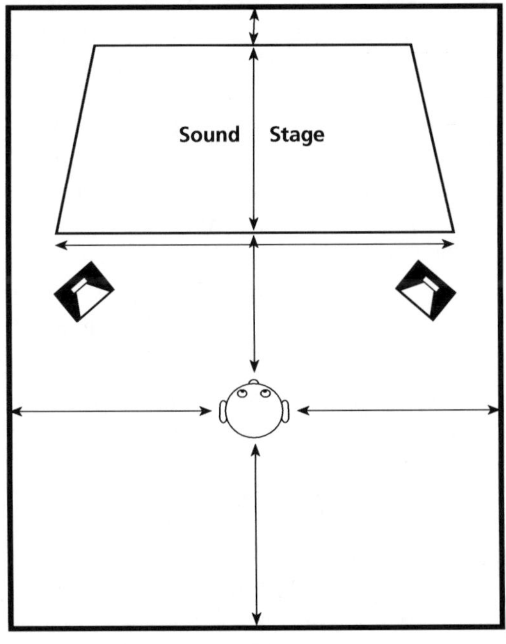


**Figure 10.1**
The sound stage within the perceived performance environment.

The listener will also calculate their location with respect to the front of the sound stage.

## Sound Stage and Imaging

Within this perceived performance environment is a two-dimensional area (horizontal plane and distance) where the performance is occurring—the *sound stage*. The sound stage is the area where the sound sources are perceived to be collectively located, as a single ensemble. The listener will unconsciously group all sources into a single performance area. The performance (that is the recording) will thus emanate from a single location.

The area of the sound stage may be any size. The size of the sound stage may appear to be anything from a small, well-defined area (an infinitesimally small world), to a space occupying an area extending from immediately in front of the listener to a location (spanning a great distance) well beyond our sight line (perhaps conceived as being an area beyond the size of anything possible on Earth) and, within the horizontal plane, filling an area stretching beyond the stereo array.

The sound stage may be located at any distance from the listener. The placement of the front edge of the sound stage may be immediately in front of the listener, or at any conceivable distance from the listener.

The perceived distances of the sound sources from the listener determine the depth of the sound stage. The sound source that is perceived as nearest to the listener will mark the front edge of the sound stage. The sound source that is perceived as being furthest from the listener will define the back of the sound stage and will also help to establish the rear wall of the perceived performance environment. The perceived location of the rear boundary (wall) will be determined by the relationship of the furthest sound source to its own host environment. The rear wall of the perceived performance environment may be located immediately behind the furthest sound source, or some space may exist between the furthest sound source and the rear wall of the sound stage/perceived performance environment.

All sound sources will occupy their own location in the sound stage. Two sound sources cannot be conceived as occupying the same physical location. Our sensibilities will not allow this to occur. It is possible for different sound sources to occupy significantly different locations within the sound stage, anywhere between the two boundaries. *Imaging* is the perceived location of the individual sound sources within the two perceived dimensions of the sound stage (see Figure 10.2). Sources are located within the sound stage by their angle (on the horizontal plane) and distance from the listener.

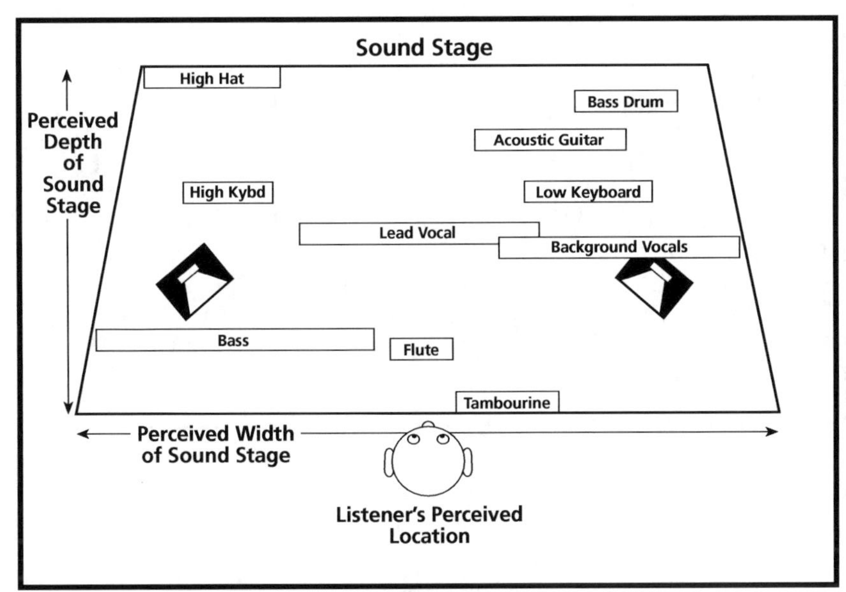


**Figure 10.2** Imaging of sound sources within the sound stage.

## Environments of Individual Sources

The placement of each source within its own environment influences imaging. The characteristics of the unique performance environments of each sound source might enrich source width and distance cues, and enhance the dimensions of the sound stage. In current music productions, it is common for each instrument (sound source) to be placed in its own host environment. This host environment of the individual sound source (a perceived physical space) is further imagined to exist within the perceived performance environment of the recording (a perceived physical space). This creates an illusion of a *space* existing *within* another *space*.

> **Listen**
> to tracks 42–44
> for narrow and wide guitar phantom images, and a narrow image broadened by reverberation.

The environments of the sound sources and the overall program may be of any size. The acoustical characteristics of just about any space may be simulated by modern technology. The sound sources may be processed so that the cues of any acoustical environment may be added to the individual sound source, to any group of sound sources, or to the entire program. Not only is it possible to simulate the acoustical characteristics of known, physical spaces, it is possible to devise environment programs that simulate open air environments (under any variety of conditions) and

Figure 10.3
Space within space.

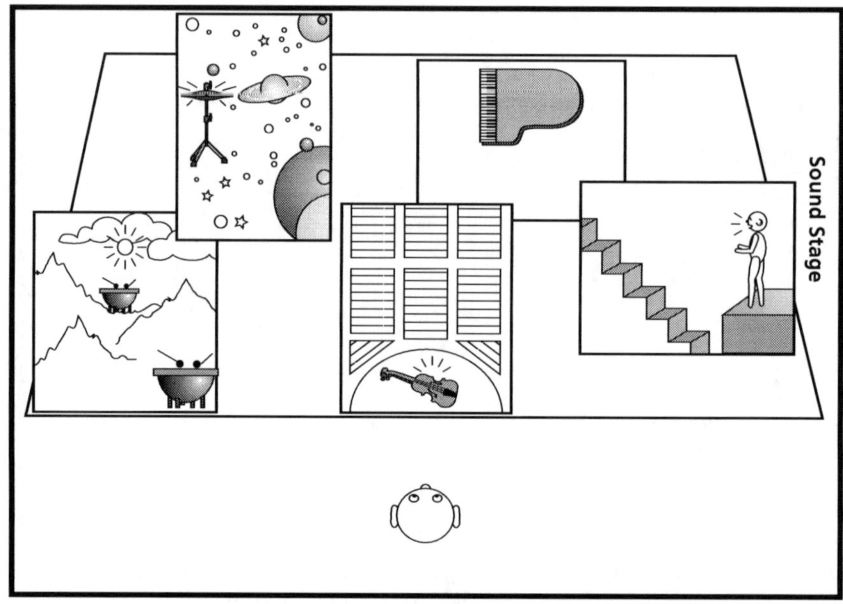


programs that provide cues that are acoustically impossible within our known world of physical realities.

Figure 10.3 presents an easily accomplished set of environmental relationships, with individual sound sources appearing in very different and unique environments:

- Timpani placed in an open-air environment
- A stringed instrument placed in a large concert hall
- A vocalist performing in a small performance hall
- A piano sounding in a small room
- A cymbal appearing to exist in a very unnatural (perhaps otherworldly or outer space), remarkably large environment.

These many simulated acoustical environments are perceived as existing within the overall space of the perceived performance environment. The spaces of the individual sources are within the space of the perceived performance environment. It is possible for more than one source to be placed within an environment. Sources contained within the same environment may have considerably different distance locations, or they may be similar.

The environments of the sound sources and the overall program may be in any size relationship to one another. The environment of a sound source may have the characteristics of a physically large space, and the perceived performance environment may have the characteristics of a much smaller physical environment. This is a common relationship, and the reverse is also possible (though more difficult to achieve). The spaces of the individual sound sources are understood (by the listener) to exist within the all-encompassing perceived performance environment, no matter the perceived physical dimensions of the spaces involved.

**Listen**

to tracks 45–47

for a number of different space-within-space relationships.

The spaces of the individual sound sources are subordinate spaces that exist within the overall space of the recording. A further possibility (not commonly used, at present) exists for subordinate spaces to appear within other subordinate spaces, within the perceived performance environment. *Space within space* is a hierarchy of environments existing within other environments. Its creative applications have not been fully exploited in current music production practices.

The characteristics of the perceived performance environment function as a reference for determining the characteristics of the individual environments of the individual sound sources. All of the environments of a recording will have common characteristics that are created by the characteristics of the perceived performance environment (as discussed above). These characteristics provide a reference for determining the unique characteristics of the individual performance environments of the individual sound sources.

The characteristics of the perceived performance environment also function as a frame of reference for the listener in determining the distance locations of the individual sound sources within the sound stage.

### Distance in Recordings

*Distance* is perceived as a definition of timbral detail. It is further calculated in relation to the characteristics of the environment in which the sound exists, as well as the perceived location of the sound source and the listener within that environment. The listener will perceive the distance of the sound source as it is sounding within its unique environment. The listener will then unconsciously transfer that distance to the perceived performance environment, combining any perceived distance of the source's environment from the listener's location in the perceived performance environment.

The actual distance-location placement of the sound source within the sound stage is determined by (1) the distance between the sound source and the perceived location of the listener within the individual source's environment combined with (2) the perceived distance of that environment from the listener's location in the perceived performance environment. All this information blends into a single impression of distance. Through this process, sound sources (with and within their environments) are conceived at specific distances from the listener. This is all accomplished subliminally.

The placement of sounds at a distance and at an angle from the listener (imaging) takes place at the perspective level of the individual sound source.

### Directional Location

Sound sources (with their individual environments and conceived distance locations) will be located at an angle from the listener. Directional location is used differently in stereo and surround formats.

The *stereo location* will place sound sources on the sound stage, within the stereo loudspeaker array, at an angle of direction from the listener. The size of sound-source images can be narrow and a precisely defined point in space, or it can occupy an area between two boundaries. Sources that occupy an area may be of any reproducible width and may be located at any reproducible location within the stereo array. Further, under certain production practices, it is possible for sound sources to appear to occupy two separate locations or areas within the stereo array.

## Stereo-Sound Location

Sound location is evaluated within the stereo array to determine the location and size of the images of the sound sources. These cues will hold significant information for understanding the mix of the piece and may contribute significantly to shaping the musical ideas themselves. Phantom images may also change locations or size during a piece of music. These changes may be sudden or gradual. The changes may be prominent or subtle.

The *stereo-sound location graph* will plot the locations of all sound sources against the timeline of the work. The graph (see Figure 10.4) portrays the direction of sources from the listener and the size of the phantom images.

Left and right loudspeaker locations and the center position are identified on the graph. The actual boundaries of the vertical axis extend slightly beyond the loudspeaker locations (up to 15°). Placing the sound source's location within the L–R speaker-location boundaries represents source angle from the listener. Precise degree-increments of angle can be

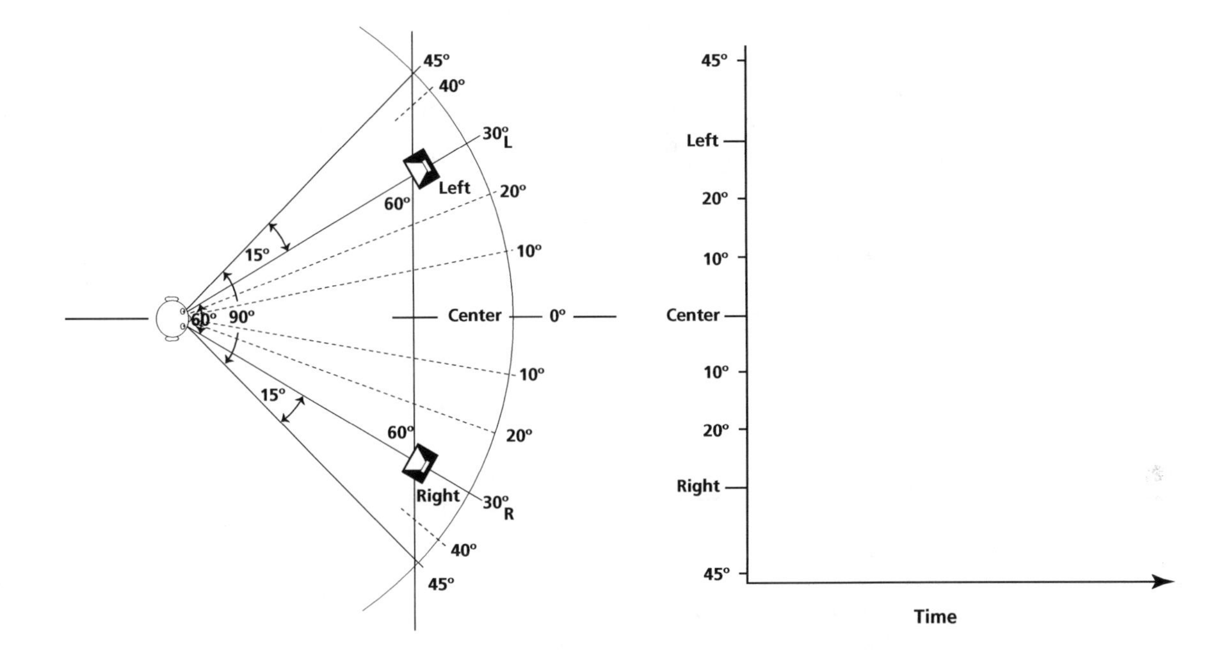


**Figure 10.4**
Calculation of degree-increments for location and the *Y*-axis of the stereo-sound location graph.

incorporated into the graph; these may be unnecessary, depending on the nature of the recording.

The *stereo-sound location graph* incorporates:

1. The left and right loudspeaker locations, a designation for the center of the stereo array, and space slightly beyond the two loudspeaker locations as the *Y*-axis; degree divisions may be added to the *Y*-axis for greater detail if the music will be better understood by such divisions
2. The graph will become unclear if too many sound sources (especially many spread images) are placed on the same tier; the *Y*-axis may be broken up into any number of similar tiers (each the same as listed in Step 1) to clearly present the material on a single graph
3. The *X*-axis of the graph is dedicated to a timeline that is divided into an appropriate increment of the metric grid, or is representative of a major section of the piece (or the entire piece) for sound sources that do not change locations
4. A single line is plotted against the two axes for each sound source; the line will occupy a large, colored/shaded area in the case of the spread image
5. A key will be required to clearly relate the sound sources to the graph. It should be consistent with keys used for other similar analyses (such as musical balance and performance intensity) to allow different analyses to be easily compared.

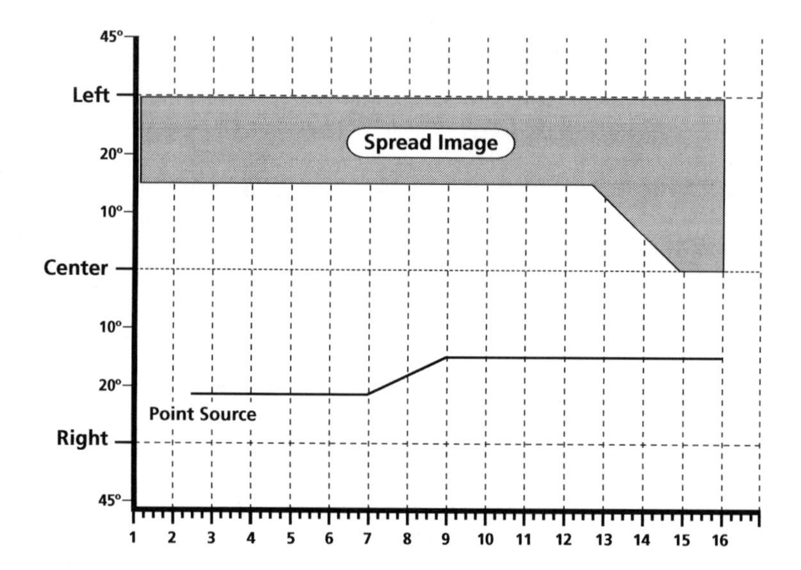


**Figure 10.5**
Stereo-sound location graph.

A sound source may occupy a specific point in the horizontal plane of the sound stage, or it may occupy an area within the sound stage. The graph will dedicate a source line to each sound source and will plot the stereo locations of each source against a common timeline.

The source line for point source images will be a clearly defined line at the location of the sound source. The source line for the spread image will occupy an area of the sound stage and will extend between the boundaries of the image itself.

It is common for stereo-sound location graphs to be multitier, placing spread images on separate tiers from point sources or providing a number of separate tiers for spread images. Figure 10.5 presents a single-tier stereo-sound location graph, containing two sound sources: one spread image and one point source. It also incorporates angle increments out from the center in degrees. The speakers should appear at 30° right and left.

Figure 10.6 presents the stereo-sound location of a number of the sound sources from The Beatles' work, "A Day in the Life." The location, size, and movements of the sound-source images directly contribute to the character and expression of the related musical materials. As an exercise, listen to the recording and notice the placement of the percussion sounds. Observe and define the stereo locations of the percussion sounds as they complement the placements of the voice, bass, piano, guitar, and maracas to balance the sound stage.

Locations and image size do not often change within sections of a work. Changes are most likely to occur between sections of a piece, or at

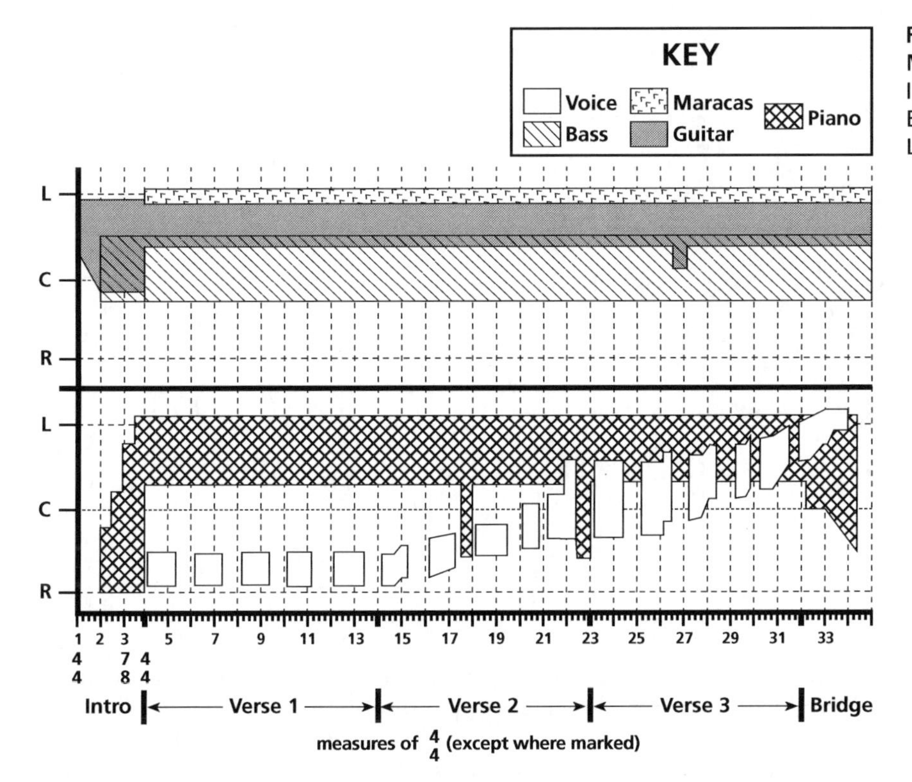


**Figure 10.6**
Multitier stereo-sound location graph—The Beatles' "A Day in the Life."

repetitions of ideas or sections (where changes in the mix often occur). The listener should, however, never assume changes do not occur. Gradual or subtle changes in source locations and size are present in many pieces and sometimes in pieces where such events are not expected. The changes in source size and location from *Abbey Road* works discussed in Chapter 2 will not be noticed unless the listener is willing to focus on this artistic element and is prepared to hear these changes.

The reader should work through the Stereo Location Exercise (10.1) at the end of this chapter to refine this skill.

## Distance Location

*Distance localization* and stereo localization combine to provide imaging for sound sources. Figure 10.7 is an empty stereo-sound stage onto which sound sources are imagined to be located. Placing sounds on this empty sound stage will allow the listener to make quick, initial observations regarding stereo imaging. These observations can then lead to the more detailed evaluations of stereo location and distance location. It is important to note that these location observations will relate to specific

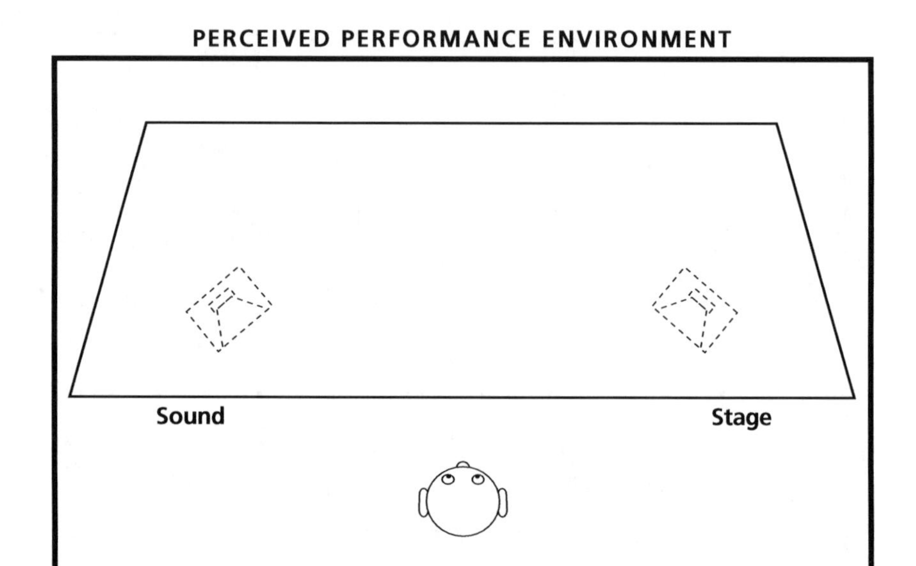


**Figure 10.7**
Empty sound stage.

moments in time or to sections of a work (specific periods of time). While location changes cannot be written on this figure, it is useful for initial observations and for graphing sources that do not change. In production work, this figure can be helpful in constructing and planning mixes, and for keeping track of parts.

Distance is the perceived location of the sound source from the listener. It is a location where the listener envisions the sound to be placed along the depth of the sound stage. We hear sounds as occupying a specific position of distance from our location. Sounds do not occupy distance areas. A source environment may provide an area of depth to the image, but the source will be heard as located at a precise point within that environment. The environment and source fuse into a single sound impression that will occupy an area of distance, with the source localized at a specific position in it.

The perceived source locations nearest to and furthest from the listener establish the front and rear boundaries of the sound stage. The front edge of the sound stage may be immediately in front of the listener, or at any distance. The depth of the sound stage will be heard as a single dimension of the area that contains all sound sources.

## Understanding Distance Location

The reader must approach distance location carefully. Distance cues are often not accurately perceived. Other artistic elements are often confused with distance. Further, humans mostly rely on sight to calculate distance and are not normally called upon to focus on aural distance cues.

Distance is *not* loudness. In nature, distant sounds are often softer than near sounds. This is not necessarily the case in recording production. Loudness does not directly contribute to distance localization in audio recordings. At times loudness and distance cues are associated in recordings, but this is often not so—especially in multitrack and synthesized productions. A "fade out" can be accomplished without causing source distances to increase. Conversely, a fade out may cause sound sources to be perceived as increasing in distance. The distance increase will be the result of a diminishing level of timbral detail; it will not be the result of decreasing dynamic level.

Very often, people will describe a sound as being "out in front," implying a closer distance. The sound may actually be louder than other sounds or may stand out of the musical texture because of the prominence of some other aspect of its sound quality. Much potential exists for confusing distance with dynamic levels.

Distance is *not* determined by or the result of the amount of reverberation placed on a sound source. In nature, distant sounds are often composed of a high proportion of reverberant energy in relation to direct sound. Reverberant energy does play a role in distance localization, but not so prominent a role that it can be used as a primary reference. Reverberant energy is most important as an attribute of environmental characteristics and in placing a sound source at a distance within the individual source environment. Humans perceive distance within environments, through time and amplitude information extracted from processing the many reflections of the direct sound. The ratio of direct to reflected sound influences distance location. Thus, reverberation contributes to our localization of distance, but it is *not* the primary determinant of distance location, in and of itself.

Distance is *not* the perceived distance of the microphone to the sound source that was present during the recording process. The only exception to this statement occurs when the initial recording is performed with a single stereo pair of microphones, and no signal processing is performed on the overall program. Microphone-to-sound-source distance is a contributor to the timbral characteristics of the sound source. Microphone-to-sound-source distance will determine the amount of definition of the sound source's timbre (how much timbral detail is present in the sound) captured by the recording process. It will also determine the amount of the sound of the initial recording environment that has become part of the sound source's timbre. Generally, the closer the microphone to the sound source, the greater the definition of timbral components captured during the recording process. This will provide distance-localization information, such that very close microphone placement will cause the image to be perceived very close to the listener (if no timbral modifications or signal processing is performed in the mixing process). The sound quality may be significantly altered in the mix, significantly altering microphone-to-sound-source distance cues. Microphone-to-sound-source distance contributes to the overall sound quality of the source's timbre. It contributes to our localization of distance through definition of timbral detail, but it is *not* a primary determinant of distance localization.

Distance location *is* primarily the result of timbral information and detail. Timbre differences between the sound source, as it is remembered in an unaltered state, and the sound, as it exists in the recording primary, determine distance localization. The listener is aware of how timbres are altered over various distances. It is through the perception of these changes that we identify the distance of a source from our listening location. Humans are unable to estimate the physical distance (meters, feet, etc.) of a sound source from their location. We perceive distances in relative terms and compare locations to one another. In our everyday activities we mostly rely on sight to make distance judgments.

Our ability to focus on the "sound" of distance has not been encouraged by our real-life experiences, and will take focused effort to develop.

We rely on timbral definition for most of our distance judgments and use the ratio of direct to reverberant sound to a far lesser degree. The extent to which we rely on either factor depends on the particular context. Distance localization is a complex process, relying on many variables that are inconsistent between environments.

The listener knows the sound qualities of sound sources within the area immediately around them. We have a sense of occupying an area, encompassing a space immediately around us. This area serves as a reference from which we judge "near" and "far." Within this area, sounds have no changes in timbre, or perhaps heightened detail of timbre. All characteristics of timbral content are present, and sounds will have more detailed definition the closer they are to the listener. The overall sound quality of the sound source may be somewhat altered by the characteristics of the host environment, but the level of detail present in the timbre causes the listener to perceive the source as being within their immediate area of *proximity*. "Proximity" is the space that immediately surrounds the listener, and is the listener's own personal space. Sounds in proximity are perceived as being close, as occurring within the area the listener occupies. The actual size or area of proximity may be perceived as being rather large or very small, depending on the context of the material and the perspective of the listener.

The listener knows the sound qualities of sound sources at "near" distances. We conceive "near" as being immediately outside of the

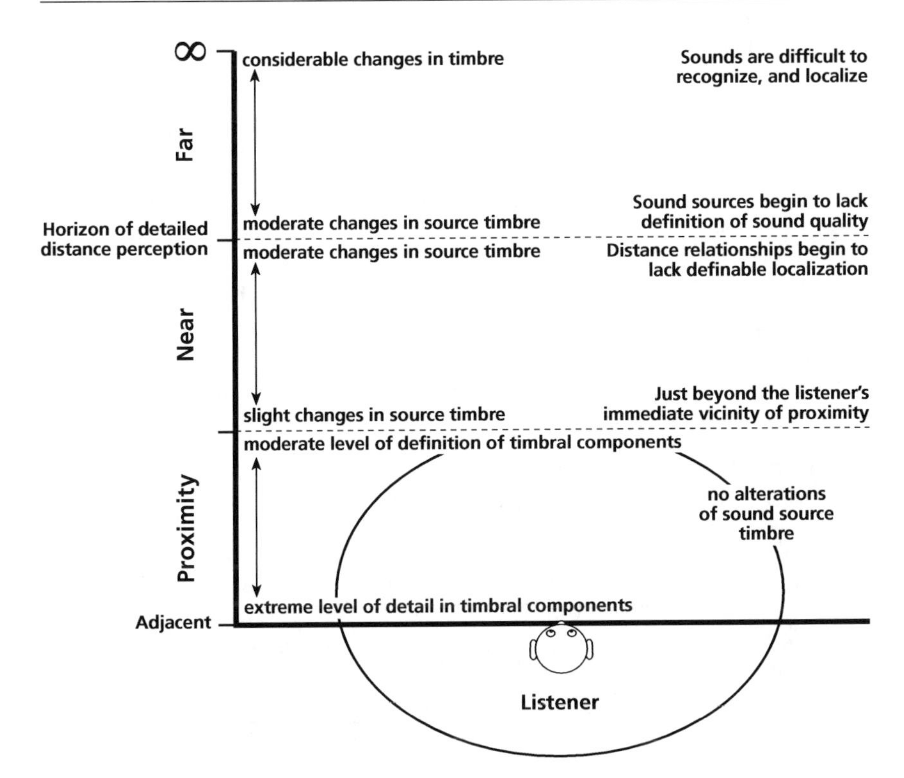


**Figure 10.8** Continuum for designating distance location.

area that we perceive ourselves as occupying. Throughout this "near" area, the listener is able to localize the sound's distance with detail and accuracy. Timbres are very slightly altered in the closest of sounds considered near and are moderately altered in the furthest of sounds considered near. An area will exist between these two boundaries where sounds are readily compared as being closer or farther than other similar sounds. Sounds cease to be considered near when the listener begins to have difficulty localizing distances in detail.

"Far" sound sources lack some definition of sound quality. The closest of far sounds will have moderate alterations to sound quality, with little or no definition. Few low-amplitude partials will be present, and amplitude and frequency attack transients will be difficult to detect. The furthest of far sounds will have considerable alterations to sound quality; the sounds will lack all definition. The furthest far sounds may even be difficult to recognize. A wide area exists between these two boundaries of "far." It could conceivably be quite large, perhaps stretching to infinity, or to the horizon of what we can imagine as possible. "Far" sounds are difficult to place in specific distance locations; they tend to be more difficult to place precisely, but can be localized quite readily by comparing them with other sounds.

## Evaluating Distance Location

In evaluating distance location we initially perceive distance cues of the sound source at the perspective of the source within its own host environment. We then intuitively transfer that information to the perspective of the sound stage. There, the sound source's degree of timbre definition, the perceived distance of the sound source within its host environment, and the perceived distance of the source's host environment from the perceived location of the listener blend instinctively into a single perception of distance location. This process will determine the actual perceived distance of the sound source, at the perspective of imaging. Fortunately, this all happens quite naturally, intuitively, and without attention.

Again, the definition of, or the amount of, timbral detail present plays the central role in determining perceived distance.

The boundaries for distance location extend from "adjacent" to the listener to a distance of "infinity." Adjacent is that point in space that is immediately next to the space the listener is occupying. It should be conceived literally as the next molecule from the listener's ear, as a sound may be localized at that location.

The continuum for distance localization consists of three areas. The areas represent conceptual distance, not physically measurable distance increments. Distance is thus judged as a concept of space between the sound source and the listener. An area of proximity surrounds the listener. This area serves as a reference for judging near and far distances. Human experience of the nature of sound is used as a reference to conceptualize the amount of space (distance) between the source and the listener.

The three areas of the continuum are:

1. An area of "proximity," the space that the listener perceives as their own area, is the area immediately surrounding the listener that may be extended in size to be conceived as the size of a small to moderately sized room. The listener will perceive the proximity area as being their own immediate space
2. A "near" area is the area immediately outside of the space that the listener perceives themselves as occupying, extending to a horizon where the listener begins to have difficulty localizing distances in detail
3. A "far" area, beginning where perception dictates that space ceases to be "near"; where detailed examination of the sound is difficult, and extending to where sounds are almost impossible to recognize. Extreme far sound sources contain very little definition of sound quality.

These three areas are not of equal physical size. The amount of physical distance contained in the conceptual area of proximity will be considerably different than the physical distance encompassed by the conceptual area of far. All three areas of the continuum occupy a similar amount of conceptual space, but represent significantly different amounts of physical area. The vertical axis of the distance location graph will clearly divide the three areas.

**Listen**

to tracks 39–41

for a single cello performance that is placed in proximity, near, and far distance locations.

The size of the three areas may be adjusted between appearances of the distance location graph. The amount of vertical space occupied by the areas may be adjusted to best suit the material being graphed, with certain areas being widened in certain contexts and narrowed in others. The far area may even be omitted in certain graphs. The area of proximity should always be included (although if necessary it may be narrowed to occupy less vertical space), to clearly present the conceptual distance between the perceived location of the listener and the front edge of the sound stage.

Sound sources will be placed on the graph (1) by evaluating the definition of the sound quality of each sound source (the amount of detail present in the timbre of the sound source), and incorporating information on the ratio of direct to reverberant sound and the quality of the reverberant sound as appropriate, and (2) by directly comparing the sound source to the perceived distance locations of the other sound sources present in the musical context.

The individual listener's knowledge of timbre and environmental characteristics, and their ability to recognize the sound source, are variables that may cause the listener to inaccurately estimate distance. For example, a very close tamboura may sound like a far sound to a person who does not know the sound of a tamboura. As the life experience of listeners varies, so does an individual's ability to identify the distance relationships of sounds.

During initial studies, distance judgments may be difficult to conceive and perceive. Distance is, however, a central concern of sound-source imaging and thus of music production. Skill in this area can be refined and should become highly developed.

The *distance location graph* incorporates:

1. Continuum from adjacent through infinity (divided into three areas) as the *Y*-axis
2. The *X*-axis of the graph dedicated to a timeline divided into an appropriate increment of the metric grid

3. A single line plotted against the two axes for each sound source
4. A key that is required to clearly relate the sound sources to the graph. The key should be consistent with keys used for other similar evaluations (such as musical balance or stereo location) to allow different elements to be easily compared.

The locations of all sources are plotted as single lines. Sources are precisely located at a specific distance from the listener. No two sources can be at the same distance level unless they are at clearly different lateral locations (placement of phantom image in stereo or surround formats).

Sound sources may change distance locations (often subtly) in real time or within sections of a work. Marked changes are most likely to occur between sections of a piece, at entrances or exits of individual sound sources, or at repetitions of ideas or sections (where changes in the mix often occur). The listener should, however, never assume changes do not occur. Gradual changes in distance are present in many pieces. At times a sound will become unmasked by the exit of another instrument from the mix, and will move closer to the listener with its increased timbral detail.

The reader is encouraged to work through the Distance-Location Exercise (Exercise 10.2) at the end of the chapter.

Figure 10.9 is a distance location graph from The Beatles' "A Day in the Life." While the vocal line has a significant percentage of reverberant sound, its timbral detail brings it to be located in the rear third of the proximity area. The other sound sources have widely varied distance

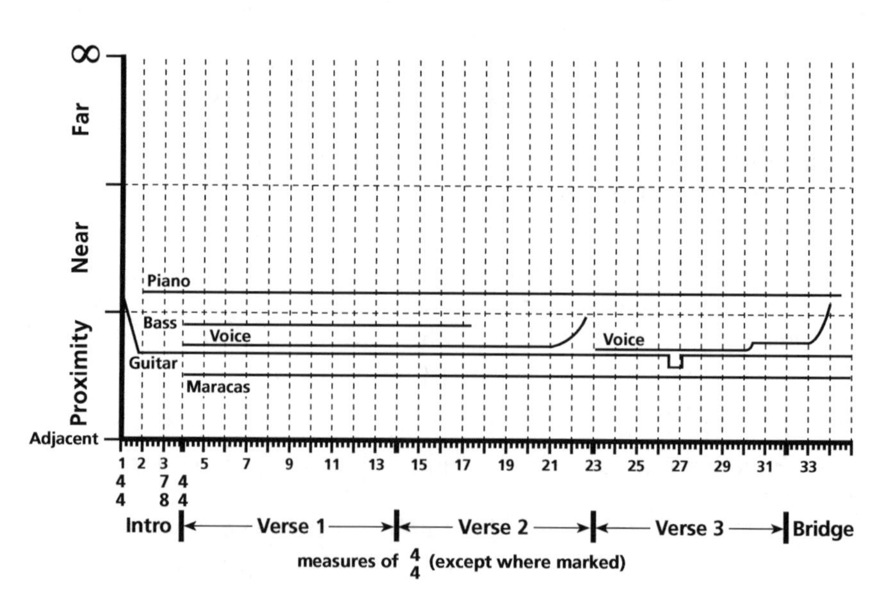


Figure 10.9 Distance location graph—The Beatles' "A Day in the Life."

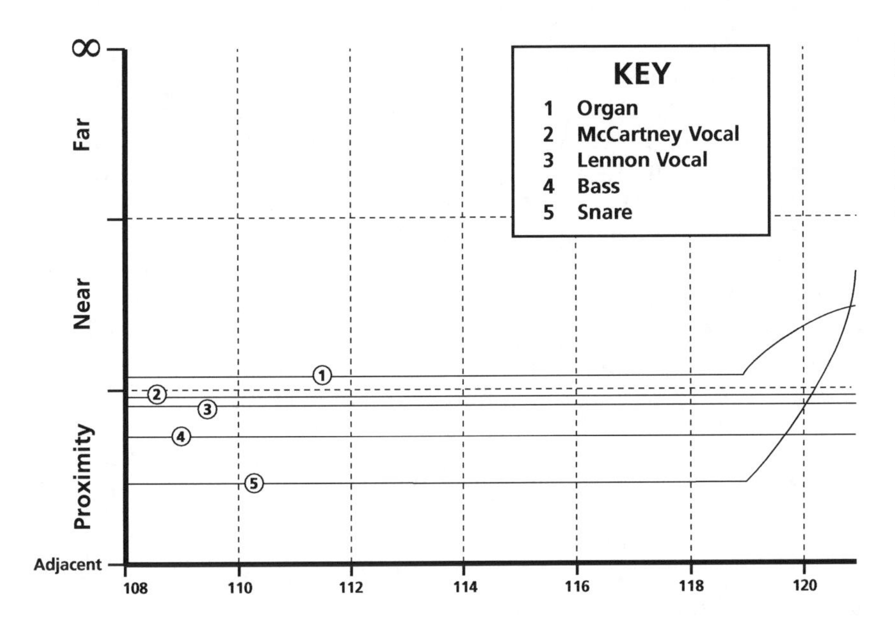


**Figure 10.10** Distance location graph of the fade out from The Beatles' "Lucy in the Sky with Diamonds" (*Yellow Submarine*, 1999).

locations, giving the sound stage great depth. The guitar, maracas, and bass are also in the proximity area, and the piano is at the front of the near area. As an additional exercise, the reader should place the percussion sounds in an appropriate distance location. Notice the great conceptual distances between the various instruments of the drum set, as some sounds are located well into the far area.

The fadeout of "Lucy in the Sky with Diamonds" is graphed in Figure 10.10. The vocal lines and bass do not change distance locations with diminishing loudness. The snare drum and organ are quite different, and do change distance locations as their loudness levels decrease. It is interesting to note that the speed and amounts of perceived distance-location changes are different for the two sounds, as they decrease in loudness almost in parallel.

## Environmental Characteristics

The characteristics of the sound source's host environment are important in shaping four qualities of the recording: (1) the overall quality of the sound source, (2) the perceived performance environment, (3) space within space, and (4) the imaging of the sound stage.

The environmental characteristics of the entire program (the perceived performance environment) shape the illusion of a space in which a performance is occurring. The characteristics of this envisioned performance environment will greatly influence the conceptual setting for the artistic message of the work.

Environmental characteristics of both the host environments of the individual sound sources and the perceived performance environment play significant roles in music production. These artistic elements have the potential to provide significant information for enhancing and communicating the musical message of the piece of music. Currently, they are most often used in supportive roles. They are coupled with sound quality in defining the unique characters of individual sounds (enhancing their sound quality and providing each sound with a sense of depth). Environmental characteristics are used as a separate element in creating depth of sound stage, in providing a resource for space within space, in creating the illusion of the perceived performance environment, and in giving breadth and depth to phantom images.

The recordist needs to be able to recognize the characteristics of the environments within which sound sources exist, and the characteristics of the perceived performance environment. This will lead to an understanding of how they influence the recording, and will bring the recordist to more effectively craft a sound stage within an envisioned environment that best suits their project.

## Evaluating Environmental Characteristics

A composite sound of environmental characteristics occurs with the sounding of a sound source within the environment. The sound source and the environment interact and fuse to create a single, composite sound—a new overall sound quality. To understand the influence of the host environment on the sound source in making this composite sound, the environment and the sound source need to be evaluated separately.

We perceive environmental characteristics as an overall sound quality that is composed of a number of component parts. As global sound qualities, environmental characteristics are conceptually similar to timbre. The evaluation of timbre (sound quality) and environmental characteristics will be similar, in that both will seek to describe the states and activities of the physical components of sound. While we might recognize large halls, small halls, and other spaces as having common environmental characteristics, each environment is unique. To meaningfully communicate information about environmental characteristics, we must define the environment by its unique sound characteristics. These characteristics can only be objectively described through discussing the levels and activities of the component parts of the environment's sound.

Environmental characteristics appear as alterations to the sound source's timbre, created by the interaction of the sound source and the environment. Therefore, to evaluate the characteristics of the environment, we need to define the changes that have occurred in the sound source's timbre after being sounded within the environment. Evaluating environmental characteristics will engage activities that are contrary to our natural tendency to fuse the environment's sound with the source's sound. Care in focusing on the correct perspective and aspects of sound will be required.

Environmental characteristics are determined by comparing our memory of the sound source's timbre outside of the host environment to the sound source's timbre within the host environment. We must go through this comparison process carefully, scanning the composite sound for information and then comparing that information with our previous experiences with the timbre of the sound source (at times considering how the source appeared within other environments). Differences in the spectrum and spectral envelope of the sound source we remember, and as heard in the host environment, form the basis for determining most environmental characteristics.

If we do not recognize the sound source (timbre) or have no prior knowledge of the sound source, we will be at a disadvantage in calculating the characteristics of the host environment. We will have no point of reference in determining how the environment has altered the timbre of the original sound source. We must rely on our knowledge of what we believe to be similar sounds, to calculate estimations of the characteristics of the environment. This may or may not turn out to be accurate.

Louder sounds of shorter duration can also be sought for evaluation. They expose much time and spectrum information, especially after the direct sound has stopped. Sounds that are exposed, with few or no simultaneous sounds, will most readily present the sound characteristics of the environments.

In evaluating environmental characteristics, we are seeking to define the characteristics of the environment itself. We must make certain we are *not* identifying characteristics of the sound source and must make certain we are *not* identifying characteristics of the sound source within the environment. The characteristics of the environment can be reduced to three specific component parts. These characteristics are what must be determined by identifying the differences between the sound quality of the sound source itself and the sound quality of the sound source within the environment.

The component parts of environmental characteristics are (1) the reflection envelope, (2) the spectrum, and (3) the spectral envelope. The environmental characteristics graph (Figure 10.11) allows for the detailed evaluation of these three components.

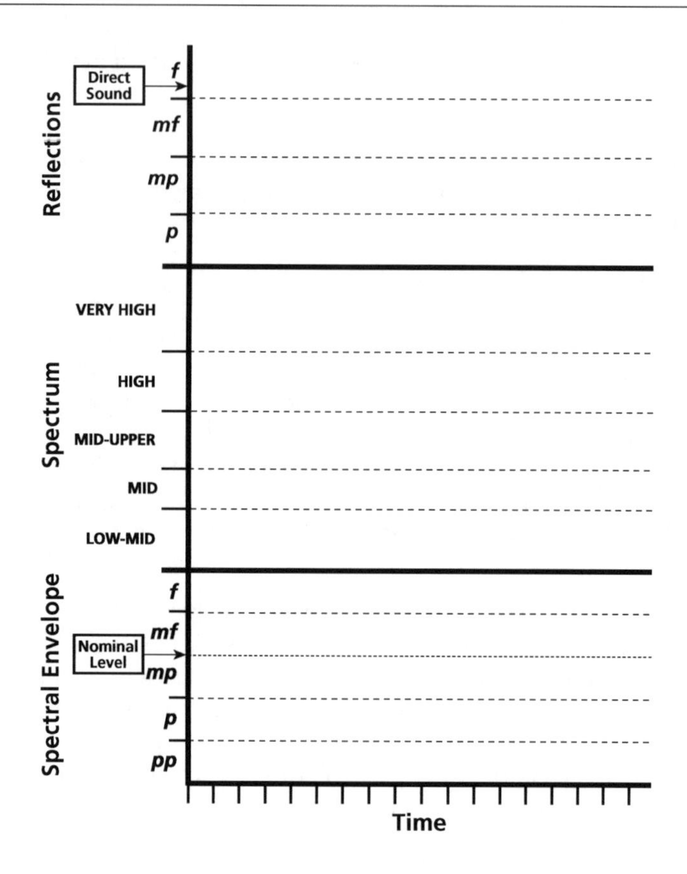


**Figure 10.11** Environmental characteristics graph.

### *Reflection Envelope*

The *reflection envelope* is made up of the amplitudes of (1) the initial reflections and (2) the reverberant energy of the environment throughout the duration of the environment's sound. This envelope is composed of many reiterations of the sound source. The reiterations vary in dynamic level and in time spacing between one another (time density).

The spacing of the reiterations of the sound source will be dramatically different over the duration of the sound of the environment (for example, the spacing of the early reflections will be considerably different from the spacing of reflections near the end of the reverberant energy). This portion of the graph can clearly show the time of arrival of the reflected sounds to the listener location, the density of the arrival times of the reflected sounds, and the amplitude of those arrivals in relation to the amplitude of the direct sound.

The amplitude of the direct sound is included to serve as a reference level for the calculation of the dynamic levels of the reflected sound. The reader should use their developing skills at pattern recognition to extract time information from the sound of the environment. The reflections portion of the graph will show the following information. The reader should organize their listening to the time elements of the environmental characteristics to recognize:

- Patterns of reflections created by spacings in time (rhythms of reflections)
- Patterns of reflections created by dynamics (rhythms of loudness levels)
- Spacing of reflections in the early time field
- Dynamic contour of the entire reflections portion
- Density (number and spacings of reflections) of reverberant sound
- Dynamic relationships between the direct sound, individual reflections (of the early time field), and the reverberant sound
- Dynamic contour shapes within the reverberant sound.

An isolated appearance of the sound source in the host environment needs to be available for all of the time and reflection-amplitude information to be accurately evaluated. This is especially true for the decay of the reverberant energy and the spacing of the early reflections (which are important parts of the environment's sound quality). A short (staccato) sound will allow the reflection information to be most audible, since it will not have to compete with the sound of the source itself.

An exercise to develop the reader's skill in this area appears at the end of this chapter. The Reflections and Reverberation Exercise (Exercise 10.3) helps readers understand and recognize this characteristic of environmental sounds, even if they never wish to perform an evaluation as detailed as the reflection envelope of the environmental characteristics graph. The exercise will lead the reader to important observations that can lead to a better understanding and use of environments.

When listening to sounds in actual music recordings, hearing these characteristics is a much greater challenge—and may at times be impossible. Without the opportunity to hear the environment's complete presentation, information related to the reverberant energy might never be completely audible. The reader should try to find several appearances of a sound source where it can be heard alone, without other sound sources, and where it is playing short durations.

Many hearings of the sound in a wide variety of presentations will be necessary to compile an accurate evaluation of all of the time characteristics of the environment.

### *Environment Spectrum and Spectral Envelope*

The spectrum of the reverberant sound and the initial reflections is a collection of all frequencies or bandwidths of pitch areas that are emphasized and de-emphasized by the environment itself. This spectrum is comprised of only those frequencies that are altered by the environment. The environment may emphasize or de-emphasize bandwidths of frequencies or specific frequencies. Often the spectrum of the environment will only contain a small number (three to seven) of prominent frequencies or frequency bands/pitch areas that are either emphasized or attenuated (de-emphasized).

These frequencies are determined through a careful evaluation of many appearances of the sound source in the environment, by listening to the way the sound source's timbre is changed by the environment over a wide range of pitch levels. Some appearances of the sound source will not have frequency information in certain frequency areas that are emphasized or de-emphasized by the environment. The reader will need to scan many pitch levels of the sound source to determine the spectral content and the spectral envelope of its environment.

The spectral envelope of the environment illustrates how the frequencies that are emphasized and de-emphasized by the environment (spectrum) vary in loudness level over the duration of the sound of the environment. The spectral envelope and spectrum portions of the graph are coordinated to present different activity of the same sound components (as with sound-quality evaluation).

A *nominal level* is used as a reference for plotting the dynamic contours of the spectral components. The nominal level will vary in loudness/amplitude over the sound's duration. The nominal level *is* the dynamic envelope of the environment, where the sound source's frequency components are unaltered dynamically. The dynamic envelope of the environment changes over time. It is the dynamic contour that is outlined by the reflections envelope. This dynamic envelope is represented as a fixed, steady-state level on the spectral-envelope portion of the environmental characteristics graph.

**Listen**

to tracks 41, 44, 45, 45, and 47

for individual sounds and entire mixes with strong environmental characteristics.

The nominal level is placed at the dynamic level that divides mezzo forte and mezzo piano. Frequencies or pitch areas that are emphasized by the environment will be plotted as activity above the nominal level. Frequencies or pitch areas that are attenuated by the environment will be plotted as activity below the nominal level.

An exercise to help develop skill in hearing the environmental characteristics spectrum and spectral envelope appears at the end of this chapter. In a similar way to the Reflections and Reverberation Exercise, this exercise (Exercise 10.4) will increase the listener's ability to recognize these important aspects of environmental characteristics. Undertaking the detailed task of creating an environmental characteristics graph is not necessary to develop the skills needed to describe the spectrum and spectral envelope of an environment. This exercise will, however, aid the listener in developing the skills to make such observations accurately and with as much detail as the audio professional's position requires.

## Environmental Characteristics Graph

The *environmental characteristics graph* allows for a detailed evaluation of the reflection envelope, the spectrum, and the spectral envelope (see Figure 10.11). Creating environmental characteristics graphs will greatly assist in understanding the nuance of any environment. When created with much detail, this graph requires great skill that will be acquired over an extended period of practice and patience. Using this graph for general observations and beginning studies will also prove very helpful to the beginner and audio professional alike. Observations can be recorded for future reference and to assist in learning, understanding, and recognizing this artistic element.

The environmental characteristics graph incorporates:

1. Three tiers as the *Y*-axis: reflections (a continuum of dynamic level), spectrum (a continuum of pitch level), and spectral envelope (a continuum of dynamic level)
2. The reflections portion of the graph, comprising a vertical line at each point in time that a reflection occurs. The height of the vertical line corresponds to the amplitude of the reflection. The dynamic level of the direct sound is indicated on the vertical axis and serves as a reference for calculating the dynamic levels of the reflections. This portion of the graph presents information on the dynamic contour of the reflections of the environment and the spacings, in time, of the reflections throughout the sound of the environment
3. The spectrum portion of the graph, which is comprised of the registers previously established. Spectral components are placed against the *Y*-axis by pitch/frequency level. A single line is plotted against the two axes for each spectral component, and it will occupy a large, colored/shaded area in the case of pitch area and a narrow line in the case of a specific frequency
4. The spectral-envelope portion of the graph, which depicts the dynamic contours of the spectral components, using dynamic areas as the *Y*-axis
5. The *X*-axis of the graph, a timeline divided into an appropriate time increment (usually needing to allow millisecond increments to be observed) to clearly display the smallest change of a duration, dynamics, or pitch present in the characteristics of the environment
6. A key that is required to clearly relate the components of the spectrum and spectral-envelope tiers of the graph.

The perspective of the environmental characteristics graph will always be of either the individual sound source or of the perceived performance environment.

It is not always possible to compile a detailed evaluation of environmental characteristics. The information of the environment is often concealed by other sounds in the musical texture. Further, it is not easily separated from the sound quality of the sound source itself. The ability to recognize environmental characteristics involves much focused attention and practice. It relies (1) on a knowledge of many sound sources, (2) on an ability to evaluate sound quality of the sound source within the host environment, and (3) on an acquired skill for comparing and contrasting a previous knowledge of the sound source with the appearance of the sound source within the environment that is to be defined.

Using the graph for general evaluations of environmental characteristics is often the most feasible approach. This approach will not require as advanced a skill level as a detailed graph and will provide a good amount of significant information. These general evaluations are acceptable for most applications. They provide pertinent information quickly, but without the subtle details that are difficult and time intensive to identify.

General evaluations of environmental characteristics will include (1) the contour and beginning level of the reverb, (2) the level of the direct sound, and (3) the most prominent frequencies or frequency bands that are emphasized or attenuated. If possible, and after practice, they should include a general description of the spectral envelope and an indication of the content of the early time field.

The complexity of environmental characteristics can vary widely. Certain environments will have very few spectral differences from the original sound source. Some environments will have no reflections present between the early time field and the reverberant energy, and the reverberant energy will increase in density through a simple, additive process. Other environments may be quite sophisticated in the way they were created, with time increments of the early time field precisely calculated at different time intervals, with spectral components precisely tuned in patterns of frequencies (designed to complement the sound source), and with spectral-envelope characteristics reacting accordingly. Natural environments and those that are created can be remarkably similar or different in character.

With current technology it is possible for all perceived environmental characteristics to be changed in real time. It is also possible for the perceived environment of a sound source to be generated solely by a delay unit, by a simple reverberation unit, or by any related plug-in. Although these environments would not be perceived as natural spaces,

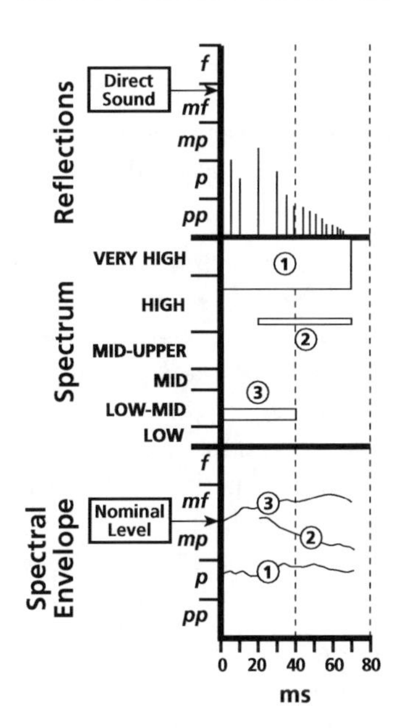


**Figure 10.12** Environmental characteristics graph of Paul McCartney's lead vocal in The Beatles' "Hey Jude" *(1)*.

the listener would proceed to imagine an environment created by the impressions of those simple characteristics. The listener simply will not perceive a sound as being void of environmental characteristics. If no environment is present, it will be imagined.

Figure 10.12 presents an environmental characteristics evaluation of McCartney's vocal from the opening of "Hey Jude" from *1*. The graph shows several alterations of spectrum and spectral envelope, and the environment's subtle time elements. The sparse texture during the beginning of the song allows the characteristics of all environments to be perceived quite clearly.

Exercise 10.5 at the end of this chapter provides guidance in learning to evaluate environmental characteristics. The reader will gain much from attempting this exercise on several different sounds from several different recordings.

## Space Within Space

The overall environment of the program provides a setting within which the subordinate environments of the individual sound sources will appear to exist. This overall environment (or perceived performance environment) is a constant that equally influences the individual environments of all sound sources. The perceived performance environment becomes part of the overall character of the recording/piece of music.

The overall environment is either (1) perceived by the listener as being a composite of the dominant, predominant, and/or common characteristics of the individual environments of the sound sources, or (2) is a set of environmental characteristics that is superimposed on the entire program.

Works will be perceived to have a single overall environment that is present throughout the piece. The listener will imagine a single space (the perceived performance environment) in which the performance (recording) occurs. When this overall environment is created by adding environmental characteristics to the entire musical texture, it is possible for the overall environment to change during the course of a work. In such instances, abrupt changes (usually at a major division of the form of a piece, such as between verse and chorus) are most common. The various environments will be perceived as having occurred within a single overall environment, even if a single environment is not present.

The perceived performance environment may be a composite of many perceived environments and environmental cues. The listener will perceive the overall performance environment in this way, if an overall environment has not been applied. In these instances, the perceived performance environment is envisioned by the listener as a result of environmental characteristics (1) that are common or complementary between the environments of the individual sound sources, (2) that are prominent characteristics of the environments of prominent sound sources (a source that presents the most important musical materials, or the loudest, nearest, or furthest sound sources, as examples), and/or (3) that are created by environmental characteristics found in both of these areas.

Within this overall environment, the individual environments of sound sources are perceived to exist. This is the illusion of space within space. If reverberation has been applied to the overall program to create a perceived performance environment, all of the recording's sound sources and their host environments will be altered by those sound characteristics.

Figure 10.13 is the perceived performance environment of The Beatles' "Hey Jude." The lead vocal, with its fused environmental characteristics of Figure 10.12, appears contained within this overall environment of the recording/performance. This space-within-space illusion is convincing in bringing the listener to accept the small space of the lead vocal contained within a midsized, natural-sounding performance space (perceived performance environment).

A complete space-within-space evaluation will include the environmental characteristics of all sound sources and the perceived performance

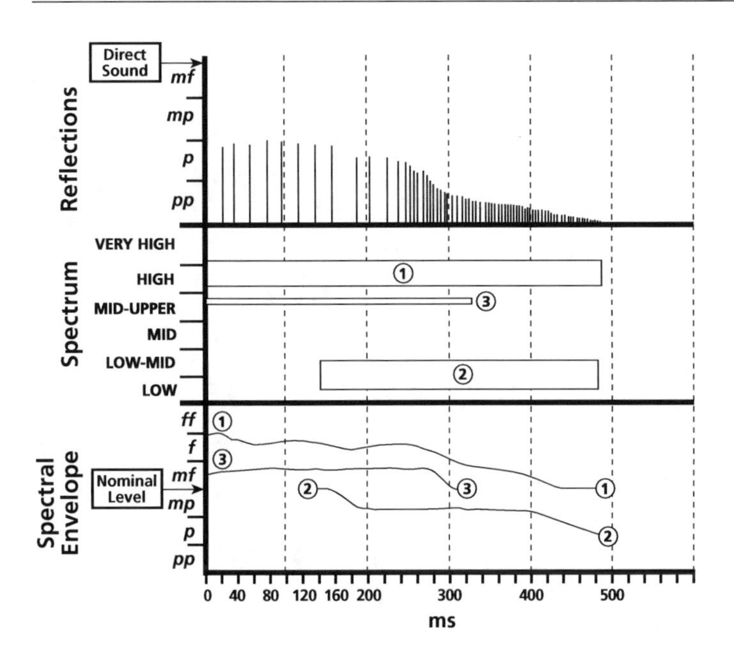


**Figure 10.13**
Perceived performance environment of The Beatles' "Hey Jude."

environment. Relationships between the environments and the perceived performance environment can then be understood and evaluated, among other possible observations. This complete evaluation might not be undertaken often, but this type of attention is often a level of focus in the final mixdown. The skill to compare environments also leads to an understanding of how environments can be used to enhance sound sources and the recording in complementary ways.

Distance location and environmental characteristics are related. The interrelationships of distance location and space within space should always be considered when evaluating a music production. They work in a complementary way to give depth to the sound stage. The two are closely interactive, and comparing the two will offer the listener many insights into the creative ideas of the recording.

Additional environments from The Beatles' recording "Hey Jude" are presented in Figures 10.14, 10.15, and 10.16. The graphs allow us to recognize the very different decay times of the three environments. When comparing the graphs to Figures 10.12 and 10.13, we can also recognize the uniqueness of all four host environments and also their relationship to the perceived performance environment. It is interesting to note that the high and very high areas are attenuated (in different bands) in the guitar and piano environments, and the spectral alterations of the tambourine environment are very subtle. The reflections are sparse in their spacing, to varying degrees, and fairly regular reflections occur in all three environments.

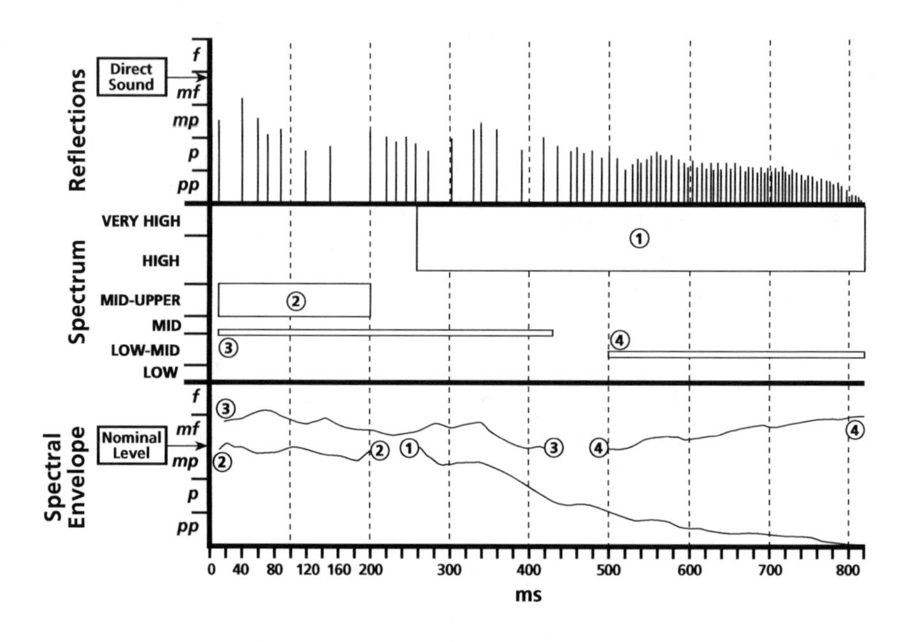


**Figure 10.14** Environmental characteristics evaluation of the first appearance of the piano in The Beatles' "Hey Jude."

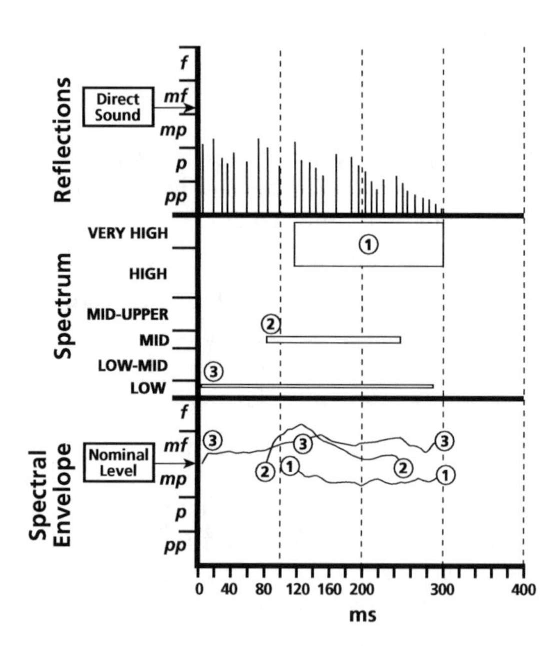


**Figure 10.15** Environmental characteristics evaluation of the first appearance of the guitar in The Beatles' "Hey Jude."

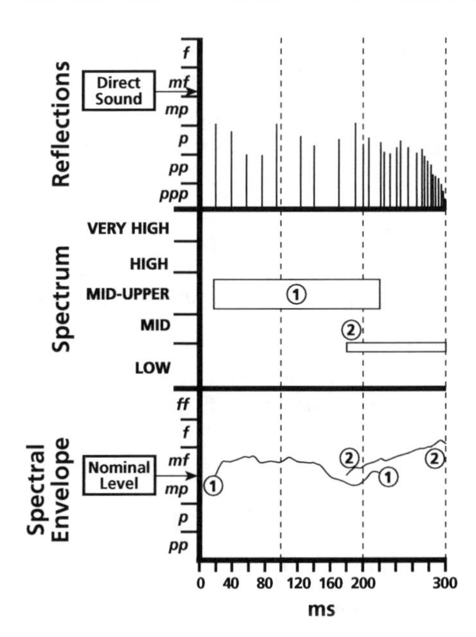


**Figure 10.16** Environmental characteristics evaluation of the first appearance of the tambourine in The Beatles' "Hey Jude."

# Exercises

## Exercise 10.1

*Stereo Location Exercise*

For this exercise, find a recording that displays significant changes in stereo location of sound sources. Plot the location of the sound source that displays the greatest amount of change, and at least two others: one spread image and one point source. Graph the sounds throughout the first two major sections of the song.

The process of determining stereo-sound location will follow this sequence:

1. During the initial hearing(s), listen to the example to establish the length of the timeline. Next, notice the presence of prominent instrumentation, note the placements and activity of their stereo location against the timeline.
2. Check the timeline for accuracy and make any alterations. Establish the sound sources (instruments and voices) that will be evaluated and sketch the presence of the sound sources against the completed timeline.

3. Notice the locations and size of the sound sources for boundaries of size, location, and any speed of changing locations or size of image. The boundaries of source locations will establish the smallest increment of the *Y*-axis required. The perspective of the graph will always be of either the individual sound source or of the overall sound stage.
4. Begin plotting the stereo location of the selected sound sources on the graph. The locations of spread images are placed within boundaries; these "edges" of the images can be difficult to locate during initial hearings, but they can be defined with precision. Focus on the source until it is defined. Work from what you know for certain (where the image *is* and where it is *not*) to gradually remove your doubt and confusion. The locations of point-source images are plotted as single lines. These sources are often easiest to precisely locate.
5. Continually compare the locations and sizes of the sound sources to one another. This will aid in defining the source locations and will keep the listener focused on the spatial relationships of the various sound sources. The evaluation is complete when the smallest significant detail has been incorporated into the graph.

As you gain experience in making these evaluations, songs with more instruments should be examined and longer sections of the works should be evaluated.

---

**Exercise 10.2**

*Distance Location Exercise*

Select a recording with at least five sound sources that exhibit significantly different distance cues. Plot the distance locations of those sources throughout the first three major sections of the work.

The process of determining distance location will follow this sequence:

1. During the initial hearing(s), establish the length of the timeline. Notice the selected sound sources and any prominent placements and activity of distance location, especially as they relate to the timeline.
2. Check the timeline for accuracy and make any alterations. Clearly identify the sound sources (instruments and voices), and sketch the presence of the sound sources against the completed timeline.
3. Make initial evaluations of distance locations. For quick, general observations, ask yourself: "Does this sound like I am playing the instrument?" That is "proximity." "Is the sound in the same small

room with me, but beyond my reach?" That is "near." Identify the general locations of the sound sources to establish boundaries of the sound stage (the location of the front and rear of the sound stage). Notice any changing distance locations and calculate any speed of changing locations. The placement of instruments against the timeline, more than the boundary of speed of changing location (which are quite rarely used), will most often establish the smallest time unit required in the graph to accurately show the smallest significant change of location. The amount of activity in each area will establish the amount of Y-axis space required. The perspective of the graph will always be of either the individual sound source or the overall sound stage.

4. Begin plotting the distance of the selected sound sources. Sound sources will be placed on the graph by (1) evaluating the timbre definition of each sound source by focusing on the amount of detail present, while being aware of the amount and characteristics of the reverberant sound; (2) transferring this evaluation into a distance of the source from the listening location and penciling in the sound on the distance-location continuum for reference; (3) reconsidering the definition of the timbre (Is the source in the listener's own space, or proximity? Is it near or far?), and then placing the sound in relation to the sound stage; (4) precisely locating the distance location of the sound source by comparing the sound source's location to the locations of other sound sources.
5. Once several sounds are accurately placed on the distance continuum, identifying additional distance locations is most readily accomplished by directly comparing the sound source to the perceived distance locations of the other sound sources present in the music. Use proportions of differences between the locations of three or more sound-source distances to make for more meaningful comparisons. Is sound "c" twice or one half the distance from sound "a" as "a" is from sound "b"? How does this compare to the relationship of sounds "d" and "c"? Sounds "c" and "b"? Continually compare the distance locations of the sound sources to one another. The evaluation is complete when the smallest significant detail has been incorporated into the graph.

Remain focused on the distance location of the sound sources, making certain your attention is not drawn to other aspects of sound.

As you gain experience in making these evaluations, you should examine songs with more instruments and evaluate longer sections of the works.

**Exercise 10.3**

*Reflections and Reverberation Exercise*

Find a snare drum sound that was recorded without environmental cues, such as Track 23 on the companion website. The sound will need to be repeated many times, over a period of 10 or more minutes. Establish a way to continually repeat the track, if possible. Route the sound through an appropriate reverb unit or plug-in.

1. Make the reverb unit or plug-in emphasize the items listed below one at a time, and make radical (perhaps unmusical) settings of these parameters to learn their characteristic sound qualities. Listen carefully to individual snare drum hits while adjusting the device.
2. Seek to create a pronounced early time field. Establish a clear set of two reflections and create a setting that will repeat this pattern. A recurring pattern of reflections is the result. Listen carefully, and alter the speed of the reflections and spacings of the reflections and patterns.
3. Repeat this sequence with a clear set of three, four, and then five reflections, establishing recurring patterns while gradually increasing the number of reflections and the complexity of the pattern. Listen carefully to create and recognize:

   a. Patterns of reflections created by spacings in time
   b. Patterns of reflections created by dynamics
   c. Spacing of reflections in the early time field
   d. Dynamic contour of the entire reflections portion
   e. Density (number and spacings of reflections) of reverberant sound
   f. Dynamic relationships between the direct sound, individual reflections (of the early time field), and the reverberant sound
   g. Dynamic contour shapes within the reverberant sound.

You will begin to notice and recognize that certain spacings in time have a certain consistent and unique sound quality. A "sound of time" can be understood and recognized for delay times and reverberation rates. With patience and practice, this skill can become highly refined—as many room designers will attest.

### Exercise 10.4

*Environmental Characteristics Spectrum and Spectral Envelope Exercise*

Find a recording of high-quality acoustic instrument sound and loop it in a DAW or otherwise where it can be controlled. Route the sound through an appropriate reverb unit or plug-in.

1. Establish a setting with three or more seconds of decay and with a high proportion of reverb signal (or only reverb signal).
2. Alter the frequency response, equalization, or any similar frequency-processing control on the reverb to emphasize and de-emphasize (attenuate) several specific frequencies or frequency bands.
3. Play single pitches with short durations. Listen carefully to how changes of settings alter the sound quality of the instrument. Keep track of the settings played. Repeat this process while moving through the entire frequency range(s) the unit will alter and listening (and learning) carefully.
4. In a separate process, listen carefully to pitches played throughout the instrument's range, played through an unchanging reverb setting. Notice how the qualities of some pitches are altered differently than others. Changes in the environment's spectrum and spectral envelope will occur only if the particular pitches performed have spectral energy at the frequencies being altered.

Repeat this process again after several hours, then again after several days. During these sessions try to anticipate what the modification will sound like before you listen to it. Check your memory and your recognition of many different spectrum changes. Keep returning to this exercise to become comfortable with the material.

### Exercise 10.5

*Environmental Characteristics Exercise*

Return to the work or works evaluated in the distance location exercise. Carefully select three of the five sound sources previously evaluated for distance, and perform environmental characteristics evaluations on those sounds as outlined below.

As an alternative, look for a suitable surround-sound recording with few sound sources. Identify three to five sources that have separate locations for their direct sound and environmental characteristics. This will greatly assist you in comparing the direct sound and the environment, and in isolating environmental characteristics.

While these evaluations are most easily accomplished for short-duration percussive sounds, environmental characteristics evaluation is possible for any sound source as long as the reverberant energy of the environment is exposed (not accompanied by or masked by other sound sources) after the sound source has ceased sounding.

The process of determining environmental characteristics will follow this sequence:

1. During initial hearings of the entire work, listen to each sound source to identify a location where the sound is isolated throughout the duration of the environment. Nearly always the graph's time increments on the timeline will need to show milliseconds. Estimate the length of the timeline for that presentation of each sound source.
2. Check the timeline for accuracy and make any alterations. Work in a detailed manner to establish a complete evaluation of the reflections of the sound. First, sketch the presence of the most prominent reflections against the completed timeline; then, establish the precise time placement and the dynamic levels of these prominent reflections against the timeline. Use the prominent reflections as references to fill in the remaining reflections in the early time field. After the early time field is plotted, complete the reflections portion of the graph by plotting the dynamic envelope and spacing of reflections (density) of the reverberant energy.
3. Notice the locations and size of any emphasized or de-emphasized pitch areas or frequencies. Scan the entire piece of music, listening to how the sound source is altered by the environmental characteristics, by listening to many different pitch levels. Throughout these hearings, keep track of pitch areas or specific frequencies that appear to be emphasized or de-emphasized. With a running list of observations, regularly identified pitch areas/frequencies will begin to emerge. Further hearings will allow you to more accurately identify these frequencies and pitch areas (that make up the spectrum of the environmental characteristics), and to place the presence of these frequencies or pitch areas against the timeline.
4. You will now plot the dynamic contours of the components of the spectrum against the timeline. This process is the same as the process of plotting the spectral envelope of sound-quality evaluations. Each component of the spectrum is plotted as a single line, and these components are listed in a key, so their dynamic contours may be related to the spectral-envelope tier of the graph.
5. Continually compare the dynamic levels and contours of the spectral components to one another. This will aid in remembering the nominal dynamic level (where the amplitude of the spectral components of the sound source are unaltered by the environment), will aid in keeping the dynamic levels and contours consistent between spectral

components, and will keep you focused on the relationships of the sound source and its host environment. The evaluation is complete when the smallest significant detail has been incorporated into each tier of the graph.

This evaluation can be detailed and time intensive. It is not proposed that these detailed evaluations be undertaken in normal, daily activities of audio professionals. This is a learning tool. This study will be very successful at bringing you to hear, understand, recognize, and remember these important aspects of sound. You are encouraged to return to this exercise. Once speed and accuracy improve, you should undertake evaluations of more complex environments and sounds that are partially masked.

**Exercise 10.6**

*Exercise in Determining the Environmental Characteristics of the Perceived Performance Environment*

This exercise will seek to define the environmental characteristics of a recording's perceived performance environment. A multitrack recording should be selected that contains no more than three or four sound sources, a sound stage that clearly separates the images, and an overall sound that appears to envelop the sound stage.

1. Identify the sound sources and the different environments of the piece.
2. Perform general environmental characteristics evaluations of the environments. These initial evaluations should be general in nature, seeking prominent characteristics rather than detail.
3. Compare the environments for similarities of time, amplitude, and frequency information to identify common traits between the individual environments. (1) When traits are common to all sounds, an applied, overall environment is often present. Seek to identify if the traits are present in all environments equally. If the common traits are not applied to all sources equally, a single environment has not been applied to the entire program. (2) Then you must look at other factors as well. Next, identify the predominant traits of the environments of musically significant sound sources. They also directly contribute to the characteristics of the overall environment.
4. Listen to the work again to identify an overall environment of the program. An applied overall environment will be most easily detected by its detail in spectral changes of the reverberant sound, and in the clarity of the initial reflections of the early time field. The

characteristics of these environments will be perceived by listening for detail at a close perspective of slight changes to the predominant characteristics of the environment. Overall environments that are an illusion (created by the composite and predominant characteristics of the individual sound sources) will have characteristics that are not readily apparent. The characteristics of these environments will be perceived by listening at the more distant and general perspective of the dominant characteristics of the environment.

5. Compile a detailed environmental characteristics evaluation of the perceived performance environment. The evaluation is complete when the smallest significant detail has been incorporated into each tier of the graph.

Repeat this exercise on other recordings until you have evaluated a recording with an applied overall environment and a recording with a perceived performance environment that is the perceived result of the environmental characteristics of the individual sound sources.

Once skill and confidence are improving, repeat this exercise on recordings that have more activity and with less pronounced characteristics in the perceived performance environment.

### Exercise 10.7

*Space-within-Space Exercise*

Select a multitrack recording containing a small number of sound sources (five or six). A recording with a sparse texture (few instruments sounding simultaneously) and pronounced environments on the individual sound sources will be easiest to evaluate during initial studies.

The process for determining space within space follows this sequence:

1. Identify the various environments of the piece. Some sound sources may share environments with other sound sources (at the same or different distances), and some sources may change environments several times in the piece.
2. Perform general environmental characteristics evaluations of the environments. These initial evaluations should be general in nature, seeking prominent characteristics rather than detail.
3. Compare the environments for similarities of time, amplitude, and frequency information. This observation will determine common traits between the individual environments of the sources. These common traits will signal a possible applied, overall environment if they are present in all environments equally. If the common traits are not applied to all sources equally, other factors are in play as well. Identify

the predominant traits of the environments of musically significant sound sources. They also directly contribute to the characteristics of the overall environment.

4. Listen to the work again to identify the characteristics of the overall environment of the program (the perceived performance environment). Compile a detailed environmental characteristics evaluation of the perceived performance environment.
5. Begin the master listing of environments with this environmental characteristics evaluation of the perceived performance environment.
6. Perform detailed environmental characteristics evaluations of the individual host environments of each sound source. The characteristics of the overall environment may or may not be present in these evaluations, depending on the nature of the overall environment and the nature of the individual sound sources' environments. The evaluation of each source is complete when the smallest significant detail has been incorporated into each tier of the graph.
7. Number each environment and enter the evaluation into the master listing of environments. Note on the master listing the sound source or sources that are present within the environment.

Once skill and confidence are improving, repeat this exercise on more sound sources in recordings that have more activity and with less pronounced environmental characteristics.

# 11 Evaluating the Spatial Elements of Surround Sound

Surround sound can deliver a stunning experience to the listener, in ways that are very different from stereo recordings—ways that have the potential to be significantly richer. Yet, music recordings in surround have not been a commercial success to date. Consumers have not adopted the format for music listening. Surround recordings have not made their way into the daily lives of consumers as many of us had hoped, and as has happened with surround sound for motion pictures. Many artists, recordists, and labels have stopped making and releasing surround versions of recordings. The reasons for all of this are both many and speculative. Still, artists and audiophiles and recordists alike often closely embrace surround's unique qualities and its ability to greatly enhance music and sound. Surround has qualities that are completely unique from stereo and that can communicate in ways that are vastly different from stereo. Perhaps surround will become commercially viable in the future. This potential exists, and might well be realized as Internet bandwidth continues to increase and thus might soon allow for uncompromised online delivery of surround. It is from this position of hope, and an acknowledgment of all that the format has to offer our art, that this chapter is offered.

Both the technical and artistic aspects of surround recording continue to be developed and refined. Some artistically effective and musically convincing recordings have been made, and we will surely continue to witness profound developments in how surround sound is used to deliver and enhance music. While recording practice is still being refined, the ways that surround locations can be used to mix music are beginning to emerge. We have discovered that the listener's relationship to the music and to the performance has the potential to be strikingly different than in stereo; it can be an experience entirely transformed.

This chapter will present a way to evaluate, document, and conceptualize the spatial information of surround recordings. It is expected that the ways the audio professional will need to evaluate surround recordings will continue to change as a reflection of developments in technology and in production practice. While how we document and evaluate surround will need to be adapted to reflect future developments, this chapter will provide a current and meaningful point of reference.

## Format Considerations

Much debate and deliberation occurred regarding surround formats, and some continues. Many different channel and loudspeaker combinations and placements have been proposed for surround: too many to accurately count let alone cover here. These include formats from four channels to seven or eight, most with subwoofers, and some with bipolar surround speakers. Some formats have all speakers at ear level, others have the surround speakers higher, and one format uses a wonderfully effective sixth overhead channel (providing subtle but convincing ambiance and some impressive vertical cues). And there are others. While there was once much debate about which format should become the standard, the five-channel system with a subwoofer for low-frequency material and effects has emerged from this fray of formats.

Widespread consumer adoption of the 5.1 cinema format has led to its adoption for music as well. Using the specifications of the International Telecommunications Union (ITU) Recommendation 775 (see Figure 11.1), the format has proven stable, and was created after a great deal of thought and experimentation. Further, it can accomplish almost all of what is reasonable to expect of surround (albeit with the sad loss of the overhead channel). While the widespread angle of the surround speakers makes rear phantom imaging unstable and can quickly pull images forward, the equidistant placement of all five speakers has advantages for dynamic balance and time-based considerations, and provides convincing ambiance.

Another positive aspect of this format is that it is compatible with current two-channel playback concerns. The 60° angle between the left and right speakers provides for accurate listening to stereo recordings (see Figure 11.2). It is the accepted listening relationship for accurate stereo reproduction and can therefore also be used for evaluating two-channel recordings.

This format in the ITU specification (also defined by the Audio Engineering Society) was used in making the evaluations of surround recordings that appear in this book.

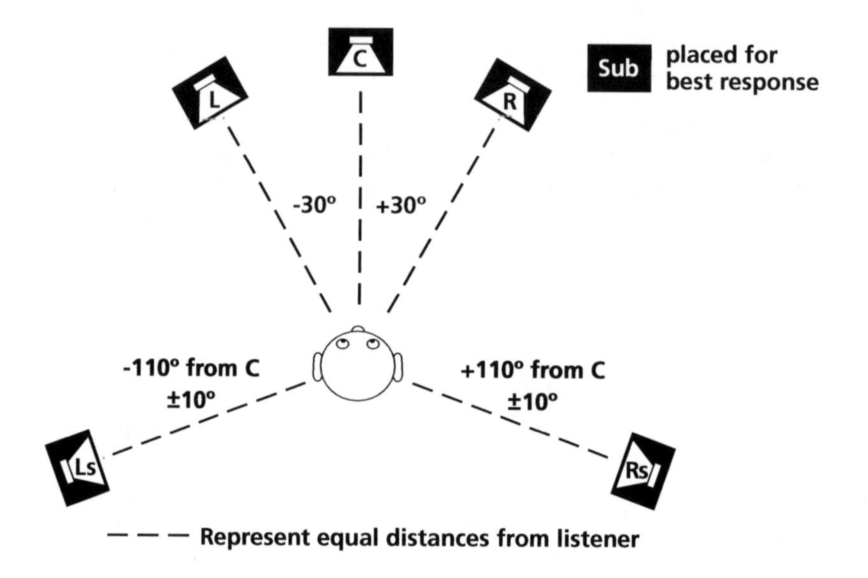


**Figure 11.1**
ITU-recommended speaker layout for surround sound.

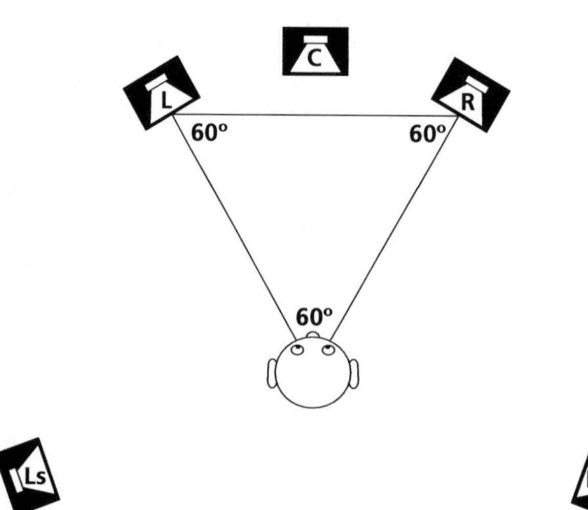


**Figure 11.2**
Two-channel sound reproduction with surround-sound loudspeaker placement.

Systems require more than correct speaker-relationship placement. Speaker placement relative to walls and the LFE subplacement are also considerations. The system must be calibrated for balance of amplitude and spectral characteristics between speakers and the listener location. Without appropriately matched sound from all speakers, recordings will not be accurate, and the listener's experience will be distorted from what was intended.

## Surround's Sound Stage and the Listener

In stereo recordings the sound stage is in front of the listener, as if the recording is a concert of sorts. The size of the stage can vary, from sources being tightly grouped in the center, to a full spread of 90° in front of the listener. The listener is an observer of the recording, a member of a conceptual audience. They may be very close or very far from the sound stage, but they will be detached from the stage of the recording, and will be an observer.

In surround, there is no telling where the sound stage will be, or the scale of its size. Every song, and potentially every moment in a piece of music, can have a unique sound stage, and might also change the listener's relationship to the music and its ensemble. This relationship of the listener to the music and its message, to the musical materials of the song, and to the sound sources/performers of the recording is both physical and conceptual. These are areas where surround can be vastly different from stereo, and deliver a vastly different experience. Surround can also enhance stereo with more realistic ambiance, or environmental characteristics, as in Figure 11.3; as such it enriches the

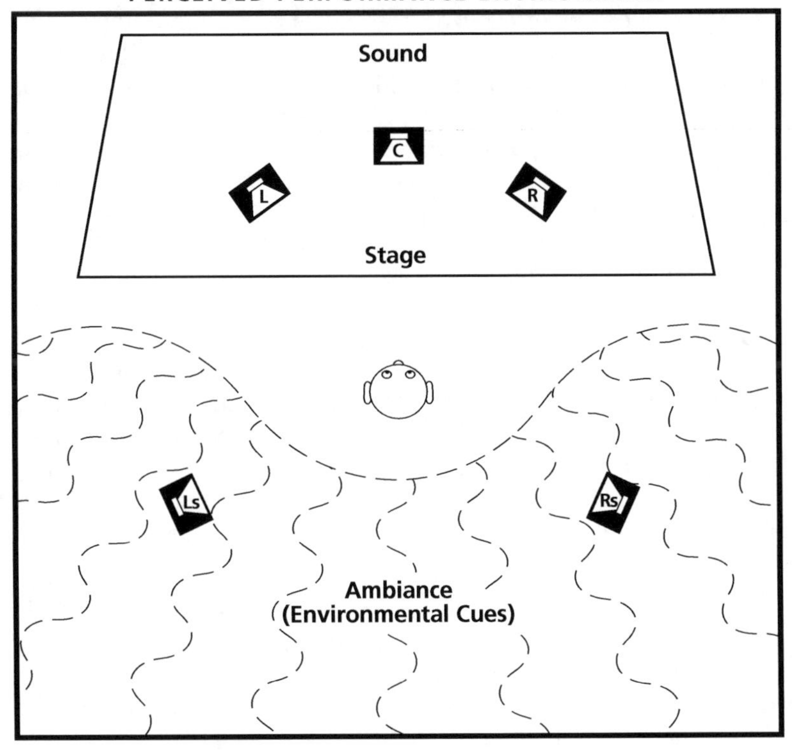


Figure 11.3
Sound stage in traditional location in front of listener, with ambiance filling the room from behind.

stereo recording without substantially changing the listener's relationship with the sound stage.

### Relationship of the Listener to the Sound Stage

The listener's relationship to the surround-sound stage plays a significant role in the character of the recording. This relationship may be examined most directly through considering: (1) the listener's location either as being located within the sound stage or detached from it as an observer, (2) if detached, the conceptual or perceived distance of the listener to the sound stage, and (3) the location of the sound stage relative to the listener.

In surround, listeners can find themselves located within the sound stage, and listeners may also find themselves surrounded by the sound stage, but detached from it. The position of listener within the sound stage is distinctly different from sound sources being merely wrapped all around the listener location. It is possible for the listener to get the impression they are within the music, within the ensemble, perhaps part of the band.

Listeners may find themselves sitting in the ensemble, and at times within the actual performance during the "I am the Walrus" version on *LOVE*. The listener's relationship to the music and sound stage shifts markedly during the song. The listener is immersed within the sound stage and the ensemble from the beginning, with instruments and voices arriving from all directions and sometimes very close. When sources begin to enter and exit, especially apparent from the start of the bridge, the listener is within the performance, with sounds bouncing around them. The motions of the sweeping sine waves in the bridge are joined by a moving cello line behind the listener; these lead to greater activities of sound movements and sudden entries of vocals throughout the verses; throughout the song effects, vocals have pronounced movement within themselves or in relation to the entries of other iterations. This immersion of the listener in the ensemble and within sound movement is brought to its height in the coda. The sound stage also has considerable depth, which brings the listener to have recognizable distance and detachment from some sources; some of the string and horn sounds are approaching the far area, while some are in proximity. This brings a sense of duality in the listener's relationship to the mix; they are an observer at times, and at times within the activity.

Surround can simulate a concert experience, sometimes with great effect. The listener can be convincingly placed within a live audience. To varying degrees, depending on the intention of the recording, the listener will experience others in the crowd and a sense of detachment from the stage. The sound stage is often in front, but not always.

The "I Want to Hold Your Hand" version from *LOVE* simulates a concert, and places the listener in the audience. This is not a typical concert recording relationship, however; the concert has a quasi-documentary soundscape, at least simulating the early1960s crowds of young women screaming. The performers are, however, very close to the listener, and not at a typical audience distance one would experience at a concert; one might stretch the imagination to place one's self at a front row seat, but there is no real ambiance from the performance space or close-up crowd presence to confirm the experience. These latter relationships are very common in surround versions of concert recordings, recordings intended to give a sense of the concert experience. The music here retains much detail and a crafted sound stage, yet provides some of the telltale characteristics of a live performance.

Separated from the sound stage, the listener can find himself or herself at any level of detachment from the music/ensemble. Levels of detachment might span extreme closeness and intimate connection, to isolation and disengagement from any personal connection. While distance cues may contribute, the level of intimacy of the music, the delivery of its performance, and its message play central roles. The significance of this detachment is heightened in surround, as there is great potential for this distance to vary considerably more than in stereo.

### Size and Position of the Sound Stage

The sound stage can be located anywhere around the listener. Its size can encompass any portion of the 360° around the listener. The position of the sound stage relative to the listener might be reduced to: (1) the extent to which the sound stage resembles traditional stereo, (2) the placement of ambiance and the perceived performance environment and of the individual sources, and (3) the presence of an additional sound stage or of sounds outside of the sound stage.

Figure 11.3 illustrates the sound stage in front, as in stereo. In this approach, the width of the stage can vary widely, from largely centered within or slightly extending beyond the 90° spread of stereo. Environment cues can be placed to the rear or to the sides of the listener, and/or added to the front channels.

The sound stage may remain in front but extended from stereo. This extended stereo may broaden the sound stage as in Figure 11.4. Sound sources are thus perceived at the sides of the listener as well as in front, yet still coalescing into a single area that includes all performers.

Ambiance may be at the rear only, blended into the sound stage, or both locations. Figure 11.4 illustrates ambiance partially blended into the sound stage, but predominantly in the rear.

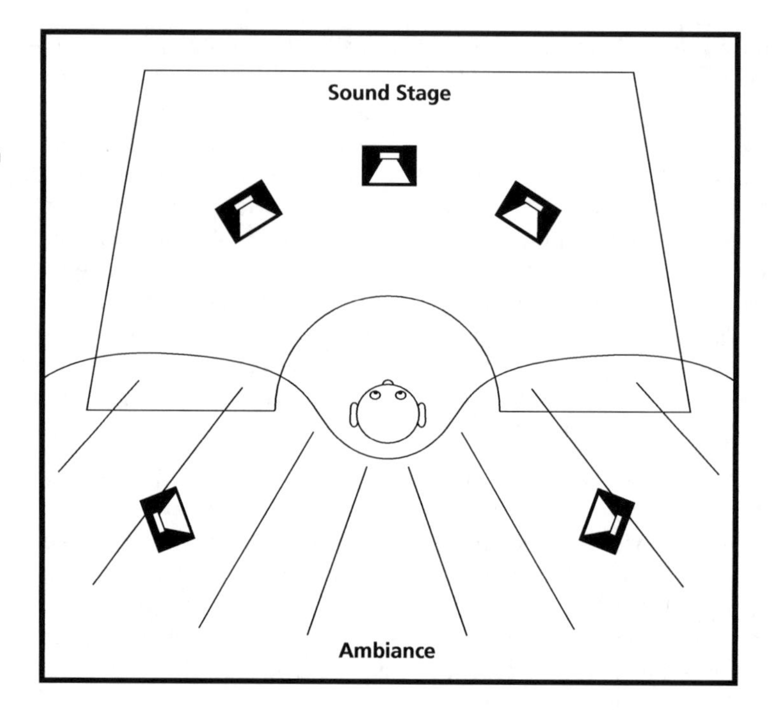


**Figure 11.4**
Sound stage extended to the sides of the listener, ambiance behind and fusing with sides of the sound stage.

Sources can appear outside of the sound stage. This is very different than stereo, where all images fuse into a single location. This situation most often occurs with a sound stage established in front of the listener, and a sound or two appearing to the rear of the listener. These rear sounds typically appear intermittently or only a few times during the work, and are clearly outside of, or detached from, the ensemble that comprises the sound stage. This technique resembles treatments of distant instruments or voices emanating from behind audiences in classical antiphonal works. The final verse, last chorus, and the coda of "Here Comes the Sun" on *LOVE* contain an isolated Moog sound in the left-rear channel that demonstrates this concept; the Moog sound is partially balanced by doubled backing vocals that appear at a lower level in the right-rear channel.

When sounds outside the front sound stage become numerous and are mostly continuous, an impression of a second sound stage can be created. Two opposing sound stages are established. The stages might have equal significance, or one may dominate. Most often the front sound stage will dominate, and have the most activity and the most significant musical materials. Sections of *LOVE*'s "Something" approximate a second, distinct sound stage behind the listener, with the rear sound stage comprised of the orchestra and low-level drums and lead vocal.

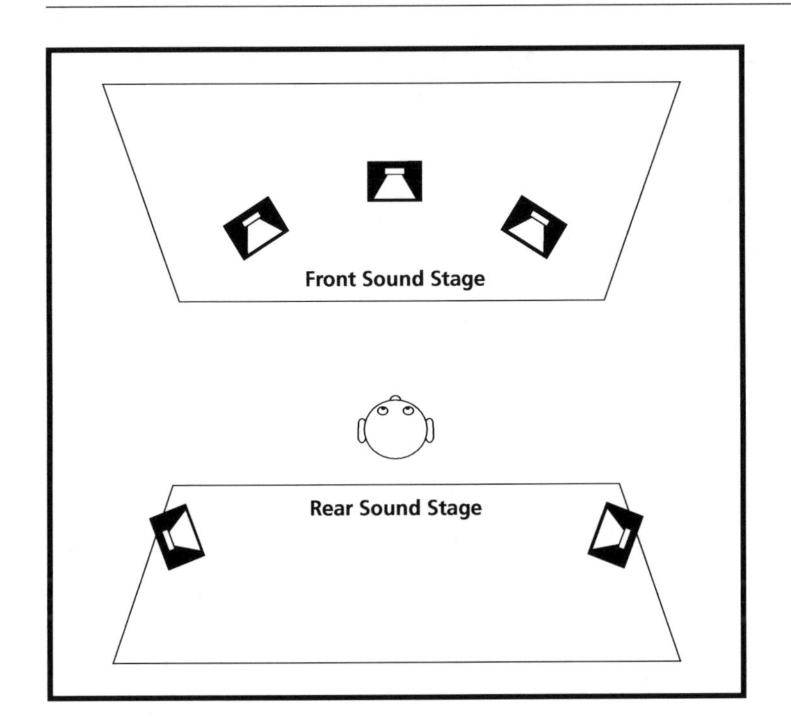


**Figure 11.5**
Surround-sound mix with two sound stages.

Finally, the sound stage may appear in the rear, behind the listener. Ambiance might be placed in front or at the sides of the listener. When the performance takes place behind the listener, it is a very different experience.

## Sound Location and Imaging

The imaging in surround sound is vastly more complicated than stereo. As discussed in prior chapters, stereo is based on phantom images between two speakers or up to 15° beyond. Phantom imaging is more complex in surround. It is composed of primary and secondary images. The format for 5.1 surround provides the opportunity for as many as 26 possible combinations of speakers. This changes phantom image placement, width, and stability greatly.

There are five primary phantom image locations existing between adjacent pairs of speakers in surround. These images tend to be the most stable and reliable between systems and playback environments.

Many secondary phantom images are possible as well. These can appear between speaker pairs that are not adjacent. These images contain inconsistencies in spectral information and are less stable. Implied are

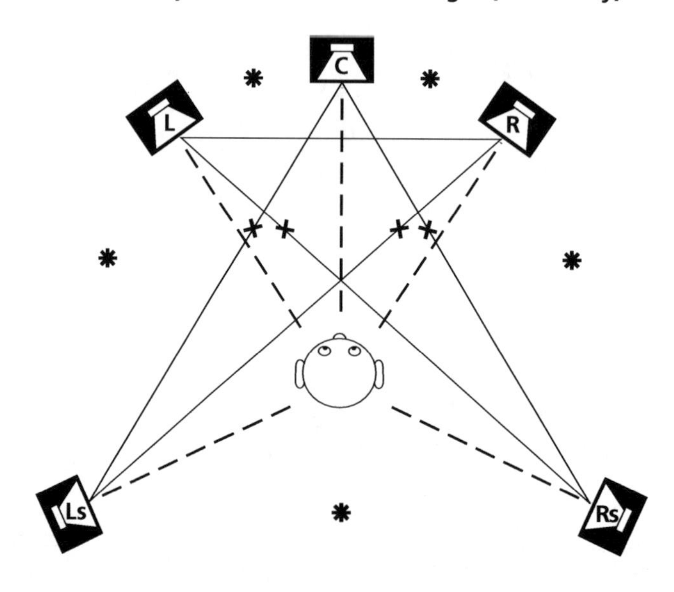


**Figure 11.6** Phantom images between pairs of speakers in surround sound.

different distance locations for these images, as the trajectories between the pairs of speakers are closer to the listener position. These closer locations do not materialize in actual practice. The distance location of these images is actually pushed away somewhat by the diminished timbral clarity of these images. This can create contradictory and confusing sonic impressions.

When we consider locations caused by various groupings of three or four loudspeakers, placement options for phantom images get even more complex.

Phantom images can be of greatly different sizes in surround. They can range from completely surrounding the listener with a spread image of enormous size, to a small and precisely defined point source. Point sources and narrow-spread images are common in current music productions, especially in the front sound field.

The center channel has changed imaging on the traditional front sound stage tremendously. Images are often more defined in their locations and narrower in size than in two-channel recordings. With the center channel, edges of phantom images tend to be clearer compared to stereo.

As with imaging in stereo, the edges of phantom images are more important to defining source placements than their centers. It is our natural tendency for the center of images to dominate our attention.

In everyday life natural sounds do not have pronounced widths in the sound images, as typically appears in reproduced phantom images. In as much as natural sounds do have size, in everyday life we are not really concerned about the size of the sound as much as we are concerned about where it is, so we can engage the object appropriately. To accurately evaluate the widths of images in recordings, we bring our attention to their edges. The edges of images allow us to understand and identify widths of images, and where and how they might overlap. With this awareness of the edges of images, we can recognize how sounds are separated in space or blended.

During imaging using non-adjacent speakers, sounds blend and fuse differently. Image placements and size can be unstable and vary quickly with any listener movement. Sounds can both separate themselves from others or become ill defined, almost transparent sonically, and largely masked.

Another unique aspect of imaging in surround is the various ways environments might appear.

The ambient sound of a perceived performance environment (PPE) can be separated from the sound stage, as in Figure 11.3, or it can be partly blended with the sound stage and filling the rear, as in Figure 11.4. Other forms are possible, such as blending throughout the 360°, perhaps equally or perhaps present more in some areas than in others.

The environments of individual sources can be separated from the source itself. In effect, the environment can become another source with its own location, width, and character. The fusion of sources and their environments that occurs in stereo may also happen in surround, but it can also be deliberately counteracted by physically separating the two sufficiently. Later, Figure 11.10 depicts a hypothetical mix placing the image of the ambiance of the lead vocal behind the listener with the vocal centered in front; note also the separate locations of the bell sound and its ambiance.

## Evaluating Location in Surround Sound

Sound location is evaluated in surround to identify the location and size of the phantom images of sound sources. These cues hold significant information for many surround productions, and may contribute significantly to shaping the musical ideas themselves. These are the cues that separate surround recordings from two-channel (stereo) recordings. Phantom images may change locations or size at any time during a piece of music. These changes may be sudden or they can be gradual, and may be prominent or subtle.

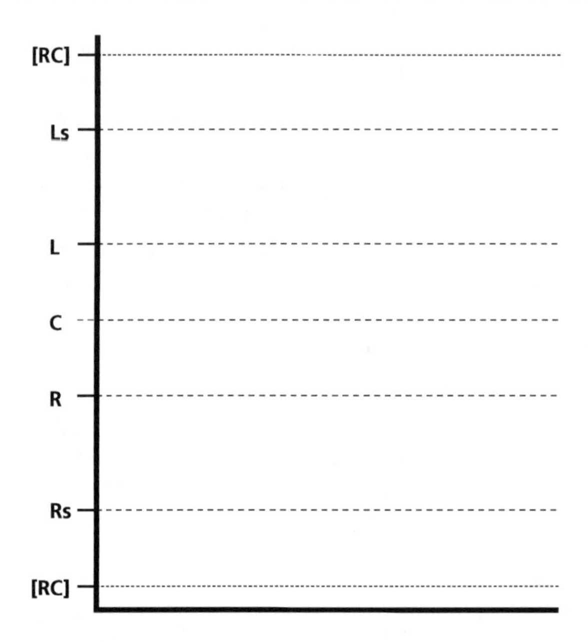


**Figure 11.7** Surround-sound location graph.

The *surround-sound location graph* will allow the reader to plot the size and locations of all sound sources against the timeline of the work. The graph can portray any lateral direction of sources from the listener and the size of the phantom images, 360° around the listener.

"Left," "right," "center," "left surround," "right surround," and "rear center" locations are identified on the graph (see Figure 11.7). The sound source's location is represented by placing a mark on the graph at the location of the phantom image. The angle of the sound source from the listener can be determined from the centerline out 180° up or down, and degree-increments of angle may or may not be incorporated into the graph, as desired. The rear-center location is placed at the very top and at the very bottom of the graph. This allows sound movement and spread images across the rear sound field to be graphed, although sometimes not as clearly as we might wish. As Figure 11.8 shows, the movement of sound sources across the rear and the locations of spread images extend off the top and bottom of the graph to wrap the source to the other rear-center location.

The surround-sound location graph incorporates:

1. "Left," "right," "center," "left surround," "right surround," and "rear center" as the *Y*-axis
2. The *X*-axis of the graph, dedicated to a timeline that is devised to follow an appropriate increment of the metric grid or is representative of a major section of the piece (or the entire piece) for sound sources that do not change locations

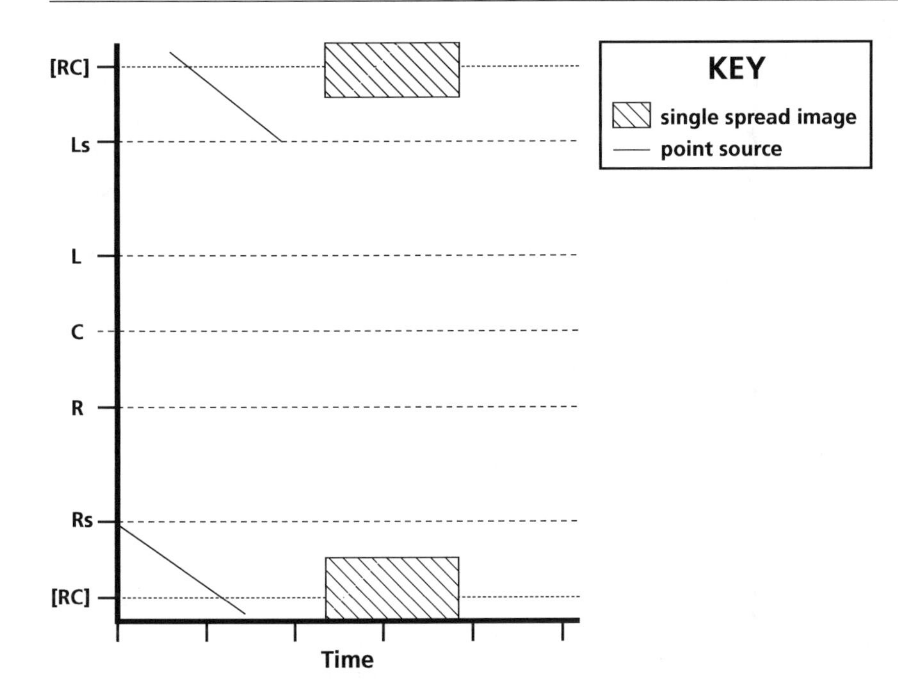


**Figure 11.8**
Rear-center sound-source movement and spread images on surround location graph.

3. A single line plotted against the two axes for each sound source and occupying a large, colored/shaded area in the case of the spread image
4. A key that is required to clearly relate the sound sources to the graph. The key should be consistent with keys used for other similar analyses (such as musical balance and performance intensity) to allow different analyses to be easily compared.

A sound source may occupy a specific point in the horizontal plane of the sound stage, or it may occupy an area within the sound stage, as was earlier found in stereo-sound location. Similarly, the surround location graph will dedicate a source line to each source against a common timeline. Point sources and spread images will be graphed in the same way as on the stereo-location graph.

A surround mix might have a sound stage behind the listener in addition to the front sound stage, and have little activity at the direct sides of the listener. Either stage may be a primary stage, with the other secondary, or they may be of equal significance. When two sound stages (front and rear) are present, the *Y*-axis presented in Figure 11.9 would more clearly show the content of that mix. Note the listener's heads are present to help the reader orient him- or herself to the graph; they would not appear on an actual working graph of a mix. The location of the rear sound stage might take some getting used to, but in time will make

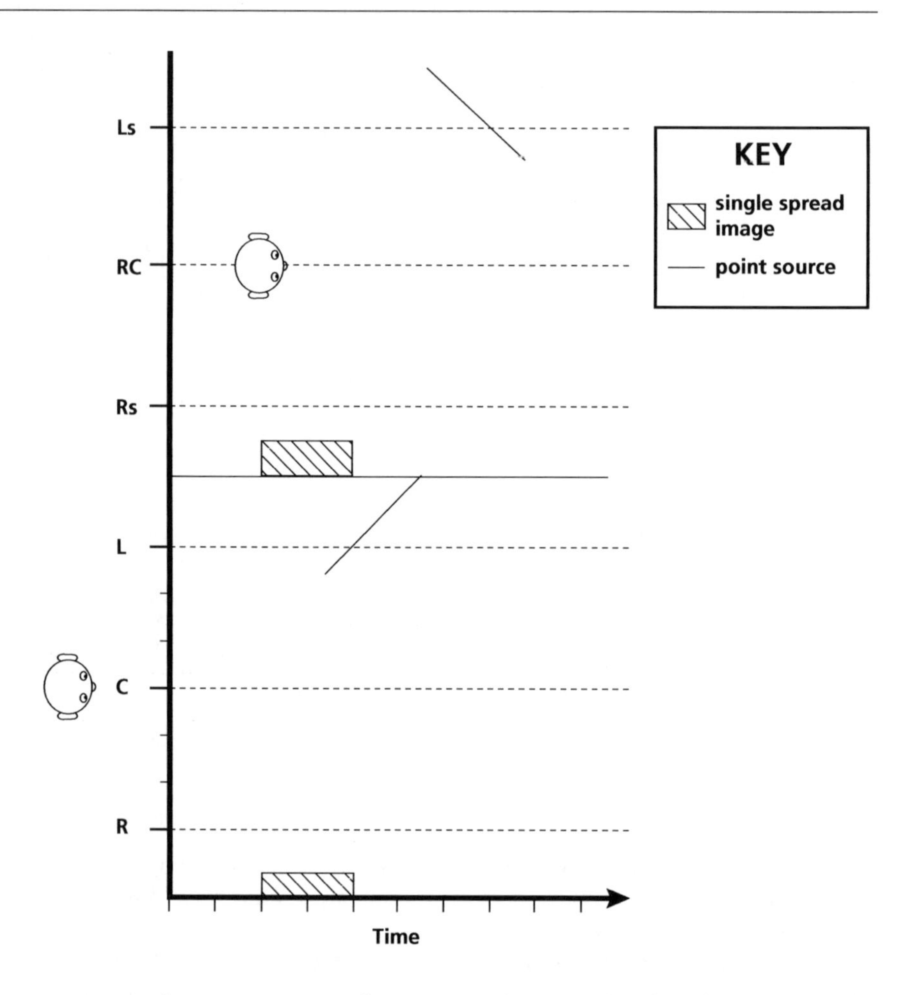


**Figure 11.9**
Alternative *Y*-axis for the surround-sound location graph.

sense; a clear stereo sound stage can be perceived and recognized in front of and behind the listener location. The sounds at the side will be cumbersome to plot; Figure 11.9 shows a spread image at the right side and a point source moving from the left front to the left rear to demonstrate how such sounds would appear. It must be noted this format might not be the best choice if many sounds of this type were present, but in recordings that emphasize the front sound stage and create a rear sound stage, this *Y*-axis would more clearly show the sound stage than would the format of Figure 11.7.

Locations and image size may be manipulated and change within a song/piece of music. Changes also often occur when instruments enter or exit the mix (due to masking). This often occurs between sections of a piece, or at repetitions of ideas or sections. These are places where changes in the mix often make the best musical sense and most often occur. The listener should never assume changes did not occur. Gradual

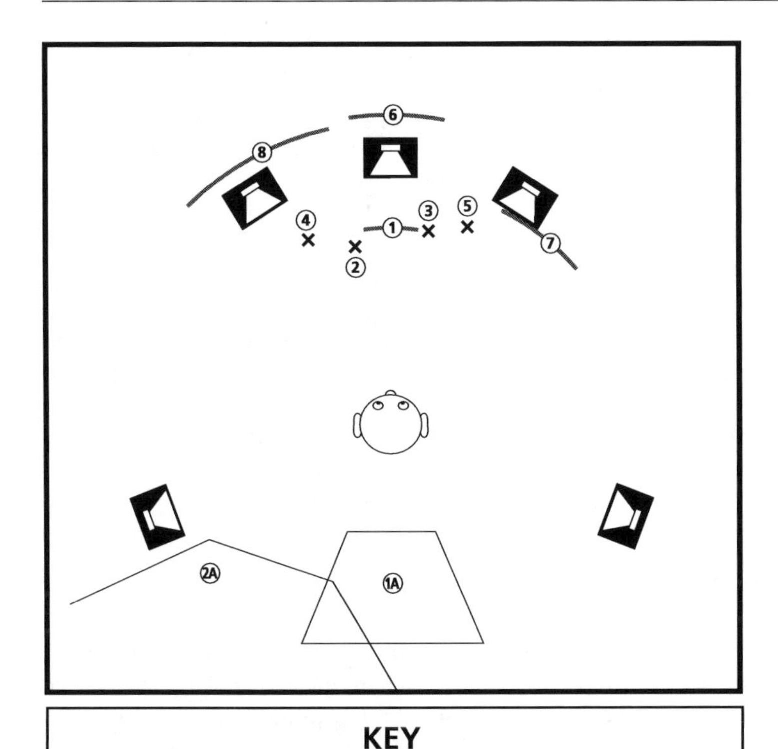


KEY

① Lead Vocal ⑩ Lead Vocal Ambiance
② Bell 1 ③ Bell 2 ④ Bell 3 ⑤ Bell 4 ②A Ambiance of all Bells
⑥ Keyboard ⑦ Acoustic Guitar ⑧ Bass
× point-source location

**Figure 11.10** Surround-sound imaging of a hypothetical mix.

changes in source locations and size are present in many pieces, and sometimes in pieces where such events are not expected.

When sound sources do not change locations in a section, a stationary graph without a timeline can appropriately be used. Figure 11.10 presents a surround location image that can be used to notate the locations of sound sources in such instances. Sizes of images and placements are clearly shown. It is also possible for the listener to sketch the sound stage—the area where the performance appears to emanate. This allows the listener to recognize when the surround speakers are being used for ambiance only.

This figure also has the advantage of allowing the listener to notate secondary sound-source phantom images created by non-adjacent pairs of loudspeakers and by groups of speakers. These images cannot be effectively incorporated into the surround-sound location graphs described above. The graph can also show separations between a sound-source phantom image and its ambiance (environmental characteristics) found in many surround productions. It can also show image location areas, containing breadth and depth. These can envelop the listener or surround the listener.

Figure 11.10 depicts the surround imaging of a hypothetical mix containing a number of sources separated from their environmental cues.

The reader will want to work through the Surround-Sound Location Exercise at the end of this chapter to refine their skill in localizing sources in surround playback. An example such as the bridge from *LOVE*'s "Being for the Benefit of Mr. Kite!" will allow one to plot moving sources surrounding the listener, as well as stationary sources, against a timeline. Most benefit will be obtained from the exercise by evaluating several different pieces of music, and by using both surround-sound location graph formats and the surround-sound imaging diagram.

## Distance Location

Distance cues, of course, remain a product of timbral detail, with some reliance on environmental cues. The enhanced presence of ambiance causes surround to more readily draw the listener into making inaccurate judgments of distance cues. Sounds can be imagined to be further away than is accurate, largely because of an awareness of extra or enhanced environment information. The listener is drawn toward ambiance and away from an awareness of timbral detail.

The depth of sound stage is extended all around the listener as well. The listener is able to perceive distance in all directions, and these cues are often strongly present in surround recordings. This provides for creative opportunities not possible in stereo, but provides the listener and the recordist with new challenges in processing sound(s).

First, we know listeners accept sound stages of great depth in the front sound field. They easily imagine they are viewing something with proportions out of their physical confines. Listeners are not prepared to perceive sounds from the side and rear in the same way. When presented with similar materials from side and rear locations, listeners can be reluctant to place sounds in or behind walls (even after they have been observed and recognized at those locations). The same cues presented in a musically different way to envelop the listener in the space may allow this greater depth to be perceived.

Second, phantom images from the side and rear are inherently filled with phase anomalies of the listening room. This can cause a lack of timbral definition and detail, and distort distance cues. Further, it is also common for surround systems to give different timbral qualities to any instrument panned across its different speaker locations. These timbre changes often translate into distance changes and blur distance location imaging.

Finally, involuntary head movement contributes to our natural localization process for acoustic sound sources. While this instinct has allowed our species to survive and evolve, it may well lead to a sense of apprehension in the listener. When presented with sounds from behind, the fight or flight instinct can be triggered, and thus distract the listener or create discomfort.

Phantom images from non-adjacent speakers complicate distance perception and depth-of-sound-stage impressions. These images exhibit peculiar characteristics. The interaural cues of sounds emanating from images produced from pairings of left- or right-surround and one or two non-adjacent front speakers place the sounds close to the listener's head, but without heightened timbral detail. This can confuse the listener and blur the sound stage—or simply provide conflicting information.

The tendency of surround to separate sources from one another spatially in the mix can provide sounds with heightened timbral detail. This, of course, will bring the sound closer to the listener. Sounds emerging from the mix or blending back into the mix with other sounds often exhibit marked changes in timbral detail, and thus marked changes in distance location.

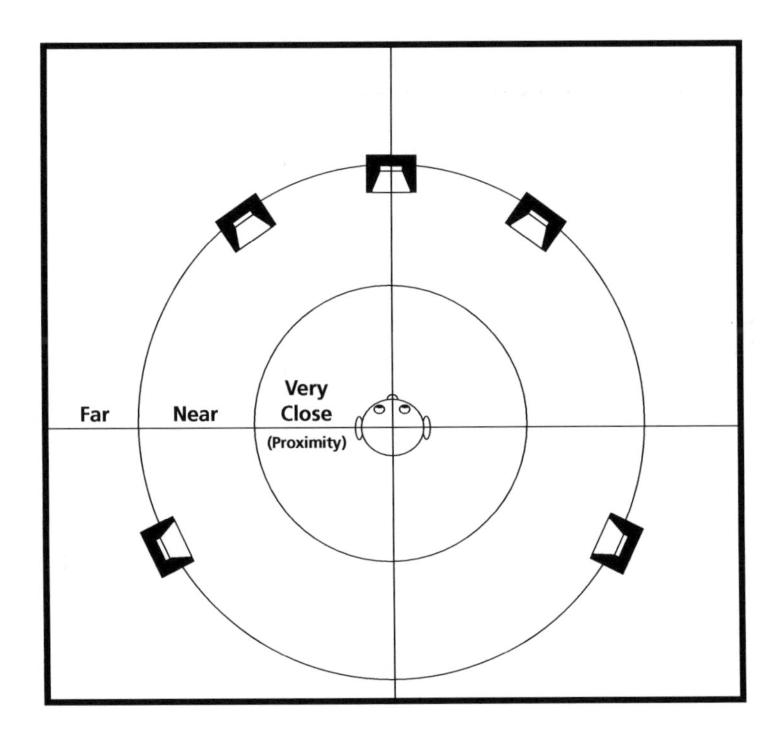


**Figure 11.11**
Diagram for imaging in surround sound with distance location shown.

Figure 11.11 is a diagram intended to be used for imaging in surround. The concentric circles allow sources to be correctly located in distance, as well as used to note the size and lateral locations of images. The distance areas presented in Chapter 10 are used here as well.

Figure 11.12 is the basic surround-sound stage of "Help!" from *LOVE*. The recreation of the mono mix experience is evident, with some important changes. Nearly all sounds are centered on the front sound stage, and ambiance is at the sides and behind. Notice the image of the acoustic guitar stretching from the front-center channels through the listener to the rear channels; this is the complicated depth and distance experience of non-adjacent speakers referenced above. The unrealistically large acoustic guitar is balanced by a lead electric guitar's descending line in the chorus, which is suddenly prominent and abruptly stretches across the entire front sound stage. It is also more prominent than the acoustic guitar except during the last verse, when the acoustic guitar is very prominent.

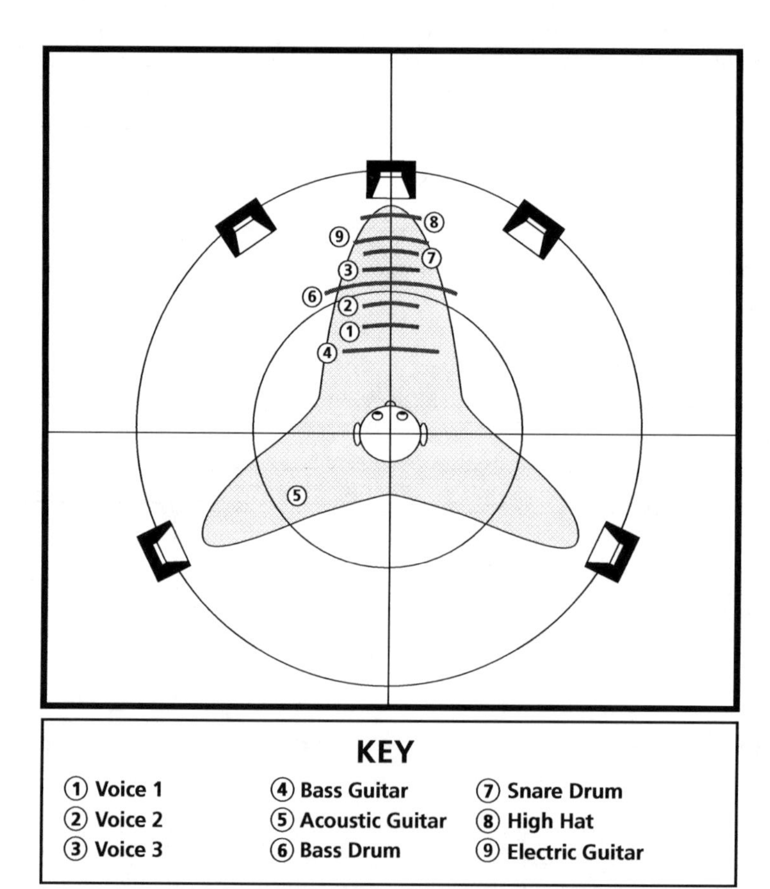


**Figure 11.12** Surround-sound stage for "Help!" *(LOVE)*.

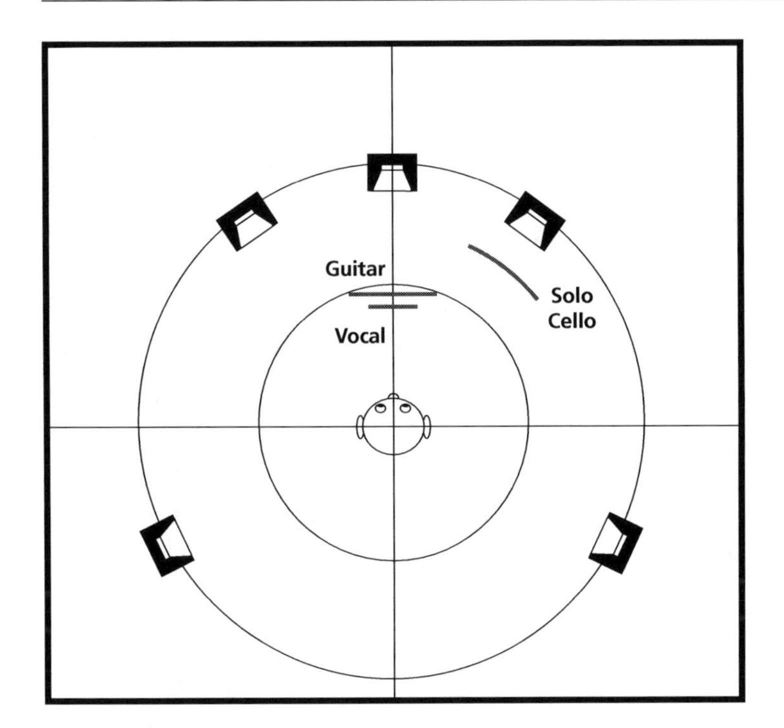


**Figure 11.13**
Surround-sound mix of "While My Guitar Gently Weeps" from *LOVE*, measures 1–24.

Figures 11.13 and 11.14 present the surround-sound stage in two sections of "While My Guitar Gently Weeps" from *LOVE*. Distance placement and image widths of sources are combined in the figures to provide a more complete impression of imaging. The diagrams clearly show the width of the sound stage changes significantly throughout the song. During the introduction, the acoustic guitar changes its position and width slightly before arriving at its final location within the front-center focused sound stage in Figure 11.13. From there the mix evolves slowly, as each of the entering string sources undergoes changes of widths and locations. Continually and gradually expanding, the sound stage grows much wider before it arrives at the stage diagramed in Figure 11.14.

The boundaries of the sound stage steadily, but gently enlarge as the music progresses, with the listener gradually becoming more and more immersed in the song as they are further and further surrounded by the musical materials and sound sources. Though subtler, distance locations change as images move and as sources enter and exit. The sound stage does not fully envelope or immerse the listener by incorporating material from the rear. A certain degree of detachment is maintained, and the listener remains an observer.

As an exercise, note the changes that occur in the sound stage between measures 24 and 57. Use additional diagrams to capture time segments

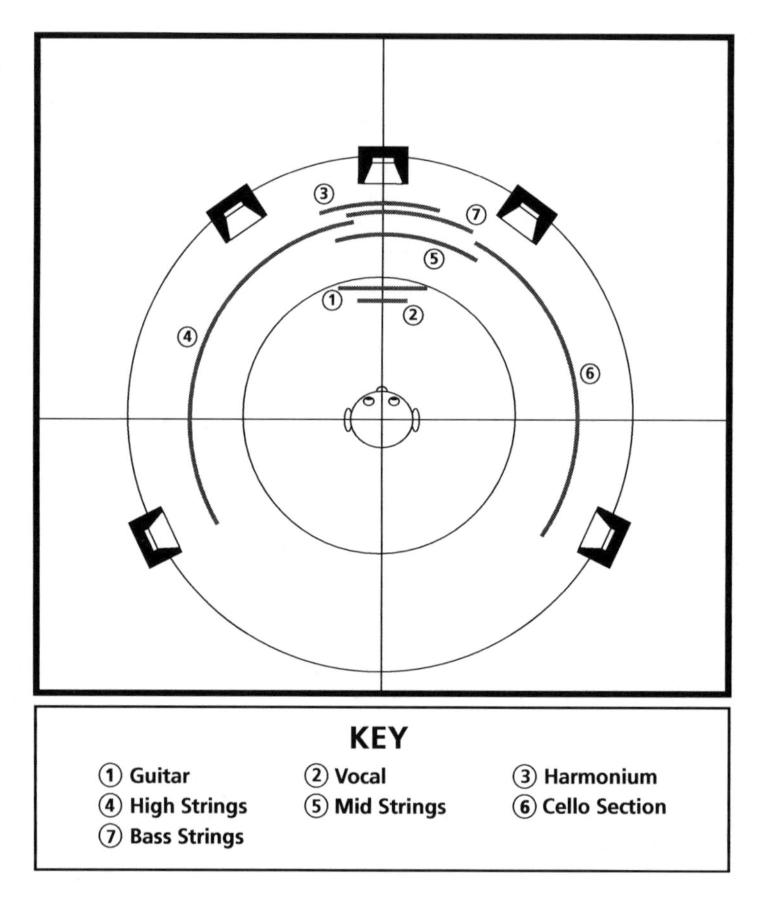


**Figure 11.14** Surround-sound mix of "While My Guitar Gently Weeps" from *LOVE*, measures 57–72.

at other important moments. You will note the differences in distance locations, lateral locations, and image size are significant. Bring attention to the edges of the images first, then focus on distance. These will become evident with some time and effort.

Listening to this surround version is a very different experience to the stereo mix. Compare the two and notice distance differences. Then note how imaging in surround is different musically to the stereo sound stage. In surround, the musical ideas are spaced further apart; they are given more room and have a degree of separation; there is less competition between sources and ideas, and in some ways there is less connection to one another.

## Sound Sources and their Environments

The sound of environments can arrive at the listener from every direction in a very natural manner in surround. This will immerse the listener in the cues and provide the life-like experience of being present in the space where the recording was made. Spaciousness can be presented by both two-channel and five-channel systems to portray a sense of space, but only surround systems can provide the sensation of being there, being present within the performance space.

At present, environmental characteristics are mostly used in ways similar to stereo recordings. The characteristics of the perceived performance environment and of individual sound sources are crafted to shape the musicality of the recordings. One exception is fully or partially immersing the listener in the ambiance of a source's host environment, while localizing the source in a specific location elsewhere (usually in the front sound field), described above.

The inherent qualities of environmental characteristics remain unchanged, with the potential exception of reflected sound arriving from the rear and side. This itself is a great difference, as the fusing of environmental characteristics and direct sound can become challenged. With sound sources and environments arriving from different directions, a variety of results such as enlarged images, unnatural effects (perhaps pleasing and suitable for the music, or perhaps not), distracting reflections, and more can occur. This is especially apparent when environment cues are sent to only a few channels and do not surround the listener. A very believable and effective host environment may also exist.

The environmental cues of individual sound sources may be perceived as separated from the direct sound, may be used to enhance imaging and space-within-space illusions, and many other alternatives exist for this new dimension. This new set of illusions makes the perceived performance environment's tendency to bind all of the spaces of the individual sound sources together even more important. The perceived performance environment (PPE) will continue to provide an important context for the recording and a critical point of reference. Indeed, the PPE is perhaps more apparent and serves a greater function in surround recordings than stereo. The PPE cues can often be heard more distinctly in surround, especially when they are separated from the sound stage, and more clearly contribute to the character of the recording. The PPE also functions more obviously in its role of providing a common, overall character within which we readily recognize all sources and their environments to reside.

## Exercises

### Surround-Sound Location Exercises

Two approaches can be used for surround-sound location. The reader should work through both approaches, as one will be more suitable to any sound material than the other. Determining which approach is most suitable will be a valuable undertaking in itself.

---

**Exercise 11.1**

*Surround Location Evaluation against a Timeline*

Find a surround recording with sources located around the listener position, but with the listener detached and at the audience perspective. Perform an evaluation of the locations of four or five sources for the first two major sections of the work.

The process of determining surround-sound location will follow this sequence:

1. During the initial hearing(s), listen to the example to establish the length of the timeline. Next, notice the presence of prominent instrumentation, with placements and activity of their surround location, against the timeline.
2. Check the timeline for accuracy and make any alterations. Establish a list of sound sources (instruments and voices) you will use for this exercise. Now, sketch the presence of the sound sources against the completed timeline.
3. Notice the locations and size of the sound sources (instruments and voices) for boundaries of size, location, and any speed of changing locations or size of image. The placement of instruments against the timeline will most often establish the smallest time unit required in the graph to accurately exhibit the smallest significant change of location. The boundaries of the sound sources' locations will establish the smallest increment of the *Y*-axis required.
4. Begin plotting the surround location of each source on the graph. The locations of spread images are placed within boundaries. The boundaries may be difficult to locate during initial hearings, but they can be defined with precision. Continue to focus on the source until it is defined. The locations of point-source images are plotted as single lines. These sources are easiest to precisely locate and are most likely to change locations in real time.
5. Continually compare the locations and sizes of the sound sources to one another. This will aid in defining the source locations and will keep you focused on the spatial relationships of the various

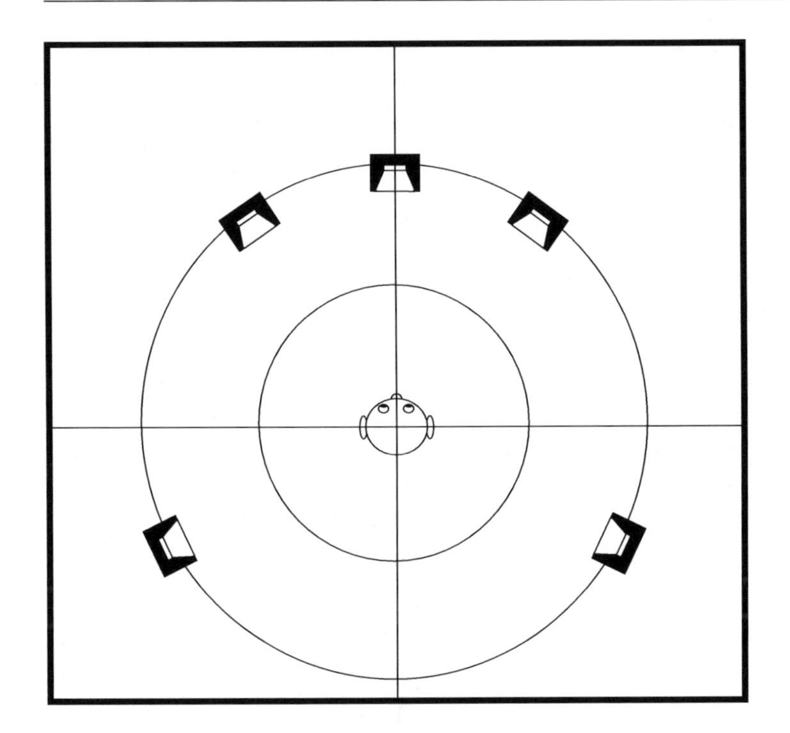

**Figure 11.15** Surround-sound imaging diagram.

sound sources. The evaluation is complete when the smallest significant detail has been incorporated into the graph.

Locating sound sources originating from behind normally causes a listener to move their head. You should consciously keep your head still and focus on the direction and size of the image as heard.

### Exercise 11.2

*Surround Location Evaluation with Imaging Diagram*

This exercise should be repeated on recordings other than those used for Exercise 11.1. Find a recording with the listener located within or surrounded by the ensemble or the performance.

When sound sources do not change locations or when sources appear to envelop the listener, a stationary-sound location diagram can be substituted for the surround location graph. Use the empty surround-sound location diagram of Figure 11.15 for these types of evaluations. The diagram represents a specific moment in time or a defined period of time that should be clearly defined.

Identify and graph several point sources and several spread-image sound sources in the song. A separate diagram will be used whenever source size or locations change. Create diagrams of several different sections of the work separately. Note and evaluate the characteristics and changes of imaging that occur in the song.

---

**Exercise 11.3**

*Incorporating Distance Evaluation into the Surround-Sound Imaging Diagram*

Figure 11.11 illustrates how the concentric circles in the diagram radiate distance from the listener and designate distance areas. Using the same recording(s) as Exercise 11.2, bring your attention to timbral detail.

Focusing on one sound source at a time, define the distance locations of each sound source. Compare the distance locations of sources, asking if "sound source a" is further or closer than "sound source b." Next, seek to define the amount of different between the two, and how they both relate to "sound source c."

Add this information to the imaging diagram, combining it with the lateral location information from Exercise 11.2 to create a complete drawing of the surround-sound stage for that section of time.

---

**Exercise 11.4**

*Source Location Separated from Environment/Ambiance*

Find a surround recording that places the environment in a different location than the direct sound of the sound source. The two should have clearly different locations, so they can be heard as separate sounds. Next, identify the edges of each image carefully to be certain you are able to keep them separate as you make your evaluation.

Next, evaluate the environmental characteristics of the ambient sound. Use the same techniques as the environmental characteristics evaluation exercise, Exercise 10.5. You will be able to directly compare the environment to the direct sound in making this evaluation. Much of the guesswork in trying to separate the environment from the source in stereo will be eliminated. Continue until your evaluation of the environmental characteristics is complete.

### Exercise 11.5

*Evaluating an Exposed Perceived Performance Environment*

Find a surround-sound recording that separates the sound stage and the ambient sound of the overall program. A recording with the two clearly separated will be most helpful for this initial exercise. The ambiance of the overall recording is the environment of the sound stage, and is the perceived performance environment (PPE).

Perform an environmental characteristics evaluation on the PPE, using the process just practiced in Exercise 11.4. Bring your attention carefully and deliberately to the dynamic contour of the reverberation, the density of the reverberation, and the frequencies emphasized or attenuated by the environment, in separate listening sessions. Seek greater detail in each listening, and move toward defining missing information one deliberate step at a time. Continue working until the evaluation is complete.

Finally, with the PPE complete, there is an opportunity to evaluate how the individual sound sources and any environments of the sources relate to the PPE. This will allow the reader to experience how those spaces contribute to the overall space of the recording, and bring a more tangible understanding to space within space. It will also allow the reader to perceive how the sound sources themselves relate to the PPE, and how they are impacted.

# 12 Complete Evaluations and Understanding Observations

The evaluations of artistic elements from previous chapters will be drawn together and compared here. Observations will be made from examining those evaluations and comparing the artistic elements to the musical materials.

Our evaluations have primarily been on three different levels of perspective: (1) the characteristics of an individual sound, (2) the relationships of individual sound sources, and (3) the overall musical texture, or overall program.

The highest level of perspective brings the listener to focus on the composite sound of a recording, or its overall texture. At this level, all sounds are summed into a single impression. Much recreational listening occurs at this level of perspective, with shifts of focus moving to text and melody—and other aspects attractive to the listener, such as beat or pulse, or the emotion and message of the song—in a random manner and at undirected times.

The overall texture is very important for the recordist as well, as its dimensions can profoundly shape the music/recording. These dimensions are the piece of music/recording's form, perceived performance environment, sound stage, reference dynamic level, program dynamic contour, and its timbral balance. The overall dimensions that provide the recording and music with its unique character must be readily recognizable and understood by the recordist, listening at the upper levels of perspective. All but one of these dimensions, timbral balance, have been thoroughly explored in previous chapters.

This chapter will examine relationships of the various dimensions of the overall texture and explore timbral balance and pitch density more deeply. The information offered by comparing evaluations of individual sources is explored next. An examination of how the artistic elements shape a recording will lead to a summary of the system for evaluating recorded/reproduced sound.

## Pitch Density and Timbral Balance

The process of evaluating pitch density is directly related to pitch-area analysis from Chapter 7. *Pitch density* is the amount and placement of pitch-related information of a single sound source within the overall pitch range of the musical texture. It is composed of the pitch area of the source's musical materials fused with its sound quality (timbre). *Timbral balance* is the "spectrum" of the overall texture that is created by the combination of the pitch densities of all of the sound sources in the music. The two are interrelated, as each source's frequency content contributes to create the overall frequency content of the recording.

### Pitch Density

The concept of pitch density allows each musical idea/sound source to be perceived as having its own pitch "placement" or "location" in the musical texture. The range of pitch that spans our hearing (and the musical texture) can be conceived as a space. Within this space, sound sources might be understood as being placed and/or layered according to the frequency/pitch area they occupy. The overall pitch range can be perceived as being divided into areas. Some of these areas will be occupied by the sound sources, or left empty; some areas might contain much pitch/frequency material, some little. The size of pitch areas and their placement are unique to each piece of music. Further, they may remain stable throughout a song, or may change at any time.

With this approach, the concepts of pitch density and timbral balance are often applied to the processes of mixing musical ideas and sounds in recording production. This is similar to the traditional concept of arranging and orchestration, where instruments are selected and combined based on their sound qualities, and the musical materials they are presenting. The recording medium and its various formats provide new twists to this traditional approach to combining sounds.

Pitch density is (1) the pitch area occupied by the musical material of the sound source with (2) a density of pitch/frequency information provided by the sound quality (timbre) of the sound source.

Musical materials create a single concept or pattern. The materials will come together in our memory and perception as a single idea comprised of a group of pitches. The material will be heard to occupy a specific pitch area. The pitch area of the musical material is defined by its boundaries—its highest and lowest pitches. Within the boundaries a bandwidth of sorts is established. The number of different pitch levels that comprise the musical material and their spacing creates a density within the pitch area.

The sound quality/timbre of the sound source will also influence the bandwidth of the pitch area and the pitch density of the musical idea. The primary pitch area of the sound source's spectrum will often be made up of the fundamental frequency of an instrument or voice, perhaps with the addition of a few prominent lower partials. A primary pitch area might also contain environment information (delayed and reverberated sound). These cues may add density to the sound without introducing new pitch information. Distortion sounds and other processing effects may also provide additional spectral information and added density to the primary pitch area.

Often, the primary pitch area of the sound source is rather narrow, often slightly more than the fundamental frequency alone. This is especially true when an instrument or voice is being performed at a moderate to low dynamic level. Strong secondary pitch areas in the sound-source timbre can be present. These provide additional pitch information that will widen the pitch area of the source (raising the upper boundary of the pitch area) and/or that can add to its density. Formant regions will often add a consistency of pitch-density information between a sound source's pitch-density events.

As the dynamic level of a sound source increases, lower partials will often become more prominent, and the width of the pitch area will tend to widen. In this way the spectrum of the sound source and its performance intensity impact pitch area.

Using this information, the highest boundary of the pitch area is determined by perceiving the spectrum of the sound source. The upper boundary of the pitch area "bandwidth" will be located at the pitch/frequency level where significant harmonics and overtones cease to be present. Typically there is approximately a 3-to-1 loudness ratio between the lowest boundary of the pitch area and the upper boundary.

The pitch area of each musical idea is determined by (1) defining the length of the idea. It is then possible to define (2) the lowest boundary of the pitch area, and (3) the highest boundary of the area. Once the bandwidth of the pitch area has been established (4), the amount of spectral information present can be observed, completing the concept of pitch density.

Most listeners can easily determine the length of the idea by simply asking, "When does this idea end, and when does the next idea performed by this instrument/voice begin?" This usually takes place at the perspective of the individual sound source. At times an instrument such as a piano may present several musical ideas simultaneously, in which case each would be understood and processed separately. At times, a group of instruments such as a brass section may present a single musical idea and be heard as a single unit; these would be grouped together.

Pitch information can be determined by examining the melodic activity (and/or harmonic activity for instruments like guitar and keyboards) of the musical idea to determine the shape and the highest and lowest pitch levels; this will establish the lowest boundary of pitch density. Adding detail pertaining to the sound qualities of the sources performing the idea provides the upper boundary and density information. The boundaries of the pitch areas of musical ideas may or may not change over time.

The pitch area of a sound source's timbre as it presents a musical idea is a composite impression that establishes a frequency band we call pitch density. Pitch density contains all of the appearances of the sound source performing all of the pitch material within the time period of the musical idea. It is the sum of all of the pitch levels and the significant sound-quality information of its sound source(s). The relative density of the idea is determined by the amount of pitch information generated by the musical idea and the spectral information of the sound source(s). As a song unfolds, a source's pitch area will change as it presents different musical materials and/or changes timbre by differing performance intensity, loudness, expression, etc. This gives the different pitch-density events different characters and characteristics that directly contribute to the music.

This process for determining pitch density is repeated for each individual musical idea. All sounds can be plotted on a single graph to allow the overall texture to be observed. This graph will represent the timbral balance of the recording/music.

### Timbral Balance

Timbral balance is the combination of all of the pitch densities of all of the recording's sounds. It is a dimension of the overall texture that conceptually represents its "spectrum." Timbral balance is the distribution and density of pitch/frequency information in the recording/music.

Evaluating timbral balance is a sound-quality evaluation of the overall texture. Here the individual sound sources that make up the overall texture can be conceived as individual spectral components. Individual sound sources are analyzed for their contributions to the sound quality of the overall program in terms of their timbres and the musical materials they present (pitch density).

All sound sources are plotted on the timbral-balance graph. The graph may take two forms, with or without a timeline. The graph may simply plot each sound source's pitch area against one another (as the pitch-area graph, in Chapter 7), and be a rather general representation of the overall texture. Most helpful in beginning studies is when the sound sources are plotted individually against the work's timeline, allowing the graph to visually represent changes in timbral balance as the work unfolds.

In either form, the timbral-balance graph contains:

- Each sound source, represented by an individual box denoting its pitch area
- The Y-axis, divided into the register designations first presented in Chapter 5
- The density of the pitch areas, which can be denoted on the graph through shadings of the boxes of the sound sources or by other descriptions.

The pitch densities of all of the sound sources may be compared to one another and to the overall pitch range of the musical texture. Timbral balance allows the pitch densities of all sound sources to be compared. Thus, the recordist is better able to understand and control the frequency content of the recording by observing the contribution of the individual sound source's pitch material to the overall musical texture and of the mix.

The timbral balance of the beginning sections of The Beatles' "Lucy in the Sky with Diamonds" (1999 *Yellow Submarine* version) appears in Figure 12.1. The work uses density and the registeral placement of sound sources and musical ideas to add definition to the musical materials and sections of the music. Timbral balance itself helps create directed motion in the music. The musical ideas are precisely placed in the texture, allowing for clarity of the musical ideas. The expansion and contraction of bandwidth of the overall pitch range and textural density of the musical ideas (and sound sources) add an extra dimension to the work and support it for its musical ideas.

While dynamic levels of sources can impact perceived timbral balance, dynamic-level information is not contained in the timbral-balance graph. Information on the dynamic levels of the sound sources can be found in the musical balance graph, however. Viewed in this way, musical balance conceptually represents the "spectral envelope" of the overall

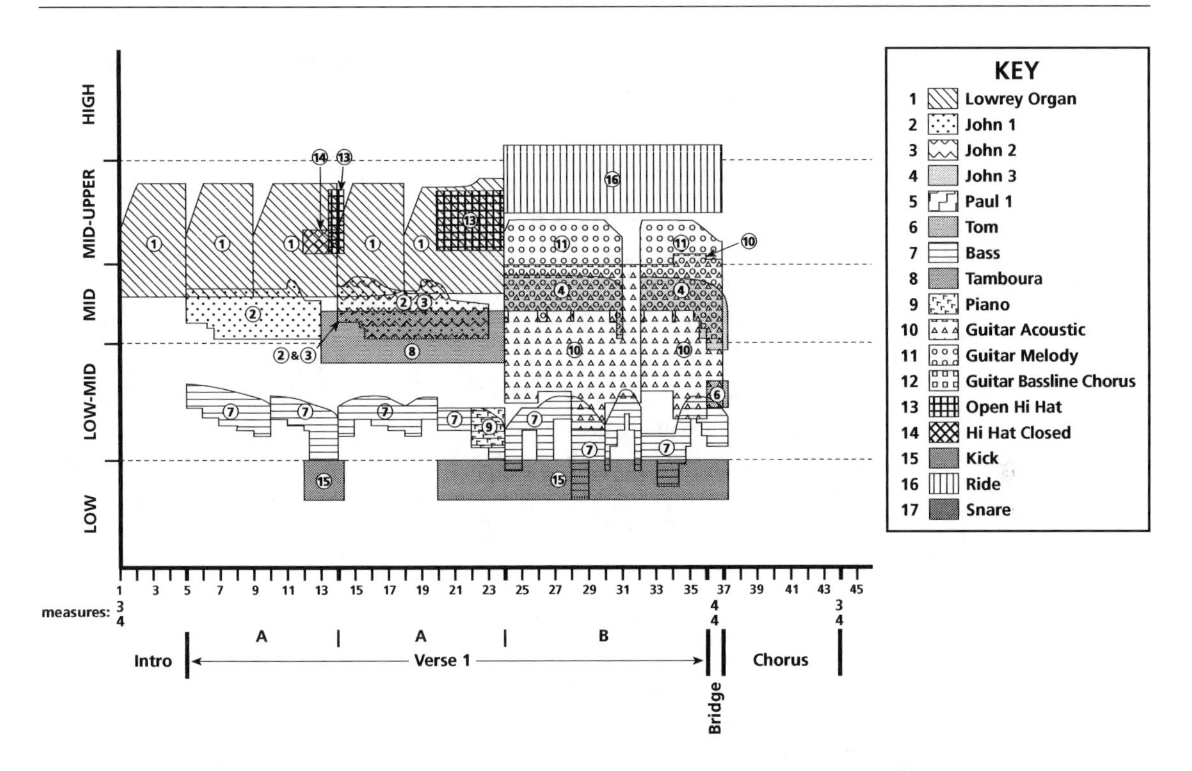


Figure 12.1 Timbral-balance graph—The Beatles' "Lucy in the Sky with Diamonds" (*Yellow Submarine*, 1999).

texture. By comparing the two graphs, the reader can understand more about how the recording made use of pitch and dynamic information to shape its overall sound quality, or timbre.

Similarly, stereo location (or surround-sound location) can impact perceived timbral balance. Timbral balance is distributed across the sound stage. The spectral information of the overall texture is balanced by location as well as by the distribution of pitch/frequency information by register. Paying close attention to this type of interaction allows us to recognize how a sound source can emerge from the timbral balance of a mix by moving it to a new location; sounds can be blended or given clarity by their placement in location and by their spectral content. By comparing the timbral-balance graph to the stereo (or surround) location graph, we can recognize how the spectrum of the recording is distributed by location. Notice how the shift of the sound stage between the introduction and chorus of "Here Comes the Sun" is reflected by a change of timbral balance; notice also how changing stereo location gives added clarity to this shift of timbral balance.

Exercises for determining pitch density of a single source and for determining a recording's timbral balance appear at the end of this chapter. The reader is encouraged to spend enough time with each exercise to feel comfortable with these concepts that are so important to the mixing process.

## The Overall Texture

The overall texture of the recording is perceived as an overall character, made up of the states and activities of all sounds and musical ideas. Pitch-register placements, rate of activities, dynamic contours, and spatial properties are all potentially important factors in defining an overall texture.

The characteristics of the overall texture provide many fundamental qualities of the music and recording. These greatly shape the music and its sound qualities, and communicate most immediately to the listener. The framework for the music and the context of the message of the recording are crafted at this level of perspective.

The characteristics of the overall texture are:

- Perceived performance environment
- Sound stage
- Reference dynamic level
- Program dynamic contour
- Timbral balance
- Form

The perceived performance environment creates a world within which the recording exists. This adds a dimension to the music recording that can substantially add to the interpretation of the music. The level of intimacy of the recording can become related to the level of intimacy of the message of the music. This element will largely be defined by how the sound stage is placed in the perceived performance environment, the listener's perceived distance from the sound stage, and the size and depth of the sound stage.

The reference dynamic level represents the intensity and expressive character of the music and recording. What the music is trying to say is translated into emotion, energy, expression, and a sense of purpose. These are reflected in recordings and performances, and are understood as a perceived performance intensity that is used as a reference dynamic level. Understanding this underlying characteristic of the music allows the recordist to calculate the relationships of materials to the inherent spirit of the song/composition.

The program dynamic contour allows us to understand the overall dynamic motion of the entire piece of music. Actual dynamic level will impact the recording process in many ways, and it also shapes the listener's experience. While program dynamic contour depicts how the work unfolds dynamically over time, this contour is often closely matched to the drama of the music. The tension and relaxation, the points of climax and repose, movement from one major idea to another, and more are contained in the contour of this sum of all dynamic information.

Timbral balance provides information on the spectrum of the entire musical texture. The movement of musical ideas through the "vertical space" of pitch provides a sense of place for musical materials and adds an important dimension to the character of the overall texture. The number of sources, their densities, and distribution throughout the pitch registers provide a density to the texture that can also shape the direction, motion, and sound quality of the overall texture. The sound quality of the overall texture is largely shaped by timbral balance. Timbral balance can be envisioned as providing spectral information—materials that are harmonically related, and those that are not, all add their unique qualities to the production as they change over time or provide a continual presence that is part of the overall sound of the recording/music.

Finally, the form of music is created by all of these characteristics, plus the text and musical ideas. This essence of the song is what reaches deeply into the listener. It resonates within the listener when the song is understood. Form is the overall concept of the piece, as understood as a multidimensional, but single idea.

Figure 12.2 presents a timeline and the structure of "Lucy in the Sky with Diamonds." This is an outline of the structural materials that contribute to the form of the piece. The shape of the music and interrelationships of parts, as well as elements of the text, can be layered into this graph and made available for evaluation.

Matching the text against the structure of the piece will allow the reader to notice recurring sections of text/music combinations—verses and choruses. The musical materials enhance nuances of the meaning of the text as they were captured and enhanced by the recording process.

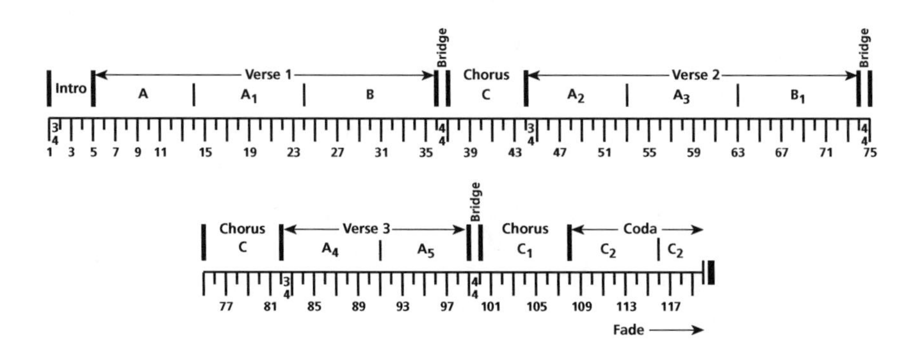


**Figure 12.2**
Timeline and structure—The Beatles' "Lucy in the Sky with Diamonds" (*Yellow Submarine*, 1999).

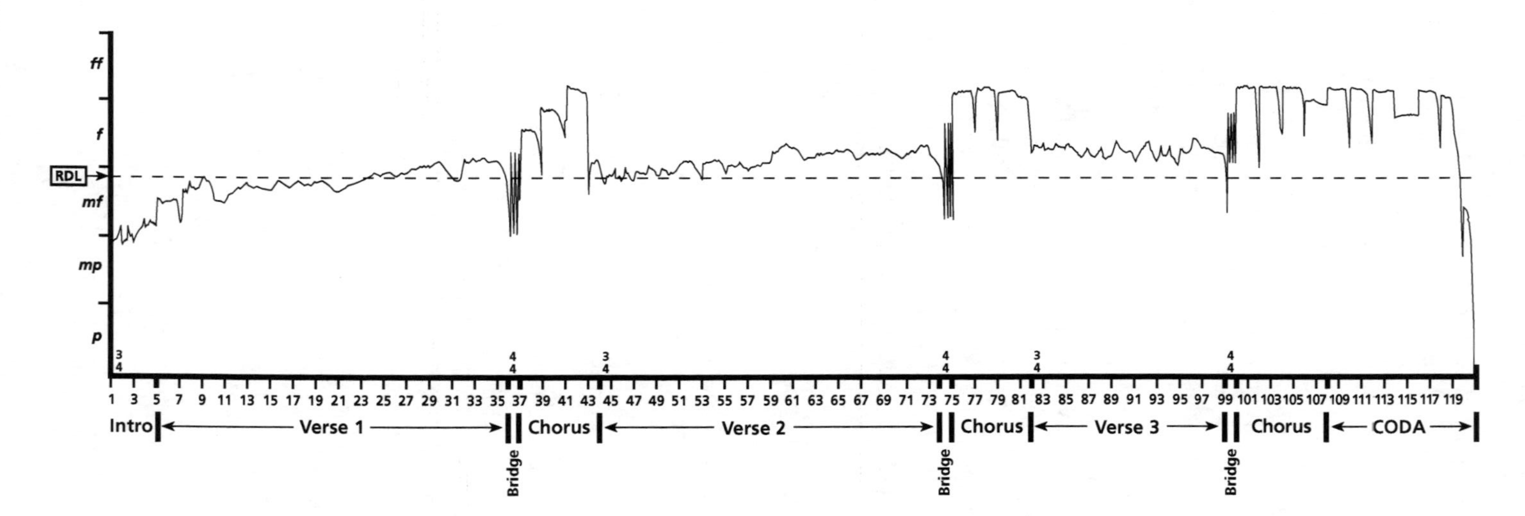


**Figure 12.3**
Program dynamic contour—The Beatles' "Lucy in the Sky with Diamonds" (*Yellow Submarine*, 1999).

Dynamic relationships between the various sections of "Lucy in the Sky with Diamonds" are present. These are clearly observed in the program dynamic contour graph (Figure 12.3). The graph represents the overall shape and dynamic motion of the song. The reader can experience how that motion relates to the song's text and sense of drama by observing the graph while listening to the recording.

The reference dynamic level of the song is also identified in Figure 12.3, as a high mezzo forte.

The timbral-balance graph of the song in Figure 12.1 allows the reader to identify and understand how the song emphasizes one pitch area for a time, then more evenly distributes pitch-density information—moving from one texture to another between sections. Evaluations of the perceived performance environment and sound stage will provide the remaining characteristics of the overall texture, allowing all to be compared and considered. Through this process, important and fundamental characteristics of the song can then be more readily understood and communicated to others.

## Relationships of the Individual Sound Sources and the Overall Texture

The mix of a piece of music/recording defines the relationships of individual sound sources to the overall texture. In the mixing process, the sound stage is crafted by giving all sound sources a distance location and an image size in stereo/surround location. Musical balance relationships are made during the mix, and relationships of musical balance with performance intensity are established. The sound quality of all of the sound sources is finalized at this stage also, as instruments receive any final signal processing to alter amplitude, time, and frequency elements to their timbre and environmental characteristics are added.

These elements crafted in the mix exist at the perspective of the individual sound source. Many important relationships exist at this level. This focus is common and important for the recordist, but is not common in recreational listening. The many ways sound relationships are shaped during mixing bring the recordist to often focus on this level of the individual sound source and also at the next higher level of perspective, where sounds can be compared by giving equal importance and attention to all sources. Learning to evaluate these elements, and to hear and recognize how these elements interact to craft the mix, is one of the most important listening skills to be developed for the recordist.

Figures 8.4 and 12.4 present two musical balance graphs of the beginning sections of "Lucy in the Sky with Diamonds." These are evaluations of two separate versions of the song. Mixing decisions brought certain

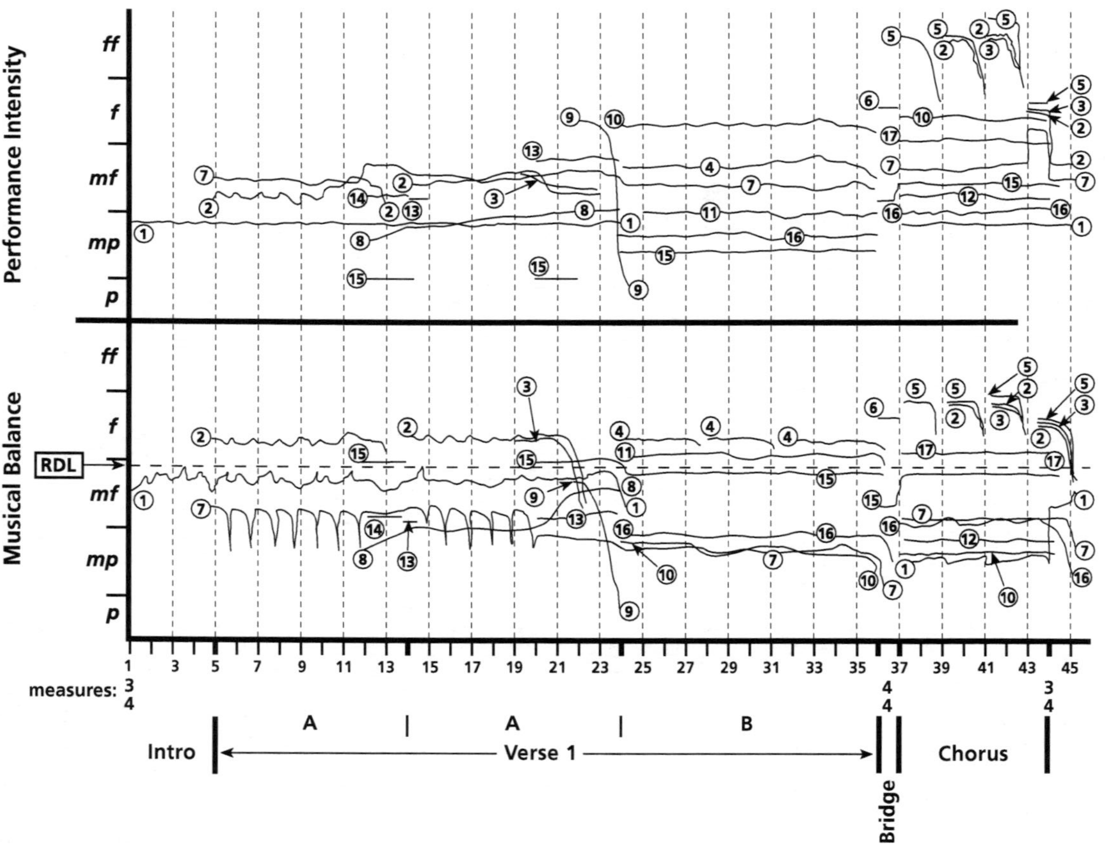


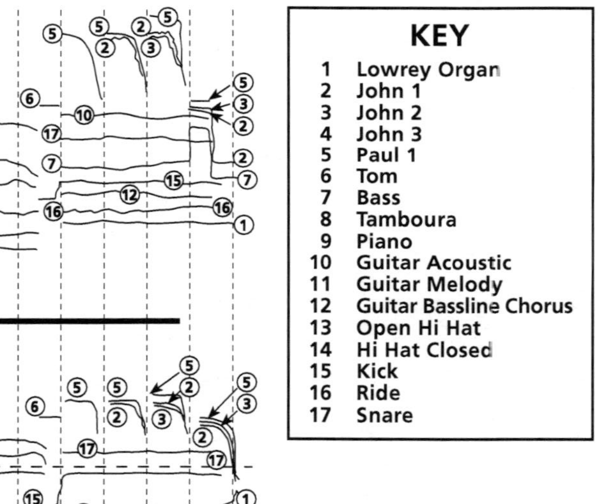


**Figure 12.4**
Musical balance versus performance-intensity graph—The Beatles' "Lucy in the Sky with Diamonds" (*Yellow Submarine*, 1999).

sounds to be at different dynamic levels in each version. Listening with a focus on several different sound sources, while comparing the two graphs to what is heard, will provide the listener with insight into these very different mixes. The performance-intensity tier of Figure 12.4 should also be examined and then compared to the final dynamic levels. With this graph the reader is able to identify how the mixing process transformed performed dynamics. This gives much insight into the performance of the tracks and their sound qualities, and the musical balance decisions that followed.

The sound stage provides each recording with many of its unique qualities. In examining the structure of sound stage, the listener will learn many things about a recording. Important among the many qualities are:

- Distribution of sources in stereo or surround location
- Size of images (lateral and depth)
- Clearly defined sound-source locations, or a highly blended texture of locations (wall of sound)
- Depth of sound stage
- Distribution of distance locations
- Location of the nearest sound source
- Sound-stage dimensions that draw the listener's attention or are absorbed into the concept of the piece
- Changes in sound-stage dimensions or source locations

Figure 12.5 presents the stereo location of sound sources in the 1999 *Yellow Submarine* version of "Lucy in the Sky with Diamonds." The phantom-image locations and sizes add definition to the sound sources and musical materials, and the width of several images changes between sections.

Comparing the graphs for timbral balance (Figure 12.1) and stereo location (Figure 12.5), it is possible to examine how pitch information ("vertical space") is distributed along the stereo sound field (lateral space). As the song begins, each source is in its own pitch area and stereo location. As sounds with similar pitch areas enter, some are given their own location to give the parts clarity. This allows them to be distinguished from others of similar pitch areas; notice, for example, the high hat open and closed, relative to the Lowrey organ in measures 12–14. Parts in the same pitch areas that are intended to blend (or fuse) are at the same or similar stereo location; this is especially evident with Lennon's lead vocal and its doubling in measures 20–22.

Comparing this stereo-location graph (Figure 12.5) to surround placements in Figures 12.6 and 12.7 allows some interesting observations. The lead vocal, Lowrey organ, and tamboura have been graphed into

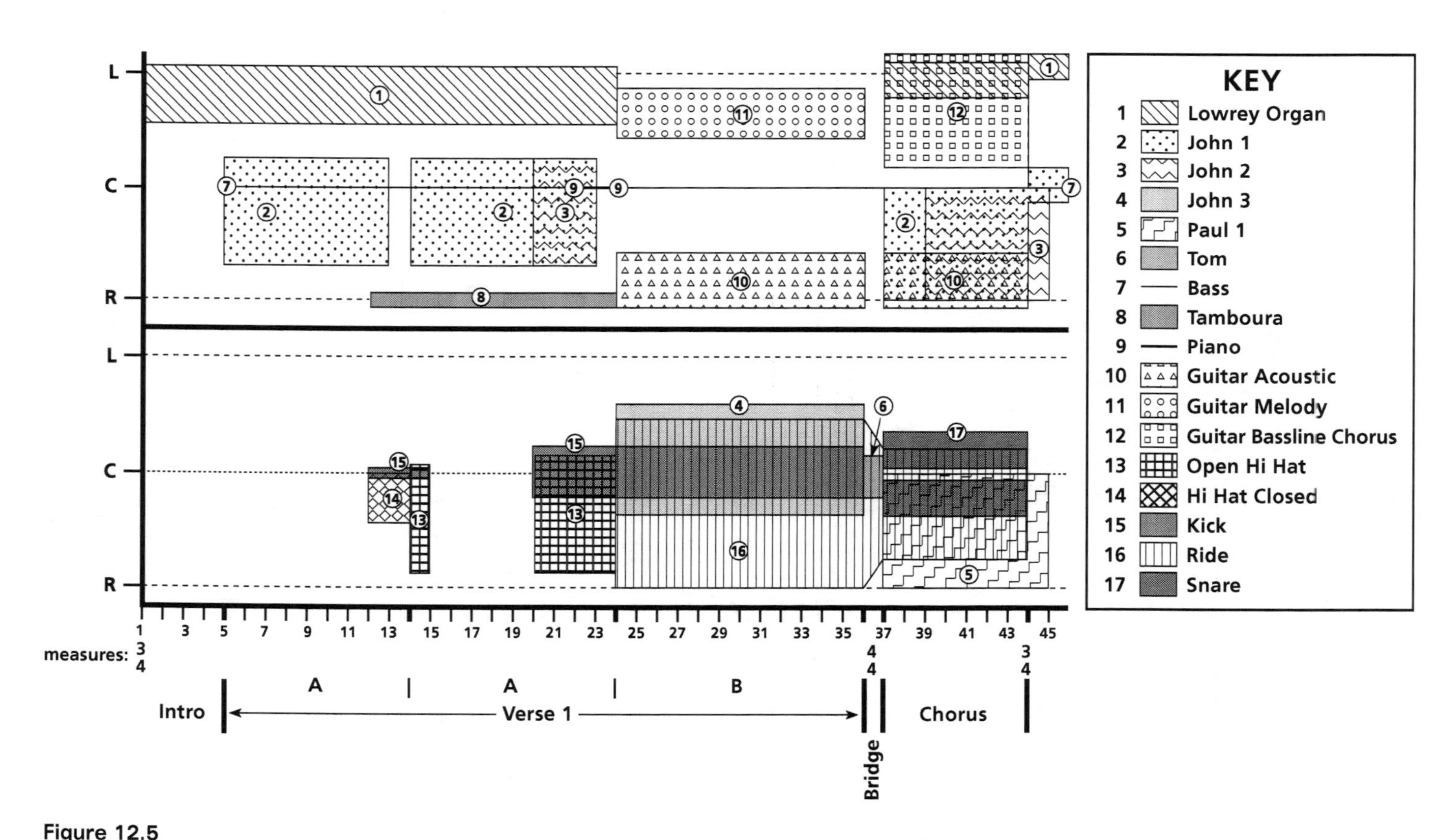


**Figure 12.5**
Stereo-location graph—The Beatles' "Lucy in the Sky with Diamonds" (*Yellow Submarine*, 1999).

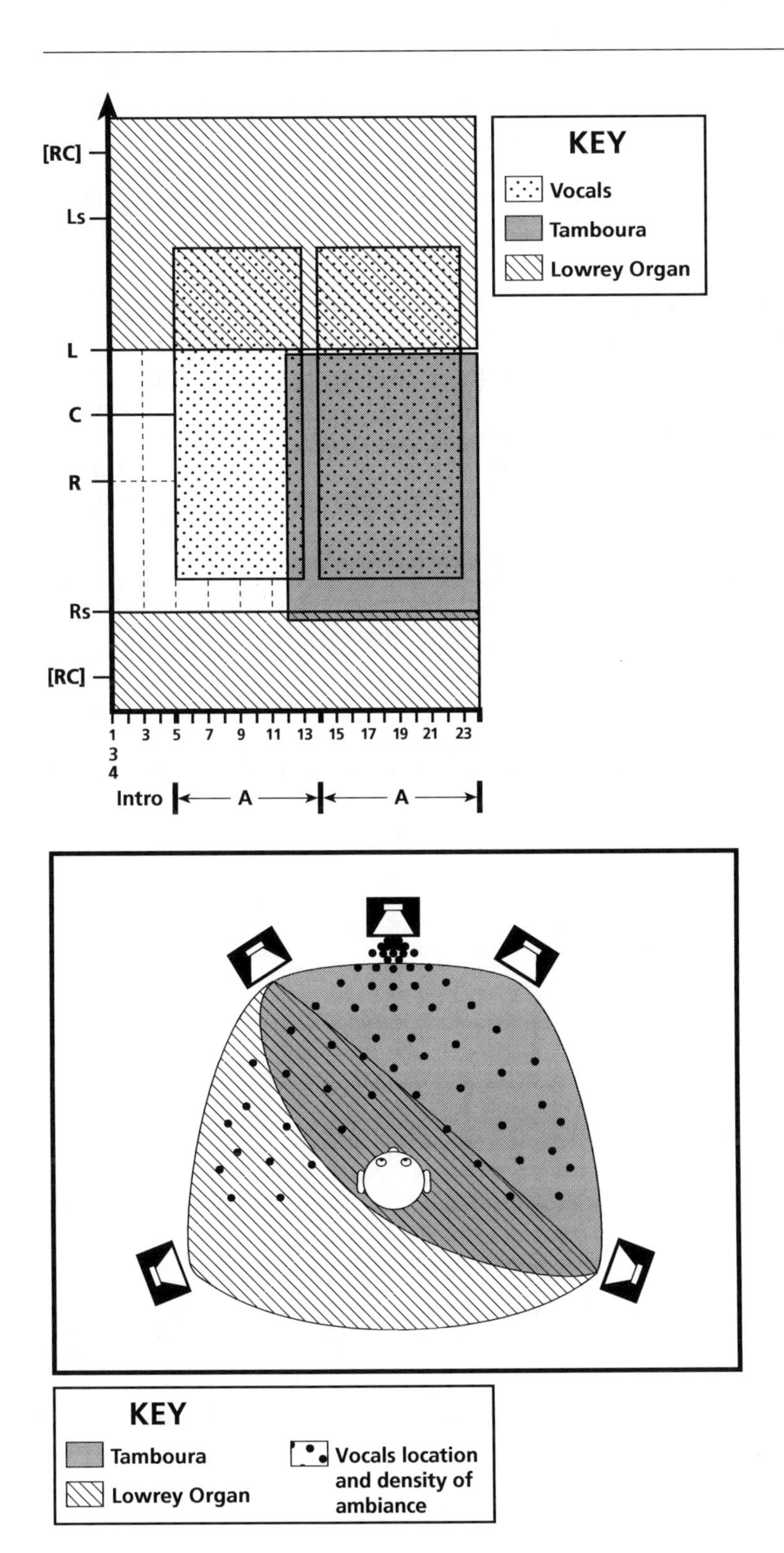


**Figure 12.6**
Surround-sound location graph—The Beatles' "Lucy in the Sky with Diamonds" (*Yellow Submarine*, 1999).

**Figure 12.7**
Surround image placements of Lowrey organ, tamboura, and vocal images—The Beatles' "Lucy in the Sky with Diamonds" (*Yellow Submarine*, 1999).

the first verse in the surround version, and both graphs need to be used to clearly define their dimensions. The image sizes and locations in a surround mix are quite different from the two-channel versions, and offer a very different experience of the song. Comparing these location graphs will allow the reader insight into what makes each mix different.

The reader is encouraged also to evaluate the stereo location of the sources in the original *Sgt. Pepper's Lonely Hearts Club Band* version of the song. Then compare the three location evaluations for similarities and differences of image location and sizes, and note when and how images change sizes or locations. In the end, consider how the different sizes and locations of the images impact the musicality of the song and how this aspect of imaging contributes to presenting the musical ideas.

Figure 12.8 is a blank distance location graph for the beginning sections of "Lucy in the Sky with Diamonds." The graph can be used to plot the distance location of sources from any of the three versions of the song —or to compare one or more selected sources from all three versions. In listening to all three versions of the song, the reader will notice some striking differences in distance location. Between the two-channel versions, sound-source changes in distance locations are more pronounced

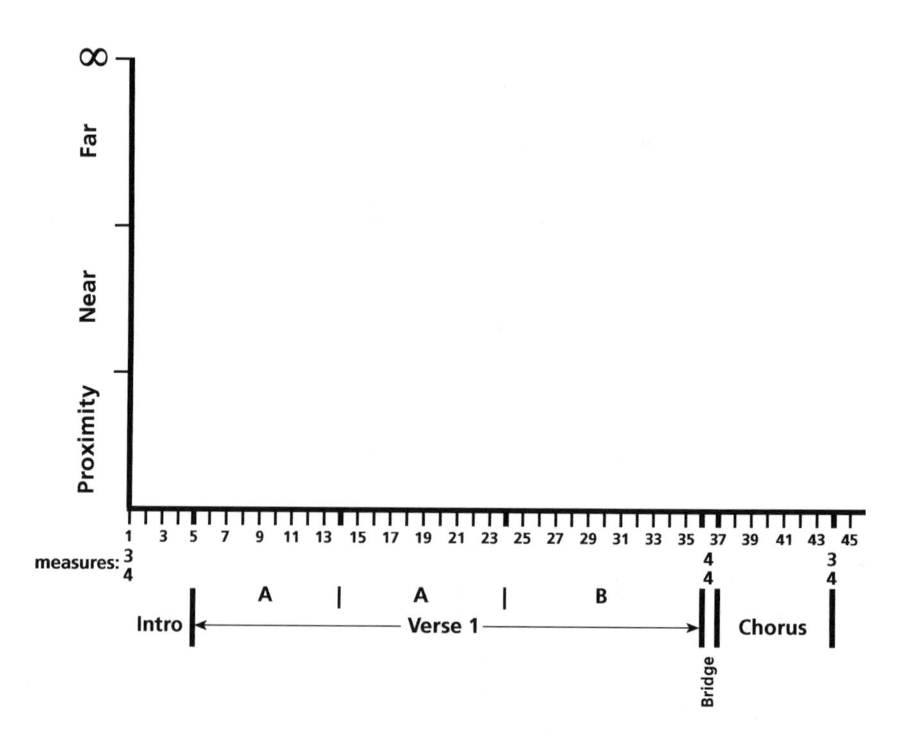


Figure 12.8
Distance location graph for The Beatles' "Lucy in the Sky with Diamonds."

in the *Yellow Submarine* version, and during the first verse the bass and lead vocal are closer than in the *Sgt. Pepper* version. Consider how distance cues enhance certain musical ideas and sound sources, and provide clarity or blending of images in various instances.

Taken together, the sound qualities of all of the sound sources jointly shape the timbral balance of the song. Sound quality also contributes fundamentally to the performance intensity of the sound source, and might contribute to shaping the song's reference dynamic level. The environmental characteristics of the source also contribute to its overall sound quality. They fuse with the source's sound quality to add new dimensions and additional sound-quality cues to the resulting composite sound. Lastly, the sound quality of each source provides important distance information, as the amount of timbral detail primarily determines distance location.

This timbral detail carries over into clarity of sound-source timbres in the mix. Sound sources can have well-defined timbres with extreme detail and clarity. Conversely, their sound qualities can be well blended with details absorbed into an overall quality. Both extremes are desirable in different musical situations. Both would place the sound at different distances, and each would cause the sound source to be heard differently in the same mix—each has the potential to be more or less prominent in a musical texture than the other.

How sound sources "sound" is an important aspect of recordings. Models of instruments and specific performers have their own characteristic sound qualities. Sound qualities are matched to musical materials and the desired expressive qualities to create a close bonding of sound quality and musical material. For instance, the opening of "Lucy in the Sky with Diamonds" would sound very different on a piano than on the Lowrey used in the recording. A listener would recognize something incorrect about the source, even before they might identify the sound quality as being different. That musical idea is forever wedded to that original sound quality. This becomes much more pronounced with vocalists. Consider the many covers of "Something" from *Abbey Road.* No matter how convincing and effective the performance of the cover, it will be heard as distinctly different from George Harrison's original.

## The Complete Evaluation

Great insight into productions can be found through an evaluation of all of the artistic elements in a particular recording/piece of music. This process can allow the listener to explore the inner workings of the sound relationships of a recorded piece of music in great depth. The listener would benefit from performing this exercise on a number of pieces, over the course of a long period of time. Not only will the listening and evaluation skills of the listener be refined, if recordings are thoughtfully chosen, these evaluations will also provide many insights into the unique production styles of certain engineers and producers, as well as an intimate understanding of the artists' work.

Complete evaluations can also bring attention and training to shifting focus while listening. Individual evaluations of individual artistic elements maintain the listener's focus at a specific level of perspective; this has been our listening practice thus far. Now comparing graphs of various elements and graphs at various levels of perspective will bring the listener practice in shifting focus with a purpose. This training in shifting focus is excellent preparation for the continually shifting focus required in producing recordings. It will also lead the listener to be able to make faster evaluations of recordings and judgments of the various qualities of recorded sound.

The project of performing a complete evaluation of a piece will be lengthy. It will take the beginner many hours of concentrated listening. The demands of this project are, however, readily justified by the value of the information and experience gained. This project will develop and refine critical- and analytical-listening skills in all areas.

For greatest benefit, an entire song should be evaluated. The listener should be mostly concerned with evaluating all of the artistic elements. An evaluation of the traditional musical materials and the text might be helpful as well, but is not necessary for most purposes. After an evaluation of the artistic elements individually, the listener should evaluate how these aspects relate to one another, and how they enhance one another.

### Elements to Be Evaluated

This complete analysis of an entire recording is strongly encouraged. Many aspects of a recording will only become evident when evaluations of several artistic elements are compared with one another, and compared to the traditional musical materials. The use of the artistic elements in communicating the musical message of the work will become much more apparent when their interrelationships are recognized.

The listener will compile a large set of data, in performing the many evaluations spanning Chapter 7 through the pitch density and timbral-balance evaluations of Chapter 12. These many evaluations will represent many different perspectives and areas of focus. Some of this information will be pertinent to understanding the musical message of the work, and some of the information will pertain to its elements of sound (such as how stereo location is used in presenting the various sound sources).

In addition, some of the information will be pertinent to appreciating the technical qualities of the recording. When evaluating recordings, we should also bring our attention to the quality of the signal, the presence of noises and distortions, performance issues, and other matters related to the integrity of the signal and the quality of the production. If desired, one can listen to seek information on "how" certain recording results were achieved.

All of this information will contribute to the audio professional's complete understanding of the piece of music, of how the piece made use of the recording medium, and of the sound qualities of the recording itself.

The sequence of evaluations that is usually most efficient in evaluating an entire work (depending upon the individual work, it may vary slightly) is:

- List all of the sound sources of the recording
- Create a timeline of the entire work
- Plot each sound source's presence against the timeline
- Define unknown sound sources and synthesized sounds through sound-quality evaluations
- Designate major divisions in the musical structure against the timeline (verse, chorus, etc.)
- Mark recurring phrases or musical materials, similarly, against the timeline; an in-depth study of traditional musical materials would be appropriate at this stage, if of interest
- Evaluate the text for its own characteristics and its relationships to the structure of the traditional musical materials, as appropriate
- Evaluate the pitch areas of unpitched sounds
- Determine reference dynamic level
- Perform a program dynamic contour evaluation
- Create a musical balance graph
- Create a performance-intensity graph
- Graph the song's stereo location or surround location
- Evaluate the work for distance location
- Perform environmental characteristics evaluations of all host environments of sound sources and the perceived performance environment
- Create a timbral-balance graph of the work
- Study these evaluations to make observations on their interrelationships and to identify the unique characteristics of the recording
- Examine the recording for technical issues, integrity of the signal, performance and recording technique flaws, and other issues with the quality of the recording

- Identify a stream of shifting focus on various elements and perspectives, after compiling a complete evaluation graph from the previous graphs
- Observe how the artistic elements work jointly with the musical materials and the text to create the recording and the song/music.

### Observing the Roles and Interaction of Elements

As we have learned, all of the elements of the recording have roles in delivering the musical ideas and message of the recording. Further, all of the elements interact and support the activity of the other elements (and potentially can detract from other elements). In examining these aspects, we learn much about the recording and how it presents the music.

The following materials may be coupled on the same graph (on separate tiers), or on similar graphs. They are all at the same perspective (at the level of the sound source):

- Performance intensity
- Musical balance
- Distance location
- Stereo location or surround location
- Pitch density (timbral balance showing individual sound sources)

Pitch density is timbral balance considered at the perspective of the individual sound source. Often when the perspective is not of the individual source, groups of instruments present a single musical idea; in this case they function as a single source. Instruments capable of playing several parts simultaneously (such as a keyboard or drum set) are divided into several separate pitch areas—each functioning as a separate sound source. In this way the timbral-balance graph's information can be directly compared to the four elements listed above.

These five artistic elements will be interrelated in nearly all recording productions. Observing their interrelationships will allow the listener to extract significant information about the music and the mix. The reader can explore this material by reviewing the graphs of "Lucy in the Sky with Diamonds" found throughout this chapter.

In making these observations, the listener will continually formulate questions about the recording and seek to find meaningful answers. The questions of how artistic elements (and all musical materials) relate to one another will center on:

- Patterns of activity within any artistic element (patterns of activity are sequences of levels within the artistic element, and rhythmic patterns created by the relationships of those levels)

- Levels of any artistic element (how high pitched, what loudness levels, etc.) and the areas they span
- Speed or rate of patterns or changes within the elements
- Interrelationships of patterns between artistic elements (do the same or similar patterns exist in more than one element?).

Music is constructed as similarities and differences of values and patterns of musical materials. This is also the way humans perceive music. People perceive patterns within music (its materials and the artistic elements). Listeners will perceive the qualities (levels and characteristics) of the elements of the music and will relate the various aspects of the music to one another.

At the same time, the listener should compare what they are hearing with what was previously heard, and to their previous experiences. Meaning and significance will be found in this information by looking for similarities and differences between the materials.

The listener should ask, "What is similar?" between two musical ideas (or artistic elements); "What is different?"; "How are they related?" These will be answered through observing the information that was collected during the many evaluations. The shapes of the lines on the various graphs may show patterns. The vertical axes of the graphs may show the extremes of the states of the materials and all of their other values.

The listener's ability to formulate meaningful questions for these evaluations will be developed over time and practice. They will be asking, "What makes this piece of music unique?"; "What makes this recording unique?"; "How is this recording constructed?"; "How is this piece of music constructed?"; "What makes this recording effective?"; "What is important and how is it presented?"; "How does [insert any element] contribute to delivering the music or musical idea(s)?" Many other, much more detailed questions will be formulated during the course of the evaluation. The listener should finally ask, "Which of these relationships are significant to the communication of the musical message; which are not?"

The use of the artistic elements in the recording can also be considered in their relationships to the traditional musical elements and materials. This brings an understanding of the importance of each musical idea, as related to the piece as a whole. Through these observations, the recordist will obtain an understanding of the significance of the artistic elements to communicating the message (or meaning) of the music. The recordist can then understand and work to control how the recording process enhances music, and how it contributes to musical ideas and the overall character of the piece. The complete evaluation graph will assist all this, and more.

## Stream of Shifting Focus and Perspective

A final step in a complete evaluation of a recording is to identify a stream of shifting focus. This will help one to recognize the various elements and perspectives that shape the recording in significant ways, as the work progresses. This will be shaped at several perspectives and will be different from any one moment in time to another.

Figure 12.9 presents a "complete evaluation graph" to assist in compiling this information. The elements at the perspective of the overall texture appear in the top tier; the three elements at the very top of this tier are qualities that do not change. They are separated by a thin line from the three qualities of the overall texture that can change over the course of the song. A heavy line separates the top tier from the bottom tier where the elements at the perspective of the individual sound source appear.

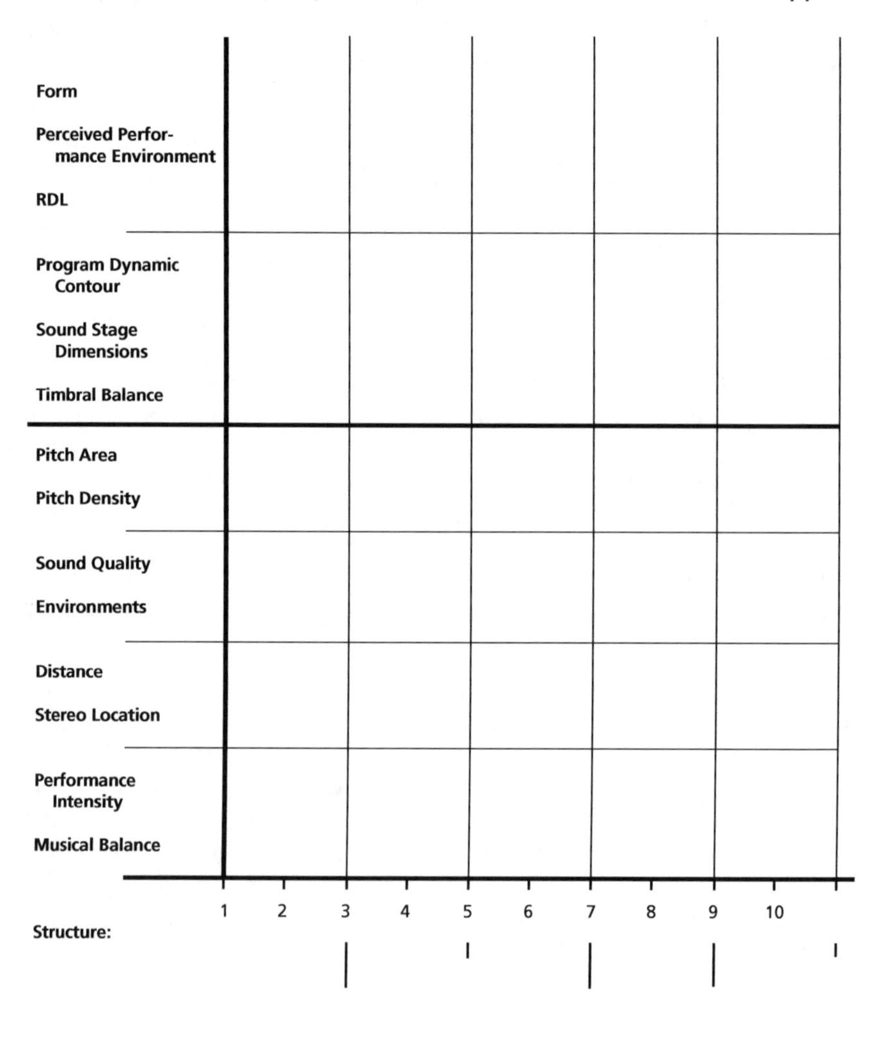


**Figure 12.9** Complete evaluation graph.

Notations drawing attention to important activities in each element are made on the graph. Notes about important characteristics and qualities of elements should also appear here. This allows for elements and perspectives to be easily compared. The timeline allows the reader to follow the elements as they unfold over the song.

This study will bring the listener's attention to important aspects of the music and the recording. If followed throughout a work, it will provide an experience of hearing the song move from one idea to another, from one element to another, from the importance of one level of perspective to another. As an example, one might hear how distance shapes a certain sound in an important way at a certain moment, and then quickly shift to an important quality of timbral balance that was created by this change in distance, moving again to hearing and understanding how the program dynamic contour was then altered, then recognizing that a certain sound source became less prominent in the mix because of this activity, and upon investigation finding that its loudness did not change but its spectral content was altered. The reader will be able to use this graph to recognize and follow the most significant events and activity at any one time, and also to recognize and follow how all other elements contribute at the same and at other levels of perspective.

This stream of information brings a new understanding to the music and the mix, to how the recording process shaped the music and how the artistic elements work jointly with the musical materials to successfully deliver a quality music recording. The role of shifting focus and shifting perspective will be better understood. Of as great importance, the reader will be able to practice the process of shifting perspectives and focus that takes place continually in recording production. It can be difficult to learn how to execute this skill in a controlled and deliberate manner, if only practiced during actual production work.

Figure 12.10 is a complete evaluation graph of the opening of The Beatles' "Here Comes the Sun." This summary of elements provides an overview of important activities, and from this we can achieve a greater understanding of the recording. The graph clearly shows that changes in instrumentation and elements of the mix coincide with the changing structure of the song. During these 15 measures, important changes take place in sound-stage dimensions, timbral balance and pitch density, distance location, stereo location, and musical balance. These all contribute to a changing context for the song that adds to its character. Attention should be drawn to these at strategic moments. A few of these are:

- The sound stage moving from fully left to the center during measure 8, followed by a sudden broadening of the sound stage in measure 9

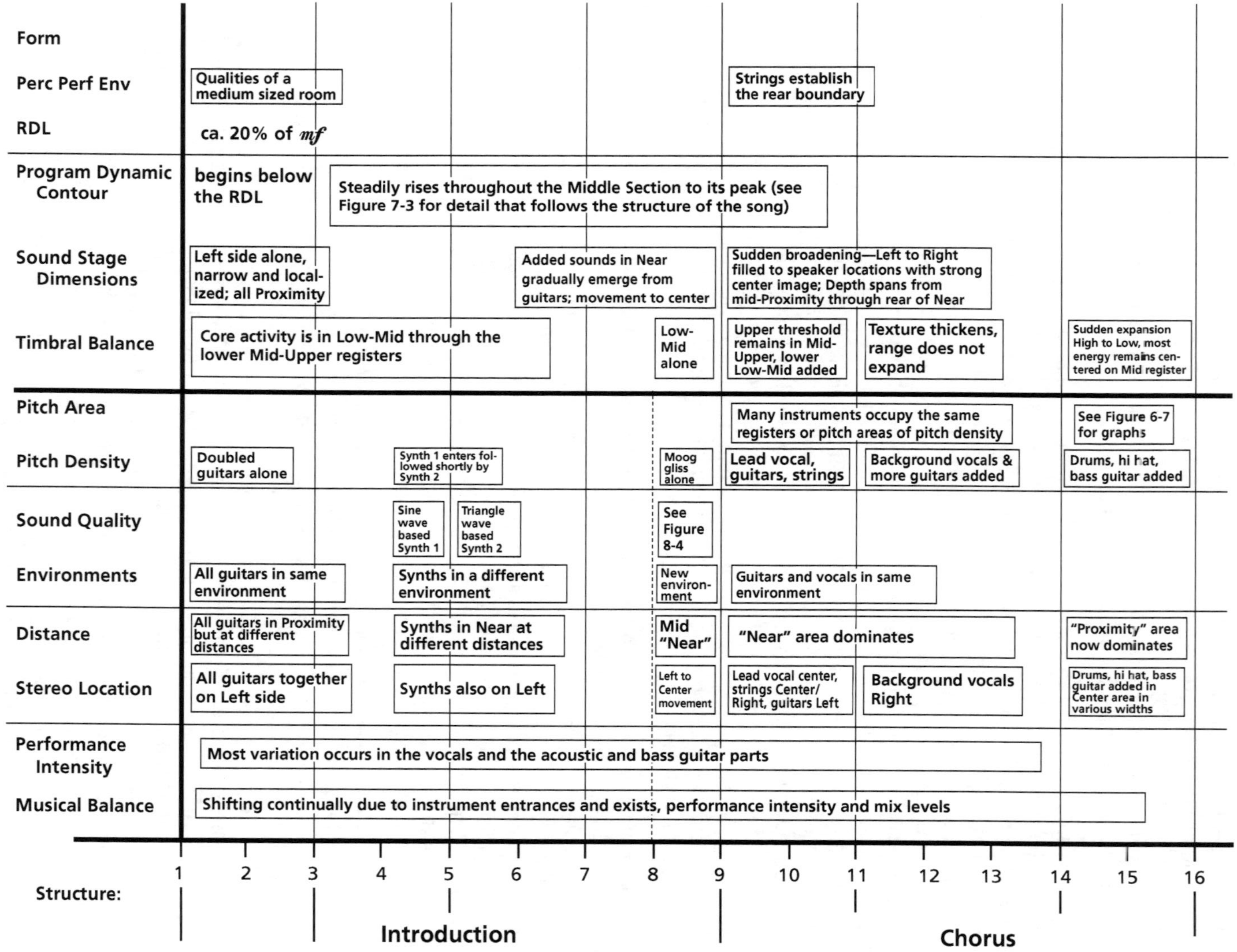

| Parameter | Entries (left to right) |
| --- | --- |
| Form | |
| Perc Perf Env | Qualities of a medium sized room; Strings establish the rear boundary |
| RDL | ca. 20% of *mf* |
| Program Dynamic Contour | begins below the RDL; Steadily rises throughout the Middle Section to its peak (see Figure 7-3 for detail that follows the structure of the song) |
| Sound Stage Dimensions | Left side alone, narrow and localized; all Proximity; Added sounds in Near gradually emerge from guitars; movement to center; Sudden broadening—Left to Right filled to speaker locations with strong center image; Depth spans from mid-Proximity through rear of Near |
| Timbral Balance | Core activity is in Low-Mid through the lower Mid-Upper registers; Low-Mid alone; Upper threshold remains in Mid-Upper, lower Low-Mid added; Texture thickens, range does not expand; Sudden expansion High to Low, most energy remains centered on Mid register |
| Pitch Area | Many instruments occupy the same registers or pitch areas of pitch density; See Figure 6-7 for graphs |
| Pitch Density | Doubled guitars alone; Synth 1 enters followed shortly by Synth 2; Moog gliss alone; Lead vocal, guitars, strings; Background vocals & more guitars added; Drums, hi hat, bass guitar added |
| Sound Quality | Sine wave based Synth 1; Triangle wave based Synth 2; See Figure 8-4 |
| Environments | All guitars in same environment; Synths in a different environment; New environment; Guitars and vocals in same environment |
| Distance | All guitars in Proximity but at different distances; Synths in Near at different distances; Mid "Near"; "Near" area dominates; "Proximity" area now dominates |
| Stereo Location | All guitars together on Left side; Synths also on Left; Left to Center movement; Lead vocal center, strings Center/Right, guitars Left; Background vocals Right; Drums, hi hat, bass guitar added in Center area in various widths |
| Performance Intensity | Most variation occurs in the vocals and the acoustic and bass guitar parts |
| Musical Balance | Shifting continually due to instrument entrances and exists, performance intensity and mix levels |

| | | | | | | | | | | | | | | | |
| --- | --- | --- | --- | --- | --- | --- | --- | --- | --- | --- | --- | --- | --- | --- | --- |
| 1 | 2 | 3 | 4 | 5 | 6 | 7 | 8 | 9 | 10 | 11 | 12 | 13 | 14 | 15 | 16 |

Structure: Introduction; Chorus

**Figure 12.10** Complete evaluation graph of the opening of The Beatles' "Here Comes the Sun."

- The different "proximity" distance locations of the doubling guitars in the first four measures, the dominance of the "near" area in measures 9 through 13, and the dominance of the "proximity" area in measures 14 and 15
- Timbral balance is focused on the "low-mid" through "mid-upper" ranges from the beginning through measure 13; the density of activity varies significantly but the bandwidth of pitch registers being used does not change markedly (except for measure 8's solo Moog glissando). This is especially noticeable in measures 9 through 13; at measure 14, the pitch areas covered suddenly expand to encompass activity from the "low" register into "high."

The reader would benefit from performing several or all of these graphs to notice the subtleties of activity. Following all elements into the next section would provide much more interesting information; as the song unfolds, more changes in the mix enhance the song profoundly. The reader might wish to refer back to Figure 7.5 for pitch areas of several percussion sounds and to Figure 8.3 for the entire program dynamic contour of "Here Comes the Sun."

This complete evaluation process will greatly assist the recordist in understanding how the artistic elements can be crafted during the recording process to enhance, shape, or create musical materials and relationships.

The recording production styles of others can be studied and learned. By understanding the sound qualities of a recording and being able to recognize what comprises those sound qualities, the sound of another recording or type of music can be emulated by the recordist in their own work as desired.

## Using Graphs for Making Evaluations and in Production Work

Graphing the artistic elements may be time consuming and at times tedious and perhaps frustrating. It is important, however, for developing listening skills and evaluation skills, especially during beginning studies. It is also valuable for in-depth looks at recordings, providing insights into the artistic aspects of the recordist's own recordings and the recordings of others.

This process of graphing the qualities of the various artistic elements is also a useful documentation tool. Graphs can be used to keep track of how a mix is being structured or how the overall texture is being crafted. Many of the graphs or diagrams can even be used to plan a mix. For example, the imaging diagram can be used to plan the distribution of instruments on the sound stage and consider distance assignments before beginning the process of mixing sound—perhaps even before selecting a microphone to begin tracking. Working professionals through beginning students will find these useful in a variety of applications.

Detailed graphs of artistic elements are not proposed for regular use in production projects. The graphs are not intended for the production process itself, though they might be of some use in certain planning and record-keeping matters. Audio professionals must be able to recognize and understand the concepts of the recording production, and hear many of the general relationships, quickly and without the aid of the graphs. The graphs are intended to develop these skills, and to provide a means for more detailed and in-depth evaluations that would take place outside of the production process.

Recordists who have developed a sophisticated auditory memory may also find these graphing systems to be useful for notating their production ideas, and for documenting recording production practices. These acts will allow them to remember and evaluate their production practices more effectively, allowing them more control of their craft.

## Summary

The pitch area that comprises the focused frequency content of individual sound sources is joined with the source's musical material to establish pitch density. This takes place at the perspective of the individual sound source and its interaction with other sources. At the highest level of perspective, overall texture, we observe timbral balance, as the frequency content of the recording is established by the frequency content of each sound source. In this way, each sound source and the musical materials they present can be envisioned as a "partial" in the "spectrum" of the overall texture's timbral balance. With timbral balance addressed, all of the dimensions of the overall texture have been explored.

Chapter 12 concludes Part Three with a discussion of how the individual elements contribute to shaping the recording. At the perspective of the individual sound source, musical materials are presented and enhanced by the artistic elements. Musical balance, distance location, image size and placement, sound quality, and other elements contribute to shaping the music and the recording. These relationships are realized in production through using pitch, dynamic, and spatial relationships in planning (composing) and executing (performing) the mix. These individual elements become part of the fabric of the overall texture, at the highest level of perspective.

There are six characteristics or dimensions of the overall texture. The three dimensions of form, perceived performance environment, and reference dynamic level are qualities that are single, global concepts that remain fixed throughout the work—they are not time dependent. The three qualities of sound stage, program dynamic contour, and timbral balance are variables; they may change from moment to moment as the song is experienced, and are the product of all of the sound sources coming together to create an overall state within the elements of spatial relationships, dynamics, and frequency content, respectively. These three qualities are time dependent and create an overall shape of the recording.

As we have seen, the recording's artistic elements and the song's musical materials and text are fundamentally linked. The recording's artistic elements contribute to the delivery of musical ideas and can help shape them; they present the musical ideas in important ways, and at times can add important dimensions to the music. The artistic elements act in concert with the musical materials and text to create the music and the recording. Part Four will explore concepts that will assist the reader in putting these ideas into practice in their own recordings.

Throughout Part Three a systematic method for evaluating the artistic elements of recordings and their impact on the recording's delivery of musical ideas has been explored. This method embraces each element separately and works toward their interaction. Materials and exercises have been ordered carefully. All of the exercises presented in the text are listed at the beginning of the book after the table of contents. The exercises are ordered to systematically develop the reader's sound evaluation and listening skills. Working through the exercises in this order will be most effective for most people. Readers with much experience and well-developed skills will still find at least a few exercises that are sufficiently advanced to test and improve their skills. Learning anything new requires effort and a willingness to reach into the unknown.

Very sophisticated listening skills are required in audio and music. Developing such skills from the beginning will take practice, patience, and perseverance. At times the listener will be told to listen to things they have never before experienced. They will have no reason to believe such sound characteristics even exist, let alone that they can be heard, recognized, and understood. Faith will be required: a willingness to be open to possibilities and leap blindly into an activity, searching the sound materials for what they have been told exists. Using the processes that were learned and the graphs that were created throughout Part Three will make this process easier and more productive.

Following the system will give the listener a refined ability in critical and analytical listening. The listener will learn to communicate effectively about sound, and will be able to apply this new language to many situations.

Among other things, Part Four will explore how these new listening skills and knowledge of sound can provide the recordist with the ability to craft high-quality and artistically inspired recordings.

## Exercises

### Exercise 12.1

*Pitch Density Exercise*

Select a recording with an instrument performing alone, or with very few other instruments, over the first several sections of the work. The instrument's musical material should show some noticeable and considerable change in pitch levels. This exercise will graph the pitch density of this single instrument over this period.

Pitch density will be graphed to show the musical material the instrument is performing and the prominent aspects of its spectrum.

The process for determining the pitch density for a single source will follow the following sequence:

1. During the initial hearing(s), establish the length of the timeline. Notice entrances and exits of the instrument against the timeline.
2. Check the timeline for accuracy and make any alterations. At the same time, work to identify the musical ideas that the instrument is presenting and note their presence against the timeline.
3. Transcribe the instrument's pitch material onto the graph, so its melodic contour is represented as a single line on the graph. This might represent the fundamental frequency of the sound-source timbre.
4. Next notice how the single melodic line falls into phrases that generate distinct, individual musical ideas. Mark the beginning and ending points of those ideas, and modify the melodic contour to show a more general outline of the line, eliminating small and fast variations of pitch level.
5. Turn focus to the spectrum of the instrument's first sound. Determine the bandwidth of the pitch area by identifying where the spectrum becomes about one third the loudness of the lowest frequency. This will determine the upper boundary of the pitch area.
6. Scan the musical material to determine changes in performance intensity. These changes will bring about changes in the instrument's spectrum that often change the bandwidth of this pitch area. Note these changes on your graph.

7. Map out the upper boundary of the pitch area by listening to the spectral information against the lowest frequency.
8. Once these boundaries are finalized, make observations on the density of spectral activity (amount of frequency and overtone information) within the defined pitch area.

**Exercise 12.2**

*Timbral Balance Exercise*

Select a recording with four to six sound sources, to graph the pitch density over the first three or four major sections of the work. The recording should contain timbral-balance changes within and between sections.

The musical texture must (1) be scanned to determine the musical ideas present. The musical ideas might be a primary melodic (vocal) line, a secondary vocal, a bass accompaniment line, a block-chord keyboard accompaniment, and any number of different rhythmic patterns in the percussion parts. They should then (2) be identified by instrument or voice performing the material, and listed in a key.

Each idea will (3) then have its pitch areas defined as a composite of its pitch material and the prominent aspects of the sound quality of the instrument(s) or voice(s) that produced the idea.

The process of determining the timbral-balance graph will use the same skills developed in Exercise 12.1. This graph will plot the pitch density of all of the musical ideas of all of the sound sources to clearly represent the complete frequency content of the recording.

The reader should perform a pitch-density evaluation of all of the sound sources in the example, singling out each source for careful evaluation. Once these are completed, this pitch-density information of all sound sources will be combined into a single timbral-balance graph.

**Exercise 12.3**

*Shifting Focus and Perspective Exercise*

Select a short song, with only a few instruments, to evaluate completely. Create all of the graphs for the overall texture and for the individual sound source, as outlined above.

Create a complete evaluations graph. Transfer the important events and characteristics of each element onto the graph, in its designated place. When an element changes, note where this change occurred and its nature.

Once the graph is completed, listen to the work again, following this graph. Highlight the most important changes that occurred. Next, observe the changes that you heard as not being as significant as others, and note how they might lend a supportive role to the other elements.

Practice changing focus and perspective in a controlled and deliberate way by listening for specific material and changes while following this graph. Next, listen to the recording with your knowledge of the important elements and changes of perspective, and when important shifts occur, try to follow these changes without the aid of the graph.

# Part Four

---

# Crafting the Mix:

## Shaping Music and Sound, and Controlling the Recording Process

# 13 The Roles of the Recordist and the Aesthetics of Recording Production

This chapter will establish some context for how the recording process shapes music. Here we will examine the roles of the recordist in creating the recording and its art, and explore some basic concerns of the aesthetics of recording production.

## The Functional Roles of the Recordist

The recordist can have many roles in making a recording. There are many steps and phases in recording projects, and the recordist may be involved in any one of those phases, some of the phases, or in all. Many recordists specialize; for instance, one person's work may be dedicated to only mastering, and another may specialize in tracking, but will mix if the project requires it. Some recordists become producers and engage projects from the spark of the idea, nurturing them all the way to release—and perhaps beyond. Today's recordist will often carry the combined roles of audio engineer and producer, sometimes equally and at times weighted in one role over the other.

The recordist may have their role change as the project progresses. It is common for one person to cover many tasks, and it is also common to hand off a task to another recordist at mid-project. Flexibility in being able to cover various roles is invaluable—as is keeping careful documentation of the project, so handing it off to another can be relatively seamless and as efficient as possible.

Recordists do not work in a vacuum. The recordist will directly interact with performers during some phases and not on others. There is very significant interaction in tracking sessions, or direct-to-two-track sessions. Working directly with artists in preproduction is common, and highly desirable preceding tracking sessions.

All artists are rarely present for entire mixing sessions; the presence of lead artists, producers, or other interested parties varies with the project. Often the mix engineer will have some significant interaction with artists and/or the producer during the process of making mix decisions, and also work a significant amount of time alone.

Mastering is often largely solitary. Interaction with clients may often not include direct contact with artists. No matter the extent, how the recordist interacts with clients and artists impacts the project and/or the performance that creates the project.

## Facilitating the Creation of Music

Among the recordist's most significant functions is to create a work environment where the magic of making art can happen. No matter what phase of the project, the recordist usually sets the tone for recording projects and can control many things, including a project's pace and how people interact. The recordist is usually largely responsible for guiding recording projects, whether or not the others involved acknowledge this or are even aware of it. Recordists are often responsible for keeping the creative process moving effectively, efficiently, and invisibly, giving guidance to the artists or giving them enough space and support so they are free to be creative.

It has often been said that the recordist should first be a psychologist. While this statement may be somewhat extreme, the recordist needs to be sensitive to interpersonal relations. How they work with performers and others will shape the project as much as their actual recording duties. And sometimes they may be needed to facilitate in helping people get along, working together productively.

The ways people normally treat one another in everyday living, and especially in standard business environments, are often nonproductive (at best) in the recording studio. Recordists should consider how they speak with the artists about the project, and how they interact with the artists socially and during the creative processes. The type of image the recordist presents to the artist will influence the artist's comfort level and ability to trust them as they work together. The recordist will typically strive to keep artists relaxed in the studio environment and focused on the project.

Recordists strive to get creative people to do their best work, while attempting to perform their own tasks at the highest standard. The process of creating art (a music recording) is an emotional roller coaster of ecstasy of what has just been discovered and anguish over not having an equally brilliant idea for "what comes next?" Time and financial constraints further stress artists (clients).

Musicians/creative people are exposed and vulnerable in the recording process. The recordist must be certain to do nothing, and not to allow anything to happen within the session environment, or by anyone else, to make artists feel exposed, unprotected, or, worse, threatened. Musicians must have the freedom to be creative around the recording studio, without feeling that their every move is watched or evaluated. Allow artists to fail privately, even if you witness the off performance, mistake, or idea that didn't "work." At times they will need to feel they are alone. At times they may need to feel they are in their own home rehearsal space, or some other "safe" location. The recordist should be sensitive to the need of the performer to feel they belong in the recording space, and are in a risk-free environment where they can take chances and know there is a safety net (aka, the delete key).

The expressive nature of performing music will often involve taking chances, stretching performance abilities to their limits or beyond, and making mistakes. These necessary activities can potentially embarrass confident (let alone less than confident) performers if they are critically judged at this vulnerable time. Performers need to be confident to perform well. The recordist is attempting to get the performers to exceed the height of their ability. Nothing should be allowed to happen that would diminish the confidence level of the performers and to take away from the trust that the performers must have in the recordist.

While evaluations have an important place in the recording process, judgments—especially those of a negative nature—rarely can be used constructively and are often destructive to the process.

### The Roles of the Recordist in the Phases of the Project

There are distinct phases to the recording project, and the recordist has a different function in each of those phases. Exactly what the role will be varies by project, and the needs and wishes of the artists, producer, and others involved. The four primary phases of a typical recording project are: preproduction, tracking, mixing, and mastering.

Preproduction is about planning the project, and making certain the project is ready to move forward in a productive, effective, and efficient way. It includes getting to know the artists and their material. The material may be complete and rehearsed, or in need of adjustments; the album may be ready, or at any stage of completion. Preproduction brings clarity to the artist's core concept of the project, with the recordist providing detail and a roadmap to realize it. A preproduction demo of a song or two can be useful to bring focus to the project.

This roadmap is a plan from this beginning stage through completion. Not only will it involve the goals, approach, and sound qualities of the finished project, it also involves money, budgeting, and making estimates. Financials will impact many decisions.

A host of logistics bridge preproduction and tracking: timeline for the project, schedules of performers and studios; selecting a tracking studio, selecting a mix studio, and deciding on a mix engineer; determining the content and orderings of sessions and how they will be run; session protocol; and so much more. All logistics relate back to what will best serve the core concept of the project, efficient and effective workflow, and keep people on track and maintain even temperaments. Notice: The recordist has yet to actually start recording.

Tracking is recording. The recordist records the music at this stage. The sound qualities of the final recording, and many of the relationships of those sounds, are impacted by the decisions made here; some of these decisions will not be able to be changed in the future. The plan established in preproduction, of what parts get recorded when and how, is executed in the studio. Any number of approaches can take place here, as parts/performers can be tracked separately or in groups, as complete songs are eventually compiled. This process can be quite involved, as parts are repeated, performances evaluated and repeated, then combined, and so much more.

After the tracks are recorded, the project is sent off for mixing. Before the project is ready to go on, tracks must be prepared for mixing. Tracks should be completely finished before delivery to the mix engineer. The project should no longer need editing, arranging, tuning, etc. Notes about effects and processing to be added and documentation of the contents and specifications of the tracks must also be complete.

The mix is what likely brought you here, to buy this book. The mix engineer selectively combines the elements recorded in tracking to shape much of the sound of the final recording. The mix brings all of the artistic elements into play and balanced according to the needs of the project and to complement the music. This balance can be dynamics, as is readily apparent; it will also be balancing the sound stage in lateral and distance imaging, placing sounds for pitch density and sound qualities, and giving space to the recording. The mix is where the sound qualities of the recording are crafted and the recording receives its performance.

The recordist as mastering engineer is responsible for catching any problems with the project, the result of tracking or mixing. Those problems that can be fixed are, and all of the songs or parts of the project are combined into a single recording. The mastering engineer also gives overall shape to the project, and can fine tune individual songs. Often working in exacting ways but in broad strokes, the mastering engineer subtly tweaks gain, compression, and equalization to craft the overall character of the recording. Mastering is the last step of the project, and completes it.

The dimensions of these phases will be further explored, and woven throughout the remaining chapters.

## The Artistic Roles of the Recordist

The recordist must have a clear idea of their role in the creative process for each project. The project may include the composer of the music, one or many performers, a conductor of an ensemble, and/or a specific recording producer. The recordist will need to identify what is expected of them toward the final artistic product, as well as the roles, contributions, and responsibilities of the others involved.

Of the many possibilities, the recordist may be functioning to capture the music as closely dictated by the composer. They may be functioning to capture, as realistically as possible, the performance of an ensemble, as precisely directed by the conductor; they may be functioning to capture the interactions and individual nuances of a group of performers, without altering the performance through the recording process; or the recordist may be functioning to precisely execute a recording producer's instructions (often in ways that transform performances). In all of these cases and many others, the recordist is allowing the artistic vision and decisions of others to be most accurately represented in the recording. The recordist's role then is to facilitate and realize the artistic ideas of others, and not to directly impose their ideas onto the project.

The recordist's role sometimes might be to offer suggestions to the creative artists or even to take an active role in the artistic decision-making processes. The role of the recordist might be active in shaping a performance of an existing work, or in creating a new piece of music. The recordist might be active in determining the sound qualities of the instruments of the recording, or in determining the sound sources themselves. Vastly different levels of participation in the artistic process are often required from one project to the next.

Among the things that are consistent, is that the recordist must be mindful of their place in the process. Their contributions may be needed, but are often not sought. Their ideas may be helpful, but might get in the way of the artist's creativity. Even when the recordist is asked what they think, the client may not really want to know. It is a delicate dance.

The process of writing a piece of music for a recording is often a collaborative effort, and this process can become even more complicated. Such songwriting may take place with many people (composer, performers, producer) or just a few (performer/composer and recordist/composer). When it takes place in the studio it can be a laborious process that seems never-ending, and is sometimes incredibly gratifying and exciting. The need for creativity is always present, but the spark can be illusive.

In many ways, the recordist functions as a creative artist and can serve the traditional roles of a composer, a conductor, and/or a performer. The recordist also shapes sounds in nontraditional ways. Recordists have unique controls over sound and live performances that allow for an additional musical voice. It is possible to compose with the equipment (instruments) of the recording studio, to shape sounds or performances through the use of recording and mixing techniques, or to create a new musical environment for someone else's musical ideas and performances.

The recording studio can be thought of as a musical instrument or a collection of musical instruments. In this way, the recordist may conduct all of the available sound sources (for example, bringing sounds into and out of the musical texture through mixing); may "perform" the musical ideas through the recording process; may alter or reshape the sounds of the sources, or "interpret" the musical ideas, in ways that are not possible acoustically; and may create (compose) new musical ideas or sounds.

## The Recording and Reality: Shaping the Recording Aesthetic

The recordist has many potential roles in shaping the recording aesthetic. The role of the recordist might be to capture a live event as accurately as possible in relation to the dimensions of that real-life experience, or the recordist might seek to alter the artistic elements of sound to enhance the quality of that real-life experience. The recordist may even seek to create a new reality or set of conditions for the existence and relationships of sounds. Reality is simulated, enhanced, or created through the recording process.

The level of similarity between the recording and the live listening experience is central to the aesthetic quality of the recording.

A recording may differ from the live listening experience by (1) the use of the artistic elements of sound in ways that cannot happen in nature, and (2) the presentation of impossible human performances and compilations of perfect performances. These may subtly enhance the impression of a live performance or sonically represent something that is outside the possibilities of physics and human performance, as potential extremes.

The aesthetic and artistic elements that most influence the life-like qualities of the recording are (1) environmental characteristics and the dimensions of the sound stage, and (2) the relationships of musical balance to the timbres of sound sources.

### Sound Stage and Environments

All of our experiences of sound are with sound as it exists in space. We conceptualize sound, especially in the context of music performances, in relation to the spaces in which the sound is heard to exist. The recording process will provide the illusion of space for the music. With extreme realism, or very little, the recording will bring listeners to associate the recording with their reality. The recording will provide the illusion of a performance space or a physical environment for the performance—this is the perceived performance environment.

As we have learned, this perceived performance environment is an illusion of a space wherein the recording can be imagined as existing during its re-performance (playback). The realistic nature of the performance of the recording will play a central role in establishing the relationship of the recording to the live listening experience. The listener will subconsciously scan the recording to establish (1) environmental characteristics, (2) an imaginary stage (sound stage), and (3) a perceived performance environment. This information allows the listener to complete the process of establishing a reality (real or imagined) for the listening experience of the recorded music performance.

These three important characteristics need to be deliberately shaped or captured to precisely determine this aspect of the recording's aesthetic. If not directly engaged, the recording will likely appear deficient in some way, though perhaps subtly.

The imaginary environments will be either the captured reality of the original performance space, an altered or enhanced reality of the original performance space, or new spaces that are created or selected for the performance through signal processing or plug-ins.

In recordings that closely match a live performance, environments of individual sound sources are typically very similar, if they are different at all. If environmental cues differ markedly between instruments, a sense of a live performance can be maintained if fusion of the source and its environment is complete, and there is an impression that they are one sound quality. Generally, having different environments for different sources will gradually pull the recording out of a "live" experience and into a "created" one, as the number increases or as they become more pronounced.

A live recording will have realistic relationships of sound-source images. Sources will be positioned laterally as if they are in a performance situation, on stage. Image widths will be proportional to one another. Distances will be very similar, with some sources located only slightly in front or behind others. As a recording deviates further from these relationships it will become increasingly "unreal."

**Listen**

to tracks 52 and 53, then 50 and 51

for the sound-stage dimensions of one mix that simulates the relationships of a live sound stage and a second mix of the same musical balance that significantly alters the sound stage to unnatural proportions and relationships. Finally compare those mixes to two stereo microphone techniques.

The perceived performance environment plays a large role in determining the overall sound quality of the recording, and its illusion/reproduction of the size of the "space" of the recording. The listener's position in relation to the sound stage (the stage-to-listener distance) plays a critical role in the impression of witnessing a live performance. A live listening experience results when the listeners find themself clearly located at a specific location within a clearly perceived overall environment. As recordings move from these relationships, the experience becomes less and less like a live concert.

### Musical Balance and Sound Quality

The interrelationships of musical balance and performance intensity are integral parts of live performances, and are easily altered by the recording process.

Recordings that attempt to capture the aesthetics of the live performance will seek to capture the musical balance of the performers as they (or the conductor) intended. The changes in the sound quality of the instruments will be precisely aligned with changes of dynamic levels in the musical balance of the ensemble and to changes in musical expression. It is important to maintain these relationships to keep the character of the live performance.

Recordings that seek to enhance the characteristics of the live performance may contain slight changes in musical balance that were not the result of the performers, but instead are the result of the recording or mixing process. These alterations will be heard as changes in dynamic levels that are not supported by changes in the sound qualities of the instrument(s). This enhancement might take place in only a few instruments, or it may be used extensively throughout the entire ensemble. This enhancement technique may be quite subtle and difficult to detect, or it may be prominent. A soloist with an orchestra is a common example of when this might occur. These enhance the performance by making it "less live."

Alterations in dynamic levels, and thus musical balance, that are not aligned with changes in performance intensities have become integral parts of music written for recordings. Multitrack mixes frequently exhibit changes in musical balance that were not caused by the performers. These changes in dynamic level, then, are inconsistent with the sound qualities of the instruments in the final recording.

**Listen**

to tracks 37 and 38

for the sound quality of the performance intensity of the instruments when they were recorded and how these coincide with the dynamics in the two mixes.

The relationship between the musical balance and the timbre of sound sources in many multitrack recordings creates a wealth of contradictions between reality and what is heard. The aesthetics of this type of recording leans toward redefining reality with each new project and is a stark contrast to the aesthetic of trying to capture the reality of the live performance.

The recordist's approach to any project should include a conscious decision on a level of realism. How will the final sound relate to real-life experiences, and how will the characteristics of sound be shaped? What is the listener intended to believe, and how can this be achieved?

## The Recording Aesthetic in Relation to the Performance Event

As we know, the recording process will shape music performances in such a way that the sound qualities and relationships of live performances may be altered. How the process alters the live listening experience is central to the aesthetics of the recording.

### Production Transparent Recordings

The recording medium is often called upon to be transparent. In these contexts, it is the function of the recording to capture the sound as accurately as possible, to capture the live performance without alteration. This type of aesthetic is common for archival recordings that function to document events. These *production transparent recordings* may or may not be sensitive to the performance environment. At times, these recordings attempt to capture the sound of the music performance without considering the artistic dimension of the relationship of the music (and musicians) and the performance space (and audience). In other instances, these recordings seek to negate any influence of the performance space on the sound of the recording.

Because these are recordings of live performances, the recordist is not involved with compiling the performance. The performance takes place in real time, and it will not be possible to back up and fix a certain section or idea. The recordist is primarily concerned with the technical aspects of the sound of the recording (critical listening) and the sound qualities of the overall program (at the highest level of perspective).

A limited number of microphones are often used in making this type of recording. Usually two microphones are used in some appropriate stereo-microphone technique, placed fairly close to the ensemble. The microphones generally are sent directly to a two-track (or surround) master, with little or no signal processing. The recordist will exercise little real-time control over the quality of the sound and over the shaping of the performance.

The recording medium can also be transparent in documenting a performance, while placing the music in a complementary relationship with the host environment of the performance. Specific pieces of music are best suited to certain environments and are most effectively perceived from certain listening distances. The artistic message of a specific piece of music will be most effectively communicated in a certain environment and with the listener at an ideal distance from the ensemble.

### Spatially Enhanced Production Transparent Recordings

*Spatially enhanced production transparent recordings* can ensure pieces of music will be perceived as having been performed in an ideal environment, with the listener located at an ideal distance from the ensemble, when listening to the recording. This approach locates the listener at the *ideal seat*, and can be accomplished without altering the performance itself and maintaining transparency of the recording process.

The recordist (often with input from a conductor or producer) will determine the type and amount of influence the acoustic performance environment will have on the final recording. Microphone selection, choice of stereo microphone array, and array placement within the performance environment are the primary determinants of the environment sound that is captured from the performance environment. Artificial reverberation units or other time processors may sensitively enhance the characteristics of the environment. The distance of the listener from the ensemble is determined primarily through microphone placement and through time processing.

This recording aesthetic attempts to present the music in the most suitable setting possible for that particular work and to simulate the listening experience in the concert hall. This recordist seeks to ensure that the sounds will be in the same spatial relationships as the live performance, that the recording process will not alter the balance of the musical parts, and that the quality of each sound source will be captured in a consistent manner. In these *live acoustic recordings*, the recording may seek to reproduce the sound of the performance space—surround-sound recordings can be used for great realism in this approach.

**Listen**

to tracks 50, 48, and 49

for a stereo microphone technique that does little to alter the character and sound qualities/relationships of the performance, a mix that establishes sounds at unnatural relationships, and a mix that adds a stereo microphone technique to the unnatural relationships in the mix of track 48.

This aesthetic can have the recordist more involved with the decision-making process in some projects than in others. This aesthetic may be used for many types of music, and may be used for live concert recording as well as session recording. While it is common in orchestral and other art music formats, it is equally appropriate for jazz or any other music recordings where the performers are refined in their sensitivity to and control of their relationships to the whole ensemble. In session recordings, some (or much) editing may be a part of this aesthetic. A consistency of sound quality and spatial relationships between all portions of the work will nearly always be sought; this is a stark contrast to many multitrack productions.

### Enhanced Performances

The recording medium may *enhance the performance* in widely varying degrees. This aesthetic might appear as a slight extension of the concept of a transparent live recording, with the recording process slightly enhancing certain musical ideas. In contrast, this aesthetic may set another extreme of being a life-like session recording that was recorded out of real time.

This aesthetic simulates a natural listening experience, by capturing or creating many of the inherent characteristics of a live, unaltered performance. The timbre and dynamic relationships, spatial cues, and editing techniques all serve to create the impression that the recording did indeed take place within reality—as an actual, live performance.

When this aesthetic is an extension of the concept of a transparent live recording, sounds are placed in the sound stage in the same relative positions as the instruments were in during the recording. The width and depth of the sound stage and image sizes are realistic, and the recording will usually have a single environmental characteristic applied to the overall program (a single soloist might be present with a slightly different environment). Dynamic changes are nearly always aligned with timbre (performance intensity) changes, though some microphone highlighting might create a limited number of dynamic changes without timbre changes. The recording process is used to slightly enhance certain musical ideas from the live performance.

This aesthetic may be used for controlled live performances (those that have been rehearsed with the recordist) or in recording sessions, for a wide variety of musical styles. Minimal miking will usually be used, often an overall stereo array with a small number of accent microphones (or stereo pairs). Accent microphones allow this aesthetic to be adaptable to stage recordings of large classical ensembles or for musical theater and opera. The recording is usually mixed directly to a two-track (or surround), with mixing decisions taking place during the rehearsals or during the recording session(s). Recording submixes or submixing related parts of the mix to a few channels as stems to a multitrack recorder or DAW is also common, but many of the decisions related to the sound of the recording are still accomplished during the recording session or rehearsals.

Recording sessions will often be composed of many takes of large and small sections of the work. As the ensemble balance is largely controlled by the performers, and the parts are not singled out (making re-recording of individual parts unavailable), ensemble problems of accuracy and sound quality often create a lengthy recording session and a large set of session takes.

The master ends up being a collection of takes, numbering from just a few to a great many. The editing of these takes becomes an integral part of the recording process. The best takes are selected based on musical and technical qualities. These are then edited together (cut and paste) to compile a *perfect performance* of the work. The master or final version represents the final performance. The goal of this approach is usually to craft the best possible (perfect) performance, interpretation, and presentation of the music.

The aesthetic of slightly enhancing the reality of the performance may also be found in session recordings that simulate natural sound relationships. Although recorded out of real time, the recordings will seek to simulate the experience of live music. Some emphasis of certain musical materials (and/or artistic elements) over others will be unavoidable in the recording process and will diminish the naturalness of the relationships of the sounds. Some recordings may simulate reality only generally, but still have a goal of providing the illusion of a naturally occurring performance—even with complete control of the multitrack recording (miking, processing, and mixing) process.

### Created Performances

Music written for the recording medium may have qualities that are significantly different from live acoustic music. It may be constructed in different ways, and it may contain additional artistic elements. Music written to be recorded, and especially music written during or through the recording process, is often composed and/or performed in layers—its performances created by compiling separately played parts.

The musical materials are often written and recorded one part at a time, or a small group of parts at a time. The recordings use close miking techniques that ensure a separation of parts (and thus allow for precise control of the individual sound source) or will physically isolate the performers/performances from one another. The parts are continuously compiled on separate tracks, with each new musical line added to the musical texture. Players often perform their parts many times; any number of versions may be recorded before the desired result is achieved. The recordist (sometimes with the aid of a producer) may be responsible for listening for performance mistakes, listening for the most interesting and successful performance, keeping track of which portion of which musical part was performed most accurately, on which take, etc., in addition to maintaining impeccable sound quality.

The final piece may be a composite of any number of performances, and it may be a controlled integration of many different musical ideas and personalities. The performances may or may not have taken place at the same studio, or during the same day (or year), and the performers may or may not have met and discussed their musical intentions.

The recording medium can create the illusion of a performance that contains characteristics that cannot exist naturally. This aesthetic has become common since the early/mid-1960s. In this "new" aesthetic, the recording medium's unique sound qualities and creative potential are used. It becomes a musical ensemble with its own set of resources for shaping a performance or creating a musical composition.

Music written to be recorded may exploit environmental characteristics, musical balance and sound-quality contradictions, sound-stage depth and width, sound-source imaging, or its other unique elements to create, define, or enhance its musical materials.

This aesthetic might purposefully create relationships that cannot exist in nature—a whisper of a vocalist might be significantly louder than a cymbal crash. This aesthetic will use the unique qualities of recorded sound in the communication of the work's musical message.

Recordings of this aesthetic might seek to create a new reality for each work or project. Unique relationships of sound are calculated and incorporated into the music. Recordists (engineers and producers) develop personal styles of the ways they shape aspects of balance, imaging, sound stage, and environment, while continuing to explore the expressive potentials of recording and the medium's relationships with reality.

Much of today's popular music falls within either the aesthetic described above or within the aesthetic of using the recording medium to enhance the illusion of a live performance. Many of the artistic considerations of the recording process are very apparent in these two aesthetics.

## Altered Realities of Music Performance

The reality of the music performance event itself is also altered by the aesthetics of the recording production. The recording provides an illusion of a live performance, and the content and qualities of the perceived music performance may vary from a slight improvement of our listening realities, to being a live performance that exists in ways that are impossible in our known world.

Recording allows a music performance to be an object that can be precisely polished by the artists, that can exist as an almost indestructible physical object and held in one's hands, and that can be owned by any number of members of the general public.

The reality of a live music performance as an experience witnessed in a fleeting instant in time, and as retained only in the memories of those who experienced the event, is significantly altered by the existence of recordings. A recording is a permanent performance of the piece of music—one that can potentially live well beyond the artist's lifetime.

Additional pressures, ideals, and aesthetics are placed on the artists responsible for any individual recording, as opposed to a live performance. A recording may transform the live listening experience: (1) by creating humanly impossible performances, (2) by providing performance conditions that are inconsistent with reality, (3) by presenting error-free and precisely crafted performances, and (4) by providing a permanent record of a music performance.

A recording is a *permanent performance* of a piece of music. It is a period of time that has been created or captured, and that may be preserved forever. The performance can be revisited (and observed at any level of detail) at any time, and any number of times, and by anybody.

Recordings can often become *definitive performances* of a piece of music. The definitive performance may be thought of as being either that of a certain artist or of the particular piece of music. An artist's performance/recording of a work might be what is widely accepted as the definitive performance (or reference) of how the work exists in its most suitable form, in relation to performance technique or to the communication of the musical message. A specific recording of a work can also serve as a definitive reference of how a work exists in its most suitable state, in relation to recording practice or to musical considerations.

Recordings not only are a means of creating an art form, they also preserve the artistic ideas of music performance and expression that do not rely on the recording process. Recordings may permanently preserve the music performances of an artist. They may provide historical documentation or archival functions by preserving the music performances of particular artists, ensembles, events, and more—even nature and its sonic landscapes.

The great contradiction of producing a recording that is a permanent record is that the recording often becomes dated. Artists develop and grow. Their musical abilities, levels of understanding, artistic sensibilities, and their musical ideas change. The permanent performance that was previously created (perhaps only a few weeks before) may no longer be representative of the artists' abilities or aesthetic opinions. It can be a snapshot of a point in time.

The recording will often represent the artist's and recordist's idea of a perfect performance of the work. Theoretically, a perfect performance of any piece of music can be produced through the recording process. The definition of the perfect performance may vary considerably between performers, but the concept of the recording itself will be similar. It will be a presentation of what the performers and producer believe to be the most appropriate interpretation of the piece of music, under the most appropriate performance conditions (instruments used, performance space, etc.).

A perfect performance will combine the artists' desired interpretation of the music (and an absence of performance inaccuracies) with an illusion of the drama of a live concert, experienced at the ideal listening location of an ideal performance environment for the ensemble and piece of music. Practical considerations of the recording process might compromise the actual quality of the recording, but the goal of the recording remains constant.

The recording may present musical ideas, containing sound qualities and relationships that are impossible to create in live performance. Musical materials may be presented in ways that are beyond the potentials of human execution. These might be rapid passages performed precisely and flawlessly, dynamics and sound-quality expressions that change levels quickly and in contradictory ways, or the use of a single human voice to perform many different parts. These are only a few of the possibilities. Humanly impossible performance techniques and relationships are easily created in recording.

The reality of what is humanly possible in a music performance is often inconsistent with the music performance of the recording. The relationships of sound sources, the characteristics of sound sources, the interrelationships of musical ideas and artistic elements, and the perceived physical performance of the musical ideas may be such that they could not take place in nature. The music performance of the recording may be such that it could not be accomplished without recording techniques and technologies. It may be impossible to recreate the music or the performance live, on stage—a situation often cited as one of the reasons why The Beatles stopped touring.

Recordings have greatly influenced our expectations of live music performance. The listening audience of a recording becomes accustomed to a recording as being the perfect performance of a piece of music. The audience may learn the subtleties of a recording quite intimately.

A particular performance of a piece of music is created or captured in a music recording. When an audience member owns a copy of the recording, or when a recording has received much exposure through media, the audience may have heard a recording many, many times. For a great many audience members, the recording becomes the definitive performance of the music. The audience will carry this knowledge of the music recording into a live music concert, and may impose unrealistic expectations onto the performers and the event.

Differences between the live performance and the known recording of a piece of music can cause unfulfilled expectations for the audience. The artists' new and different interpretation of the music, the absence of certain sound qualities from recording production, the inconsistencies of human performance, or any number of other factors might create differences between the live performance and the known recording of a piece of music. A potential exists for audiences to become less involved with the drama and excitement of the live performance of music. Audiences may attend concerts to publicly hear live performers and the music performances they have come to know well as recordings, heard privately many times. Audience members do not always accept the reality that the live performance was not the same as the studio-produced recording. A potential exists for the audience member to be dissatisfied that the definitive performance they know well was not reproduced for them at the live event.

An audience may place unrealistic expectations on the performers. Performers may be expected to perform flawlessly, or with the same version and interpretation of the work as the recording they released. An audience may expect the performing artist to provide the role of reproducing one particular performance of the work. By reacting to this audience, the performer may be restricted from allowing their interpretation of the music to evolve and change according to their growing experience, and may be restricted from creating a more exciting performance. The subtleties of artistic expression that are possible only through the artist and audience interaction, along with other unique qualities of live music performance, may be lost from an event and diminish the musical experience.

It is unrealistic to expect to hear a precise reenactment of a recording in a live concert environment. Many music recordings have been produced in such a way that a live performance of all musical parts, sounds, and relationships is not practical or possible. These are potential negative outgrowths of the audience's familiarity with certain recordings and the new listening habits afforded by readily available music performances.

The general public hears much more recorded music than live music. They are often prone to judging live performances with the expectations of a perfect, recorded performance. Human inaccuracies are sometimes not easily tolerated, and new musical and expressive interpretations of a known piece of music may be heard as simply wrong. Further, people own their own personal copies of performances. A tendency to personalize or to become attached to those performances is common. When the performance is changed, something personal ("their music") has been altered.

## Summary

The recording aesthetic is determined by the relationship of the recording to the live listening experience. The recording aesthetic is arrived at through a careful consideration of the musical material, the function of the recording (type of music recording, sound track, and advertisement), and the desired final character of the recording. The recordist's role is defined by their contributions (or lack thereof) to the process of making the creative decisions of the recording, whether making decisions or executing the ideas of others. This includes all phases of the production process: preproduction, tracking, mixing, and mastering.

The recordist's overall control of the many qualities of the final music recording is highly variable. The recordist is responsible for the overall characteristics of the recording and may be in control of (and responsible for) its most minute details, depending on the recording techniques being used. The recordist might have precise control over shaping or creating a performance, or they may be engaged in capturing the global aspects of a live performance.

The amount and the types of control used in the recording process will determine the degree of influence the recordist has on the final content of the music recording. The recording medium may be used to greatly influence the sounds being captured by the microphones, or the recording medium may shape sound much more subtly. The recording process will be used differently, depending on the aesthetic of the particular project.

The aesthetics of recording production vary with the individual, with the musical material, and with the artistic message and objectives of a certain project.

An aesthetic position or approach may be appropriate for a certain context, or it may not. An approach might enhance the artist's conception of the music, or it may not. An approach may be consistent with other considerations of the project or the music, or it may not. Perhaps consistency is not desired.

The aesthetic approach to recording production creates a conceptual context of the artistic aspects of recording. The intangible aspects of the art can then be appreciated within this context. The recordist must clearly define the aesthetic position of the recording, in order to successfully control and shape it.

How these sounds of recordings shape musical ideas and musical expression will be explored next.

# 14 The Sounds of Recordings: Shaping Musical Ideas and Musical Expression

As we have been learning, musical ideas and the materials that present them are given extra dimensions and may be thereby enhanced in recordings. The artistic elements first described in Chapter 2 provide this enhancement, and are also dimensions that can be used for musical expression. They may transform musical substance and expression in recorded music, to enrich the music and its message.

In this chapter we will explore some of the possibilities of how the sound qualities of recordings shape musical ideas. In shaping musical ideas, these elements of sound contribute to subtle qualities of musical expression and also to the substance of what gives a musical idea its shape and content. The characters and qualities of these artistic elements, how they have been used creatively in crafting mixes, and their impacts on the music and its message will be examined.

## The Artistic Elements of Sound as Musical Materials

Any and all of the artistic elements will contribute to the character of musical materials, and the recording as a whole. In their presence at all levels of perspective, they will contribute their unique qualities to the recording in various ways at different levels of detail, and at different moments of time. They will often function to support the flow of the music and the character of the melodies, rhythms, and harmonies.

**Table 14.1** Important artistic elements in music recordings

| *Artistic Elements* |
| --- |
| ***Unique to or More Prominently Shaped in Music Recordings*** |
| **Pitch Levels and Relationships**—register, range, pitch areas, pitch density, timbral balance |
| **Dynamic Levels and Relationships**—RDL, program dynamic contour, musical balance, dynamic contour/shape of musical ideas, dynamic contour of individual sounds |
| **Rhythmic Patterns and Rates of Activities**—tempo, rhythms, patterns of durations in all elements |
| **Sound Sources and Sound Quality**—timbral balance, pitch density, arranging, performance intensity, performance techniques; sound qualities and timbre of sound sources, component parts of timbre/sound quality |
| **Spatial Properties**—stereo location, surround location, phantom images, moving sources, distance location, sound-stage dimensions, imaging, environmental characteristics, perceived performance environment, space within space |

They might establish a context, or a sonic place of reference, for the materials to exist within. They also have the potential to carry the weight of musical expression. A complete list of artistic elements appears in Table 2.1. The elements unique to music recordings, or important to our discussions, are listed again here in Table 14.1.

It should be understood that in examining how the unique sounds of music recordings impact the music, we will not examine and include traditional aspects of music in our discussions. We must remember that these traditional aspects (elements) are nearly always very significant to music recordings. The qualities we are discussing are often less significant, though still hold profound importance in crafting a quality recording.

The various ways the artistic elements might appear are many; they can appear in a great many forms, and we continue to discover new ways to use them. The elements can also function in too many ways to list and define; our creativity continues to grow this number, as well.

Table 14.2 lists some important considerations for our understanding of the artistic elements, and how they are used. Each time "artistic element" appears, each item from Table 14.1 could be inserted. This provides a glimpse of just how many possibilities these aspects of sound create for shaping the music, and the experiences of our audience.

**Table 14.2** Basic questions toward evaluating the functions and activities of the artistic elements of recorded music

| Is the activity or state(s) of any individual artistic element a musical idea in itself? |
|---|
| In what way does the artistic element impact the musical material? |
| In what way does the artistic element enhance (contribute to) the musical message, musical meaning, or the delivery of the musical material? |
| Does the activity of the artistic element:<br>– Represent substantive material or ornamental embellishment?<br>– Shape the musical idea(s)?<br>– Impart character to musical materials?<br>– Impact the music directly?<br>– Shape the musical experience of the listener? |

## Substantive and Ornamental Functions

Any element can function in one of two general ways: substantive or ornamental. They can contribute to the substance of the musical idea, or they can provide an enhancement to its character.

For an element to be substantive, it will be one of the important elements for shaping a musical idea. For melody (as an example of a musical idea), the substantive elements would be pitch and rhythm, perhaps equally, or perhaps one would dominate slightly. For a drum fill, the substantive elements would often be rhythm and the timbres of the drums; dynamics might be substantive, or could be ornamental. It is not common to find one of the unique elements of recording as the primary substantive element for significant portions of a song, but it does happen.

The bridge in "A Day In The Life" is an example of timbral balance being the primary element. In that section timbral balance *is* the musical idea. The pitch densities of the sound sources combine into a gradually changing texture of timbral balance. These give shape, direction, and drama to the section. After the material returns in the song's last section, the final chord of the song pulls the listener into a different level of perspective, as we perceive the interactions of the sustained piano pitches and their spectra during the long fade; here we are experiencing timbral balance within the focused context of a single chord, with the overall shape of its decay and the subtle dynamic changes of the pitches in the chord, and the minute changes in the appearances of the partials of each pitch. To contrast, now consider Lennon's vocal during the song's first verses, as it gradually moves across the sound stage (see Figure 10.6); here stereo location, movement, and image size function as ornamental elements, as the melody and lyrics of the vocal are clearly of primary importance.

An ornamental element is one that could be omitted, and the essence of the musical idea would be unchanged. The musical material is enhanced by the ornamental element, but not substantively changed. This is an addition that elaborates, or a layer of decoration. It can often be undetected by the inattentive or untrained listener, but this does not diminish its value to the quality of the recording. The element can still be valuable in shaping the character of the music.

As an example, reconsider the Moog glissando that ends the introduction to "Here Comes the Sun." The sound not only moves in pitch, it moves in location. This movement from left to the center of the stereo field is significant to the character of the sound; when listening to the recording in a stable listening environment this movement contributes an important characteristic to the sound that is perhaps nearly as important as the pitch change. Fortunately the passage also works without the support of the changing location, thus helping the song's success in mono formats and in situations where stereo is not reproduced accurately. In a stable listening environment this movement may be heard as substantive; in others it may be heard as ornamental, if it is noticed at all (or if it is even present).

Much of what we do in crafting a recording is adding or subtracting for slight enhancement to the various aspects of the music and its performance. We will often craft the artistic elements sensitively and subtly to bring character, motion, meaning, and more. The result will be sound qualities that support the music and not detract from it. As this chapter progresses, there will be more examples.

### Prominence and Equivalence

When we listen to a mix, some sounds or musical ideas are more noticeable than others. Our attention is drawn to them. Or, perhaps they stick out. For one reason or another, they are prominent. *Prominence* is perhaps what is conspicuous, or perhaps it is the thing that draws us to hold it within the center of our attention. What is prominent is not necessarily most important or most significant.

When we learn to mix, we are learning to balance prominence.

Learning to mix, we are learning to place sounds and musical ideas in an appropriate relationship to other sounds. Loudness can make a sound prominent, but it is not the only way a sound can become prominent. A sound with some unwanted noise can be prominent; no matter how soft the sound, its distortion demands the center of our attention from our well-trained ear. A vocal line can be much softer than the other sounds in the mix, but its words and melody can make it prominent. Prominence is not related to loudness. A sound separated from others by imaging, or by pitch area, or by sound quality (to name a few) can be prominent because it is exposed and thus isolated in some way.

Prominence does not mean something is the most important or significant. It is simply the most noticeable at that moment in time, and perhaps at that level of perspective. Prominence exists at all levels of perspective. It also places sounds relative to one another, so that the snare drum of a drum set might be more prominent than the remainder of the set, but the drum sound as a whole might be much less prominent than another sound, say keyboards.

Any element can provide a sound with prominence. Any element can add dimension to any musical idea and give it prominence. This is an extension of our earlier discussion on equivalence.

We will now include "equivalence" in our thoughts of prominence. Any element can provide a sound with enhancement that will bring it prominence, or that may draw the listener's attention, even if only for a moment. It is this draw of attention from using another artistic element instead of loudness that can allow a mix to be rich and varied without relying on loudness to give significance and prominence to material.

In listening to Phil Spector's mix of "Let It Be" (from the album *Let It Be*), we hear a prominent high-hat entrance at 0:53. The instrument is not the loudest in the mix. It is prominent in its first appearance because it is a new addition to the texture; the listener is drawn to the new sound that is unlike anything that preceded it. Immediately the high hat's prominence includes its pronounced delay iterations, then its movement on the sound stage. Never is it the loudest sound, or the most important. It also is not more prominent than the lead vocal. As an exercise, listen to the song carefully and experience the high hat in relation to the piano. Is it more prominent than the piano during this passage? Listen again, and see if your perception changes. Be aware to hold both sounds as equal within your sense of perspective. As you listen a third time you may become less focused on the high-hat sound, and its prominence might begin to diminish. On fourth listening, your perception of prominence might shift again; perhaps your understanding is beginning to stabilize.

## Sound Qualities

The element of sound quality is important to recordings on a great many levels, and in a great many ways. It exists as the individual, isolated sound, and also as a musical instrument or voice performing a certain part. It exists as the changing timbre of performance intensity, and also as the expression within a musical line. It is the overall sound of the recording, and it is also timbral balance. And there is the sonic integrity of the recording, with its noise and distortion matters. There are more ways sound quality appears in our recordings, but they are mostly related to these categories.

### The Individual Sound and Sound Source

At the level of the individual sound source, the selection of instrument and the fine adjustments to the sound quality of the instrument becomes an important dimension to the musical idea. Getting "just the right sound" is an important part of what the recordist does. Shaping sound quality takes place in the decisions of preproduction and in the processes of tracking (especially microphone selection and placement), and perhaps through signal processing later during mixing.

The recordist's concerns are: How does the sound quality of the instrument or voice fit with the music? How does it appear in the mix? How does it best deliver the line? Does it provide the expressive character that is needed? Other questions and concerns certainly can be present. Answers are a matter of style and preference, and what is compatible and what is appropriate—and the ideas of the artist and the client.

The subtle aspects we learned to hear in sound-quality evaluations are now part of our creative thinking. We bring our attention to dynamic envelope and to spectral content to understand the sound, and to experience how we are capturing it and/or altering it. We shape these subtle details to shape the message and the character of the music. And these subtle aspects change slightly throughout the song as a result of performance intensity, musical expression, and how the sound is placed in the mix and combined with other sounds.

Returning to the high-hat sound in "Let It Be" (album version), notice the sound quality. Its characteristics are very exposed. Bring your attention to the unique decay of the sound caused by the many delays; listen again to get a sense of the contour and the time of the decay. Now notice the attack of the sound, and get a sense of the speed of the attack and its dynamic level. Once you have a clear impression of these, listen intently to the spectrum of the sound; there are areas of much greater activity than others. In doing this, we are beginning to understand how the sound was shaped and its unique qualities.

If we listen now to "Let It Be" on the *Let It Be . . . Naked* album, we hear this sound in a very different state. This provides a stark contrast of sound quality to the exact same sound. Bring your attention to the decay of the sound and learn its character, then compare it to the album version. Consider also the attack of this sound, then the spectrum of this sound. Notice how very different the two sound qualities are. Describing these differences objectively would be an excellent exercise in talking about sounds and sound quality. Finally, listen again to notice how those sounds relate to the other sounds in their mix for frequency content and timbral balance. The two sounds appear within a very different mix context, within a very different overall sound quality, and within a very different aesthetic related to the sound of a live performance. Taking this still further, much could be gained in refining these observations and skills by listening to the sound qualities of this high hat as found in mixes of "Let It Be" that appear on *Past Masters, Volume Two* and on *1*. The subtle and significant differences of the 2009 remastered version of the song on *Let It Be* would provide another opportunity to study this sound containing different qualities, yet again.

We have been focused on the sound of the high hat as it appears in various mixes. The sound was captured before this time during tracking. Before capturing the sound, a significant number of decisions went into shaping that sound. The recordist will be focused on the subtleties of sound quality when selecting microphones, placing microphones, deciding on other devices and technologies in the signal chain, and more. These will be explored in coming chapters.

## Performance Intensity and Musical Expression

Performance intensity is the effort, energy, and expression put into playing or singing. Sonically it is reflected in sound-quality changes to instruments and voices. Performance intensity may also contain dynamic changes, but not always.

The urgency of performance intensity can create motion and stress in music, and in a different state it can create tranquility, respose, and calm—and there are many states between. Changes can be extreme and fast, subtle and gradual, but all contribute to the character of the sound and often the substance of the music. They do much to shape expression, and also to shape the character of sounds. The timbre of performance intensity can fuse with the source's sound quality to further reinforce the impression that there is only one correct way for a line to be performed, one definitive sound for the musical idea. Intensity blends with sound quality, and blends with the musical idea, blends with its message, and establishes the expressive character of the line, and also its sense of motion.

Given all this, performance intensity is an important part of performances. It provides much character to the music, as well as contributing sound-quality content. It is also one of the aspects of performance that is difficult to continue to generate and sustain in a long recording session. Obtaining the "right" intensity for the performance can be illusive, and entire sessions can become occupied with finding just the right groove, energy, sound, expression, motion, feeling, or other illusive trait, to a lead part or a vocal, or any sound or part, or group of sounds (like a rhythm section) considered vital (and typically all sounds and parts are vital in their own way).

Typically performance intensity is captured and shaped during the tracking phase of the recording process. Awareness of performance intensity brings the recordist to focus on sound quality, and on how sound quality relates to other aspects of the music, and to the magic of musical expression.

## Timbral Balance and the Mix

Timbral balance is the sound quality of the overall mix, and more. Just as sounds are shaped by manipulation of sound quality, the mix of the overall sound is manipulated by attention to timbral balance. The sound quality of sources transforms itself into a higher level of perspective as the pitch aspects of musical ideas coupled with, or fused with, their sound qualities; this is pitch density. As we combine sources and their pitch densities, timbral balance is established at the highest levels of perspective. In this way, timbral balance is the sound quality of recordings.

Albums often have an overall "sound quality" that binds them into a single statement and a single sound quality. Typically this is related to timbral balance. This is how the recording places sounds in the frequency range, and the character of that sound. The other aspects of the overall sound (especially perceived performance environment and sound stage) contribute, but timbral balance, and the qualities of the individual sounds that comprise it, plays the dominant role.

With timbral balance, we are placing sounds in vertical, frequency space. Allan F. Moore (*Song Means,* pp. 29–44) devised the "sound box" that brings focus to this concept. Pitch and the frequency of sounds are recognized as placed on the vertical plan—not as elevation, but as a position in pitch space. Highest pitches, highest frequencies are placed in the top region of the vertical space of the sound box, and successively lower pitches/frequencies placed appropriately below. Thus, it can serve as another aid to conceptualize frequency distribution as vertical space, supplementing the timbral balance approach of the graph format found in Figure 12.1 with sound stage dimensions. In considering timbral balance we are placing sounds within that space with attention to their relationship with others, and the overall "pitch space" of the recording itself. Voices and instruments with their materials can be considered to occupy certain frequency bands (pitch areas). How these bands overlap, blend, separate, etc., all impact the nature of the mix significantly.

Here the recordist is concerned with how sounds combine, related to frequency content. Certainly the dynamics of musical balance can impact the emergence of frequency content, but our focus of timbral density is on frequency itself. In combining sounds, the fundamental decisions are: "To what extent will the sounds blend?" or "To what extent do we want the sounds to blend?" based on frequency content. The balance is between providing sounds with uncompromised clarity or completely fusing sounds together into a single sound quality. The "right" balance between these two extremes (or at either) is where artistic judgment enters, and where the recordist is shaping sound.

When sounds are in the same frequency range they will fuse more readily, and be absorbed into the timbral balance. We know this as a type of masking. As sounds differ more in frequency, they have a greater tendency toward remaining separate, and distinguishable. Emphasizing even slight portions of a sound's spectrum through equalization (EQ) can pull a sound out of the blend with another, to provide it with a bit of clarity or prominence. This is a shift in the amplitude of a portion of the spectrum: a change of loudness, dynamics. After the considerations of frequency content alone, one can turn attention to how even subtle shifts of dynamic relationships will alter frequency content.

Other dimensions of sound quality are factors as well. Attack characteristics can provide each sound with distinctive qualities that can inhibit fusion; differing sustain time and dynamic envelope contours can also differentiate sounds enough to inhibit fusion. This should bring us to remember that combining sound qualities is more than simply combining spectra. Focused attention on the characters of sounds being combined will demonstrate this.

Timbral balance works in a complementary manner with lateral location, to distribute frequency information across the width of the sound stage. The joined concepts of vertical space and lateral space allow us to envision placing sounds and perceiving sounds by their frequency/height and their lateral position. This concept envisions a "sound box" encircling frequency/height and lateral imaging/width. This provides another dimension to the mix.

Using this concept, frequency information is distributed by lateral imaging. By placing sounds by frequency content at specific locations, the recordist has another level of control over the balance of the mix.

Sounds can be separated from others with very similar frequency content by placing each in a different lateral location, adding clarity to the mix. This can keep very similar sounds from fusing or from masking.

The more distance between the images of sounds, the more they are likely to be recognized as separate, and the less likely they will fuse together or that one may be masked. And if the recordist instead seeks to fuse and blend sounds, then placing them in the same location, or in overlapping locations, facilitates this.

A multitude of possibilities begin to show themselves, as we begin to consider just how sounds can be placed in various locations by considering their frequency content. The following are offered as starting points to initiate creative thought.

**Table 14.3** Interactions of timbral balance and lateral location

| |
|---|
| Sounds with different frequencies at the same location will create layers of frequency bands, and each sound will have a degree of distinction and separation. |
| Sounds with similar frequency content in the same location will blend, mask, or fuse to some degree. |
| Sounds with overlapping images in same or different frequency bands will react similarly to the above with a less distinct result. |
| Frequency information might be distributed evenly throughout the sound stage. |
| Frequency information might dominate some parts of the sound stage and not be present in others. |
| Sounds might be placed in a small area of the sound stage, and frequency information distributed evenly, or some frequencies dominate certain locations. |

Many more options will make themselves known to the recordist. Bringing creativity into the mix process through exploring the placement and dominance of frequency activity can add much richness to the mix.

## Sound Stage

The sound stage brings to the music recording a sense of size and location. It is the area of the performance. The two dimensions of the sound stage are defined by its front edge and the rear wall, and by the furthest left and furthest right sources. Imaging and sound-stage dimensions can significantly impact the musicality of the recording, and the delivery of its message.

The front edge of the sound stage is the closest sound in the mix/music. It brings to the listener a sense of distance from the music and the performers, and from the message being communicated. Some messages in music are directed to the listener individually, or to a third person intimately; other songs speak in a more universal way. Placing the nearest sound with a noticeable level of detachment from the listener might support the material of a universal message, or a message intended to speak to a large group of people; this level of detachment would discourage a personal connection between the listener and the message that might be sought in an intimate love song, for instance.

The back wall is the rear boundary of the sound stage; locating the rear wall of the environment of the furthest sound source defines its position. It establishes the distance between the closest sound and the furthest, as well as the depth of the ensemble and sound stage. The rear wall of the sound stage establishes the horizon of the music, and the recording. Musically and dramatically, the rear wall establishes the boundary of the potential panorama of distance layers that can create distance areas for specific performers or groups of performers in the mix; this brings the performers and their musical ideas and materials different and unique perspectives, and dramatic associations for the listener. The rear wall also establishes the magnitude of the depth of the sound stage, and brings the potential of great size to the sound stage, or of perhaps placing all of the performers at a similar distance location and contracting the front-to-back depth of the sound stage to a small area.

The width of the sound stage is its lateral size. Remembering Chapters 1 and 10, the sound stage has the potential to spread up to 15° outside each loudspeaker location. The sound stage could spread to this full 90° width, or it might also be intensely focused within a small area in front of the listener. Width provides the opportunity to create a performance location for the music and its message; a wide sound stage will have a very different dramatic impact than a narrow one. Sound-stage widths of specific sizes are often coupled to various sections of the mix, where the verse will have a different width than the chorus.

Listen to the change in the width of the sound stage in "Across the Universe" (from *Let It Be*) as the song progresses between the end of the first verse and the point where the chorus is fully established; occurring from 0:34 to 0:44. The sound stage suddenly contracts again at 1:10, bringing drama to the music and the message closer to the listener. In this way, the change of the type of communication in the song is reinforced with a change in the sound stage and the mix.

Sound-stage layout and dimensions can support the perception of a live performance. When sources are located on the sound stage in similar locations and distances as a live stage configuration, the aesthetic of a live performance can be established.

## Intimacy and Connection of Distance

Of the aspects of sound controlled in recorded music, distance is among the richest and most unique in its potential for shaping the character of the recording, the music, the performance, and their message. Distance provides the listener with a virtual physical relationship to the music, to its instruments or vocals, and to the story of the song. Distance cues can add dimension to sounds, and can bring fusion; it can change the meaning of a text, and the immediacy of a line.

Distance is approached in different ways in different genres of music. It is controlled by the level/amount of timbral detail captured in the recording process that is allowed to enter into the final mix. Distance location places the listener at some location apart from the music and the performers, or with surround perhaps within the ensemble.

As examples, in recording a choir, the recordist will often seek to blend all sounds equally and place the entire ensemble within a small distance area. The goal will be to not allow single voices to emerge as being closer. An orchestral recording might provide more variation between the closest and farthest sound (the first violin and the percussion at the rear of the ensemble), yet maintain a fusion of the ensemble placed in a confined distance location and at a stable distance from the listener. Other genres of music might similarly fuse the ensemble into a single location to bring focus onto traditional musical materials or to document the performance. Other genres of music place instruments and voices separately on the sound stage (common in multitrack productions), where they might appear in individual locations or in smaller groups in similar locations (such as sometimes happens for all of the instruments of a drum set).

Here are some central contributions of distance:

- The greater the number of different distances, the more substance and potential richness is added to the mix.
- The greater the amount of distance between the nearest and further sources, the more substantial the space of the recording and its music/message.
- The closer the location of the listener to the front of the sound stage, the more intimate the relationship and the greater the connection between the listener and the music.

Significant personal connection can be achieved by placing the lead vocal close to the listener, especially if prominent and at lower performance intensity. This brings the listener to lean in comfortably to get close to the singer, to the musical line, and especially to the message being conveyed. As we have learned, this is accomplished through timbral detail; it can be simulated, also, by adding or capturing breath or mouth noises that would be too gentle or subtle to carry very far from the singer.

A detached relationship can be established by reducing timbral detail to push the location of the source away from the listener. This can change how the listener perceives the message of the song, although the musical material might not be altered. While most apparent in vocals, this can function on any instrument and the musical materials they present.

When the distance location of a sound source changes in the mix, its character changes and its physical relationship to the listener changes. The nature of these changes is important to the mix, to the expression of the line, and to the message of the music. These changes of location can occur suddenly, for example between sections of a song, or gradually, as sometimes happens in a song's fade out. Either way, the change of distance carries a significant enhancement to the musical idea and its message.

Remembering that distance appears as timbral detail, sounds are placed in distance by the amount of subtle timbral detail present in the sound. Timbral detail can be diminished in the mixing process to push sounds away from the listener; this can be accomplished through EQ and filtering techniques, and at times through adding reverberation to mask some of the detail of the direct sound. Timbral detail, however, cannot be added during the mixing process. If timbral detail is not present after tracking, it cannot be created. Therefore, in order for sounds to be placed close to the listener, the required timbral detail must be captured during the tracking, the initial recording process. Giving some significant thought to establishing the front edge of the sound stage will prove valuable in crafting a sound stage with close sources; placing sounds there cannot be manufactured later, and must be incorporated into the earliest decisions of the recording process, including selecting and placing microphones for tracking.

In multitrack recordings all sounds can be placed at a separate, identifiable distance. The distances can range from very close to the listener, to very far. Great richness of depth can be obtained by using many different distances; this can create an interesting texture and delineate sound sources. Further, a wide range from the nearest to farthest sound can be created by different distance locations, providing the mix with the potential of a significantly rich characteristic. Typically, closer sounds are perceived as more prominent or significant in musical contexts; this can be counteracted, as we witness in the lead vocal to "Something" *(Abbey Road)* sitting significantly farther away from the listener than other sound sources in the prechorus. Despite its distance and lower loudness level, it remains prominent in the mix, and of course significant.

## Image Size and Directional Location

Sounds receive prominence or enhancement through phantom image size and directional location of sounds. Thus, musical ideas are provided with another dimension with size and location, along with the instruments/voices that present them.

Sounds in the center of the sound stage are perceived as having greater significance than sounds at the sides, all things being equal. This is typical of our human condition: that something immediately before us receives our focused attention, and activities to the sides are peripheral. That said, a sound that is isolated from all others, or a sound that suddenly appears where no sound has previously been, will be much more prominent and be perceived as significant, even if only momentarily. The initial entrance of George Harrison's vocal in "Here Comes the Sun" *(Abbey Road)* is an example of a sound appearing where none has previously been placed, though the motion of the Moog glissando ending the introduction brings attention to the location where the entry occurs.

Remembering how we localize sounds will be helpful here. We use amplitude differences to localize sounds above 2.5 kHz, and time/phase differences below approximately 1 kHz. Bringing attention to time differences in production, we can bring prominence to most sounds without increasing amplitude, and thereby avoid altering musical balance and raising the overall loudness level of the program.

Sounds of greater size have strong potential to be heard as more prominent, and perceived as being more significant. Widening a sound is a way of increasing its presence in the mix without increasing amplitude. Somewhat oddly—as is often the case with art—providing a sound with contradictory cues can also make a sound more noticeable, as a point-source sound located near the edge of the sound stage might be, when all other sounds are wide and contribute to a fused texture of locations.

Sounds might be arranged on the sound stage in approximately the locations they would be in during a performance. This holds for orchestral recordings, and recordings of traditional ensembles of most kinds: wind ensembles, string quartets, jazz groups, etc. Maintaining stage or performance locations is also commonly used in imaging drum sets, from either the audience perspective or the drummer's perspective. This creates a sense of the recording simulating, or at least having some commonality with, a live performance.

Reverb can be used to widen images. This often appears as a narrower core of the image from the direct sound source with environment cues extending the width of the image to both sides. This approach brings fusion of the two into a single image. At times this can appear to have a consistent sound across its width, and in some appearances the edges may be less defined in location. The piano sound in "Let It Be" (*Past Masters, Volume Two* version) is one such exposed example; John Lennon's vocal in the first verse of "Across the Universe" (*Let It Be* version) is another.

### Placing Sounds and Clarity

As was discussed under timbral balance, by separating sounds to different lateral locations both the masking of a sound by another and the fusion of sounds can be negated to a variety of degrees. A slight separation of location can at times make a very marked difference in clarity, bringing each sound to present its qualities in considerably greater detail; in other instances a more drastic separation might be required. Generally, the more alike sounds are, a greater amount of space is required between the sounds to bring them out of fusion.

Sounds of markedly different pitch areas (for percussions sounds) or areas of dominant spectral activity (for pitched sounds) can be placed at the same or similar locations, and remain uncompromised by the presence of the other. The result will be two or more sounds present in the same location but that are readily distinguished. This often happens in the center of the sound stage, where the lead vocal, bass, kick drum, and perhaps additional instruments (such as piano) are placed in typical productions. This placement of sound sources, so that pitch densities of lines and pitch areas of instruments do not compete with one another, is common; a "sound box" analysis of the pitch content of sounds at various locations on the sound stage will bring the reader to discover this.

In this way, frequency information is crafted along the width of the sound stage, by deliberately placing sounds in space according to complementary spectral content. The placement of instruments according to their timbre and frequency information is balanced to suit the mix and the music. This creates a dimension of the overall sound of the recording that is characteristic of specific albums, types of music, and of a producer's personal style—consider the sound qualities in George Martin's single version of "Let It Be" in contrast with the influence of Phil Spector's wall-of-sound-influenced approach to the song on the album *Let It Be*.

The use of placement to blend or fuse instruments can be equally effective to the music, as is the use of placement to bring clarity to instruments and musical lines, or perhaps prominence to their parts. It is a matter of intention and of what best serves the music. Certainly both approaches can be used to any degree within any recording.

## Environments

Environments often serve to enhance the sound qualities of instruments and voices. They also add another dimension to the character of sounds, and can enhance aspects of the spatiality—width and depth of images. The perceived performance environment (PPE), as the environment of the recording and its performance, has great potential to shape the sound and character of the recording, and the message of the music.

### Crafting a World for the Song: Perceived Performance Environment

In deliberately shaping the PPE, the recordist is engaging a potentially significant dimension. The perceived performance environment (PPE) creates a unique "reality space" or world for the piece of music performed within it. This creates a context for the song/piece of music, and enhances its qualities and meaning. The PPE adds a new "concept" to the sound of the recording—a sense of space where the performance/recording exists—and it adds sound-quality characteristics as well as time cues.

This "room" or "world" provides the song with an appropriate setting for the musical materials, the performance, and the concept of the piece. How the size and qualities of the PPE enhance the recording is as individual as the song itself. A large sense of space for the recording might bring a sense of universality that the song relates to many people and situations, or it might bring a sense of emptiness or aloneness; certainly many other possible scenarios exist. Such a large space can work supportively in many ways, and it may well be inappropriate for a particular piece. The recordist's attention here can greatly enhance the effectiveness of the music and the recording.

The potential expanse of the PPE is highly variable, from incredibly small to unrealistically enormous. The sound qualities of the PPE can add marked timbre changes to the overall sound, or be very subtle. Of course any variations between the extremes are possible, and are most likely to be found in practice.

The PPE is crafted in several ways. If the recordist does not deliberately seek to shape the PPE, the listener will conceive its dimensions on his or her own. The dimensions will be imagined by listener through an intuitive awareness of what to them are most important, then extracting characteristics from those most prominent sounds, from the furthest sound's environment, and from the sounds at the front edge of the sound stage. This is common, but does have potential subjectivity to the result. As it is based on the individual perceptions of individual listeners (and their sensitivity and attention), it can vary somewhat between individuals. The recordist loses some control over the final result in this approach.

By carefully crafting the environments of the most prominent sources and most significant musical materials and most extreme boundaries, the recordist can establish a clearer sense of the overall space of the recording for the listener. This involves consciously establishing the perceived physical boundaries of the PPE, which are the nearest and furthest edges and width, and observing and crafting the sound quality (especially spectral qualities) of the most significant environments. With the characteristics of the PPE remembered as its sound-quality components (dynamic contour, spectrum, and spectral envelope) and time elements (early time-field iterations and reverberation contour and density), the recordist can more directly control crafting the PPE.

Shaping the PPE can also take the form of processing the final mix, applying an environment to the song or the entire recording prior to mastering. In some musical styles, this is common—as in classical recordings. In others, it is not at all common—such as many multitrack recordings. Applying a program to the entire mix has advantages of consistency of sound throughout the recording, and clearly defining the PPE characteristics. Its disadvantages are that the consequences of applying a program to the overall mix are often subtle changes to the sound qualities of the recording, which are unanticipated and undesirable. For instance, early time-field reflections can cause certain instrument attacks to be altered, and for their performances to be less precise and their timbres to be less defined—thus changing their distance location; these reflections can also cause some sounds to fuse with others and mask. When even subtle equalization changes are applied to the overall environment, the mix can change markedly. When using this approach, it is imperative to listen very intently to the details of the entire recording after applying the environment.

### Source Environments: Fusion of Musical Ideas, Sound Qualities of Sources, and Individual Environments

In many productions different environmental characteristics are added to the direct sound of individual instruments or groups of instruments. These characteristics bring the listener to recognize the sound's placement in a performance space that may be unique to that sound. Still, the listener is not brought to process by this information directly. Instead, there is a fusion that occurs, where the musical materials being presented are blended with the sound quality of the instrument/voice and the characteristics of the environment. A rich experience results, which carries many implications for the recordist.

Each instrument or voice will be given a new conceptual context of placement within a space. They also receive sound-quality transformation, where reverberation and early time-field information are incorporated into their overall sound, some directional cues and depth are applied, and its spectrum is modified. Here it may be helpful to consider environmental characteristics related to frequency response as being "formants" of sorts, where no matter the pitch materials of the musical line, the environment will be emphasizing or attenuating certain frequencies. With the characteristics related to time, we understand the reflections of reverberant sound and early time field as a dynamic shape as well as a complex, recursive rhythmic pattern(s). These provide identifiable character to the environment, even as it is fused with the source's sound quality.

As previously noted, environment cues can provide instruments/voices with enhancement of spaciousness as well. An individual line might appear to have an increased or diminished image size because of the nature of its environment. This can carry with it a sense of significance. This might also allow a change in the level of prominence for the sound, to be more or less prominent; here, sounds appearing in larger environments might have a tendency toward prominence over sounds in smaller environments.

Environments have the potential to mask portions of the source's timbre. When this occurs, a decrease of timbral detail is often established. The masking of timbral detail causes a sound to appear more distant. This is not a universal situation: Merely applying environmental cues, even radical ones, does not in and of itself change distance location. A change of timbral detail must be present.

When mixing in surround, there is an additional option that has come into practice. Direct sound and environment sounds are sometimes separated, with each emanating from a different location. When the instrument or voice is coming from one location and the environment from another, a surreal experience can be created whereby the listener is no longer processing space in known ways. It is often not possible for the two sounds to fuse in the listener's perception. This brings a new experience that might be interesting in a mix where it enhances a sound or musical idea effectively; it might also be unconvincing and distracting to the listener—thus diminishing the experience and the recording. This production technique warrants careful attention when being used.

### Space within Space

Space within space brings shape to musical expression by establishing relationships of individual environments to the perceived performance environment. These relationships assist in placing individual instruments/voices and their musical ideas within the context of the song as a whole.

In this way, meaning can be enhanced, drama may be supported, and musical ideas or instruments' sound qualities may be embellished or ornamented—all framed by the "world" that was created for/by the recording: the PPE.

In multitrack productions, and many multi-microphone productions, there is an inherent state of sub-environments existing within the PPE. Whether these sub-environments are captured by the recording process or are applied to sound sources through processing or plug-ins, the instrument/voice along with its fused environment are placed on the sound stage, within the perceived performance environment. This space-within-space relationship has the potential to be used to creative advantage, to bring new dimensions to spatial placement and relationships of sound sources.

Space-within-space relationships can appear more clearly in certain surround recordings. Sound sources, within their environments, are isolated in some surround recordings. In these instances the qualities of the PPE can be more directly compared to an individual sound within its environment. Listening intently, we can note the common aspects of the acoustic guitar sound to the PPE in "While My Guitar Gently Weeps" from *LOVE*, as both reverberation time and emphasized frequencies are readily heard in the guitar introduction and carry throughout the song as part of the PPE. Next, notice the environmental characteristics of the strings and compare those to the PPE; with sounds separated in space, the characteristics of individual environments are easier to detect and evaluate.

Space within space allows us to reconcile very different types of spaces in a recording, by placing those spaces within an overall PPE. The PPE provides a framework or a reference for the environments of sound sources to be understood, and it allows them to remain separate in the perception of the listener. This allows different environmental characteristics of different sounds to remain separate, and not fuse or blend, or meld into one impression of an environment.

With this awareness of relationships of environments, the recordist may deliberately place environments onto sounds in ways that complement or contrast with other sounds, interact with other environments, allow instruments and their musical ideas to remain readily distinguishable or blend with others, and more. These all may be accomplished bearing in mind that they have a relationship at a higher level of perspective, where they all are considered as an element of perception and conception of the song's PPE.

## Noises, Distortions, and Unwanted Sounds

While located near the end of this discussion, noises and unwanted sounds are not an after thought. Instead, our awareness of distortion and detrimental sound permeates our listening process. We attend to all elements of sound for the undesirable, at all times and at all levels of perspective.

The aspects of sound quality that allow us to engage the subtle details of shaping sound also allow us to work with signal integrity. The technical qualities of the recording are also common sound qualities, or noises, that appear at the smallest and the highest levels of perspective.

Noises and unwanted sounds can be created by performers or by their instruments, by the mouth sounds of singers or by their dangling necklaces and bracelets. They can be air ducts and door slams, or squeaky chairs. External sounds, such as traffic or airplanes, can also invade recordings. These sounds often are quickly noticeable, but they can sometimes be subtle.

Listen carefully to the *Let It Be* mix of "Let It Be" again. This time bring your attention to the possibility of unwanted sounds and noises. Recognize that these sounds could be soft or partially masked by other sounds. In the first few minutes of the song, the careful listener will find noises others would have chosen to edit out of the tracks before mixing. Try to identify the noise sound at 0:09. Once you have identified the sound, listen to the remastered *Let It Be* (2009) version, to experience a significantly diminished version of the noise. Turning to the *Past Masters, Volume Two* version, bring your awareness to the possibilities of vocal plosive consonants, clicks, and bumping sounds; you will find some if you are listening correctly. Continuing into the second verse, at 1:07 you should readily identify a very prominent unwanted sound in the right channel.

Distortions appear in a great many forms in all aspects of sound and at all levels of perspective. These appear subtly and in very pronounced ways. They can be related to equipment settings, such as the clipping of a mix bus or the pumping of a compressor or limiter, or the functioning of a piece of equipment, such as latency in a DAW or phase incoherence in a loudspeaker. They can be the result of the mix process or of the tracking process, of editing or of file transfers and formats. They can impact a small portion of a single sound, or the overall sound of the entire recording.

Everything at every moment provides opportunity for distortion. The prepared and skilled recordist will have learned their sounds and be able to anticipate the potentials for their appearance.

## Musical Balance

Musical balance is typically the center of attention for the beginning recordist, sitting down to their first mix. There is a natural tendency to gravitate to manipulating loudness as the means to create prominence in the recording's balance of sounds; mixing-console layout and design tend to reinforce this. To moderate this tendency to immediately reach to alter loudness, and to open the reader's thoughts to other elements of sound, musical balance is deliberately at the end of our examination of how we shape the mix.

Balance is creating the appropriate proportions of sounds, the balance of some sounds being more prominent than others in appropriate proportions. Appropriate for the piece of music, appropriate for the instruments and musical ideas, appropriate at that moment. Musical balance is balancing loudness levels of sound sources.

The dynamic shapes of musical ideas, and the changing dynamic relationships of musical parts, are part of the performances that are captured while recording tracks. Musical balance provides opportunities for creative expression in live performance, and these carry into recorded music—where they can be exaggerated or diminished. During performance, dynamics are altered and shaped by players. Their changes in loudness levels are among the staples of musical expression, and they can add direction to the musical idea and the character of the line.

Drama is often created in the crossing of dynamics between musical parts. A competition for dominance in the texture, or simply by exchanging the position of being loudest, can add significant dimension to the music. The subtle aspects of musical ideas gradually or suddenly changing loudness levels, by large or small degrees, are important parts of the music. These are aided and honored in the recordist's work in shaping musical balance.

It is important to understand that proportions of loudness bring all of the other elements into play. For example, making a sound softer will impact its sound quality, as low-level components of the sound may be lost; it may also impact the perceived spectrum, as partials may become more attenuated than the remainder of the sound at certain levels (see Figure 1.13). Other outcomes as sounds become softer are also entirely possible, especially related to distance, when timbral detail is no longer audible, and imaging, when the edges of phantom images may become altered.

Increasing the loudness of a source even slightly will often have consequences well beyond making the individual source more prominent. Increased loudness of a source can disrupt imaging and timbral balance when comparing "balances" within other elements. Increasing the loudness of a source can have a significant impact on the overall level of the program. When changing a level at a certain moment in time, that setting will continue throughout the remainder of the program, unless it is changed; further on in the song a dramatic rise in performed dynamics, or in the increase of spectra energy because of increased performance intensity, can cause the level to become out of proportion through being too high, and an adjustment may be needed, or it may become so high it creates clipping of the channel (or of the overall program).

Musical balance does not need to be altered to change the balance of prominence between sounds. Remembering this, we can avoid some of these unwanted results of altering loudness. We can seek to make sounds more or less prominent through other elements, and to engage in shaping all of the elements of sound to support the mix and to shape the music.

We will now move into planning a recording project, in all of its dimensions, in the next chapter. Planning and preproduction is a direct move toward tracking, which begins the shaping of the recording and crafting of the mix.

# 15 Preproduction and Preliminary Stages: Embracing Reality and Defining the Materials of the Project

The preliminary stages of the project will define the musical ideas of the project, many qualities about the music and the recording, and move toward making the proper production decisions to best capture or realize those ideas. The realities of artistic, technical, logistical, and conceptual concerns will all be considered and addressed.

A recording project will start with identifying the music to be recorded. This may seem obvious, but it is not so simple as the statement implies. In an album project, this will often include how the songs in the album relate to one another, and perhaps some thoughts of song sequencing. Some songs will fit the project, and others might not. A song might emerge as a central theme or an anchor to the project. An overall concept for the album might begin to emerge; certainly an overall impression and sound quality will take shape during these preliminary stages. This overall quality is often not articulated between those involved, but it is certainly present within their basic understanding of what the project is trying to communicate or achieve musically—its sound.

This overall quality will become focused while making decisions on what is trying to be accomplished musically—how to best deliver the story of the text and music. While the songwriter/composer, artists, producer, and others are involved in the project and make musical decisions, the recordist needs to place those musical ideas into the appropriate context of the recording. The recording process will impact the music and its message. How this will occur, and the aesthetic of the recording, will start to take shape at this beginning stage.

The recording process will shape the overall quality of the project and songs by establishing its overall texture. Remember, this is composed of:

- Form
- Perceived performance environment
- Pitch density and timbral balance
- Program dynamic contour
- Reference dynamic level
- Sound stage (lateral sizes and locations, and distance locations of sources)
- Listener-to-sound-stage distance and relationship (level of intimacy or connection)
- Environments of sound sources
- Musical balance of sources
- Sound quality

At the preliminary stages, these elements will begin to take shape. Decisions will be made that will influence and determine how these elements will appear in the final recording. Some of these decisions will not be able to be reversed after tracking has begun.

The recordist will need to consider the project from a variety of perspectives during these preliminary stages. The highest level of perspective will often be a suitable starting point: the overall concept and overall shape of the composition, how the song will progress dynamically as it unfolds, its reference dynamic (intensity) level, etc., as listed above. These considerations will take shape as they are built from the materials and relationships that exist at lower levels of perspective. Bringing focused attention to them now will guide the crafting of the most subtle sound qualities of the lower levels of perspective that combine to shape the overall sound.

The sound qualities and the relationships of groups of sound sources and of individual sound sources will shape these overall qualities. For instance, as the recording is being planned, thoughts of how the lead vocal should "sound" will emerge. The recordist will formulate a sound in their imagination that they are trying to obtain; this sound will have the dimensions of timbre/sound quality, which should include thoughts of environmental characteristics, and it will have performance intensity, a sense of the required level of connection to the listener (distance location) to best communicate the message of the text, a breadth (size) and location of the voice on the sound stage. These decisions will directly impact the song's sound stage, pitch density, reference dynamic level, perceived performance environment, expression, intensity, etc. As other sources or groups of sources are defined, the overall qualities are further refined.

The sound qualities of all of the sound sources can be captured or crafted in many ways. Explored below, synthesis techniques and microphone selection and techniques contribute greatly to this process. The inherent sound qualities of all devices and processes of the signal chain also contribute to crafting the sound quality of individual sources, and the overall recording, and need to be understood and recognized.

Finally, the playback system and the listening environment can alter how the sound qualities of the recording are heard. Inaccurate recordings can easily result. At this preliminary stage of the project, it is important to recognize the characteristics of one's monitor system. Accurate sound reproduction is needed in order to control the recording process, ensure the quality of the signal, and to accurately recognize and understand the sound qualities and relationships that are being crafted.

The preliminary, preproduction stage will consider, and work to define:

- The music and its desired qualities
- The recording aesthetic to best achieve those qualities
- How the recording process will contribute to shaping the music
- How the recording process will unfold
- The equipment and technologies most suitable to the desired sound qualities of the project.

Preproduction is about planning and preparing: selecting music, rehearsing, arranging and scoring; working with artists to decide how the recording will be made (all performing simultaneously, overdubbing, click track, etc.). Decisions in the studio are much more expensive, and creativity on the clock is inhibited, so preparing artists, especially debut artists, for the special environment, procedures, and time concerns of the studio is important. Preproduction brings knowledge of how the project will progress and an overall vision of the project; details will be worked out as the production progresses.

## Sound Sources as Artistic Resources, and the Choice of Timbres

The music of a project and the sound sources (voices and instruments) to perform the music are usually determined for the recordist. Clients usually dictate what is recorded and who performs. Many subtle decisions on the selection of sound sources and their qualities are, however, often made during the preliminary stages of the recording process (and during the production process itself). Sound sources are selected, created, or shaped, and in all instances, their selection is an important decision in shaping the sound of a recording.

Selecting qualities of sound sources is significant, as sound sources deliver the materials of the music production. They are vehicles for presenting musical ideas. They must be selected and shaped carefully and with attention to their anticipated roles in the production. Sound sources are often coupled with the musical ideas themselves; the musical idea and the sound quality of the source are often melded into a single artistic impression.

In selecting sound sources, decisions are being made as to the most appropriate timbre to present the musical materials. In doing so, the sound source is considered in relation to:

- Suitability of its sound quality to the musical ideas
- Potential to deliver the required creative expression
- Pitch-area information for anticipated placement in relation to timbral balance
- How it might appear in musical balance, on the sound stage, and with environmental characteristics.

In making these projections and evaluating sources, the recordist will listen at a variety of perspectives. Focus will shift from overall qualities of the source, to its dynamic envelope, spectral content, and spectral envelope, to perhaps the subtle characteristics of spectrum changes that might appear when the sound is combined with other sounds, as examples. Sound-quality evaluations performed quickly to make general observations will assist this process greatly.

### Performers as Sound Sources

Recordists utilize the sound qualities of the performers as well as the instruments themselves. Individual performers may be selected because of their unique sound qualities. Conversely, the musical materials themselves are often molded to best suit the qualities of the performer and her/his sound. Individual performers, themselves, are unique sound sources. This is especially true of vocalists, who have a unique singing voice and style, as well as a unique speaking voice.

Accomplished instrumental performers who have developed their own style(s) of playing, or who are skilled in performance techniques, are also sought for their unique sound qualities. Individual performers often bring their own creative ideas and special performance talents to a project, and considerably aid the defining of the sound qualities of the sound sources.

The act of selecting particular performers for a recording, or best utilizing the performers involved, is important for defining the sound quality of the sound source down to the minutest detail. Since the recording is a permanent performance of the piece of music, getting the right artists for this performance is an important consideration in determining the sound qualities of the sound sources.

### Creating Sound Sources and Sound Qualities

The sound manipulation and generation techniques of sound synthesis allow sound sources to be created with the design of new timbres. New sources can be invented with new sound qualities. Some synthesis techniques generate sound in its entirety, and others build on preexisting sounds. Many approaches to synthesis exist, and can be studied elsewhere.

The creation of sound sources allows the recordist great freedom in shaping sound qualities. The recordist will be functioning as a *sound designer*, whose goal is to create a sound (with a sound quality) that will most effectively present the musical materials and ideas of the music. Sound sources must precisely suit the contexts of the song.

While an examination of the sound-synthesis process is out of the scope of this writing, it is important for the recordist to be aware of the many creative options afforded by sound synthesis. The study of sound synthesis from the perspective of building timbres will greatly assist the recordist in understanding the components of sound, and how the components of sound may be used as artistic elements. This process will also help develop skill in recognizing the subtle qualities of spectrum, spectral envelope, and dynamic envelope.

It is important to note that by inventing sound sources, the recordist will be presenting the audience with unfamiliar "instruments." The sources (new instruments) may be performing significant musical material, and the reality of the performance will have been altered out of the direct experience of the listener. The recordist will create a new reality of sound relationships, or might emphasize known sound relationships to reestablish known experiences. These relationships need to be accomplished in such a way as to support the musical materials and ideas of the recording. Adding environmental characteristics to synthesized sounds can provide life and realism to these sounds—and pull them into the experiences of the listener.

### Nonmusical Sources

Nonmusical concepts often find a place in a music project. As sound sources, speech and special effects require special consideration.

With speech as a sound source, the quality of the voice needs to complement the meaning of the text and to complement the other sounds in the musical texture. The voice is carefully crafted for the appropriateness of its sound quality, and thus its dramatic or theatrical impact, to the meaning of the text to be recited.

Special effects are sound sources that are used to elicit associated responses or thoughts from the listener. The associated thoughts generated by the special effects are typically not directly related to the context of the particular piece of music. Special effects can pull the listener out of the context of the piece of music, to engage external concepts or ideas. A horn sound occurring in a piece of music is an effect, used to elicit the mental image of an automobile. The same sound used as part of the musical material, used to complement the musical ideas of the work, would not be a special effect, but rather a musical sound source.

## Microphones: The Aesthetic Decisions of Capturing Timbres

Planning microphone selections and placements is an important aspect of preproduction.

The sound qualities of instruments and voices are shaped, and can be significantly transformed, when captured by a microphone. The selection of a particular microphone, placing the microphone at a particular location, and how these complement the characteristics of the particular sound source, will determine the actual sound quality of a recorded sound.

A specific microphone will be selected to make a certain recording of a sound source, because of the ways its performance characteristics complement the sound characteristics of the sound source. This interaction of the characteristics of the microphone and of the sound source allows the recordist to obtain the desired sound quality of the recorded sound.

The recordist needs a clear idea of the sound quality sought. With this, the recordist will determine which microphone is most appropriate by comparing the characteristics of the sound quality of the sound source and the performance characteristics of the microphone, to their vision of the final sound quality sought. Perhaps no other decision shapes sound quality more than microphone selection and placement.

A sound source's sound quality (dynamic envelope, spectrum, and spectral envelope), potentials for distance location (level of timbral detail), and perhaps environmental characteristics (that might be captured from the performance space) are all determined in this process. In future processes, some aspects can be altered (such as adding a compressor to alter the sound's dynamic envelope), some elements accentuated (such as additional reverberation), and some elements deleted or attenuated (such as filtering out high frequencies). Some qualities must be captured during the initial recording that cannot be added later—for instance, a high degree of timbral detail must be captured by appropriate microphone selection and placement during the initial recording if the source is to be at a very close distance to the listener in the final mix, as this detail cannot be created later. The distance cues of the final sound stage are significantly determined by the timbral detail captured by the microphones used.

Microphone placement can be as critical as microphone selection in shaping sound. Where a microphone is placed is carefully calculated against the performance characteristics of the microphone and how the instrument/voice produces its sound. Often placement options are impacted by the recording environment and unwanted sound qualities, or by isolating the microphone from sound sources the recordist does not wish to record.

No single microphone will be the "best microphone" for every sound source or for the same source for every piece of music. The microphones selected for recording the same sound source may vary widely depending on the above circumstances, the desired sound, and what microphones are available.

## Performance Characteristics

All microphones can be evaluated by their *performance characteristics*. These characteristics are information on how the microphone will consistently respond to sound. Thus, through these characteristics, the recordist can anticipate how the microphone will transform the sound quality of the sound source while it is being recorded.

As the microphone alters the sound source, it has the potential to contribute positively in shaping the artistic elements. If the recordist is in control of the process of selecting the appropriate microphone for the sound source and conditions of the recording, the selection and applications of microphones can be a resource for artistic expression. The artistic elements of the recording can be captured (recorded) in the desired form, and the microphone and its placement will become part of the artistic decision-making process.

A number of microphone performance characteristics are significant and prominent in shaping the sound quality, which is of central concern in determining the artistic results of selecting these microphones for certain sound sources. These microphone performance characteristics are:

- Frequency response
- Directional sensitivity
- Transient response
- Distance sensitivity.

Subtle transformations of the source's sound qualities take place during this process. They can be experienced by comparing the sound of the instrument in the recording room to the sound of the signal in the control room. To hear these differences, focus on dynamic envelope, spectrum, spectral envelope, and proportions of direct-to-reverberant sound for changes. These will reveal many of these microphone performance characteristics. The recordist will ultimately learn the "sound" of specific microphones, and learn how to use them to their best advantage.

#### *Frequency Response*

*Frequency response* is a measure of how the microphone responds to the same sound level at different frequencies. Amplitude differences at various frequencies can appear at the microphone's output and define the sensitivity of the microphone to frequency.

The frequency response of a microphone often has frequency bands that the microphone accentuates or attenuates. The matching of a sound source with similar frequency characteristics may or may not provide the recordist with the desired sound. The microphone may cause accentuation of certain characteristics of the sound source, and perhaps a microphone with somewhat opposite frequency characteristics to the sound source will be a more appropriate choice. Again, this decision is dependent upon the final, desired sound. Figure 15.1 presents a frequency response for a hypothetical microphone that slightly emphasizes the frequency band from approximately 2 kHz to 5.5 kHz, and attenuates frequencies below 100 Hz and above 12 kHz.

Microphone frequency response adds new formant regions to the sound source. The frequencies emphasized and/or attenuated by the microphone act on all sounds equally, regardless of pitch level. Microphone frequency response directly shapes, contributes to, and captures the spectrum of the sound source.

#### *Directional Sensitivity*

Microphones do not uniformly capture sounds that arrive from different angles to its diaphragm. The *directional sensitivity* of a microphone is its sensitivity to sounds arriving at various angles to the diaphragm. The *polar pattern* of a microphone depicts the sensitivity of a microphone to sounds at various frequencies in front, in back, and to the sides, and the actual pattern is spherical around the microphone (Figure 15.2). Directional response measures the microphone's sensitivity to sounds

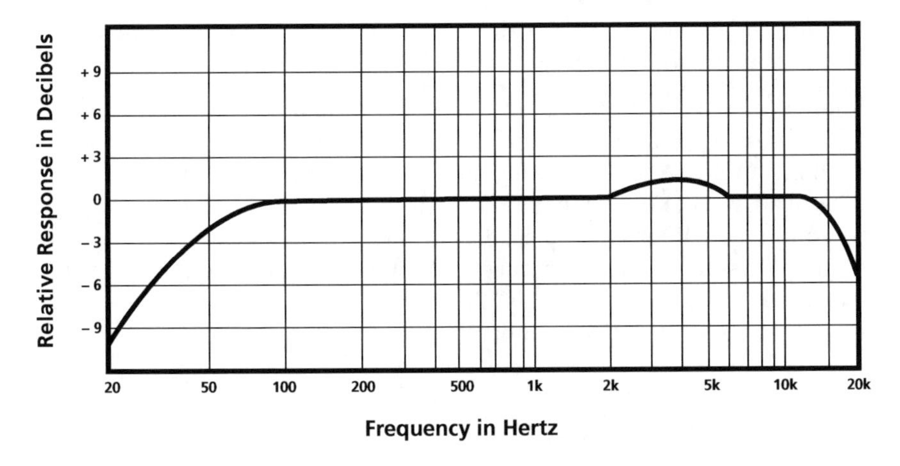


Figure 15.1
Frequency response of a hypothetical microphone.

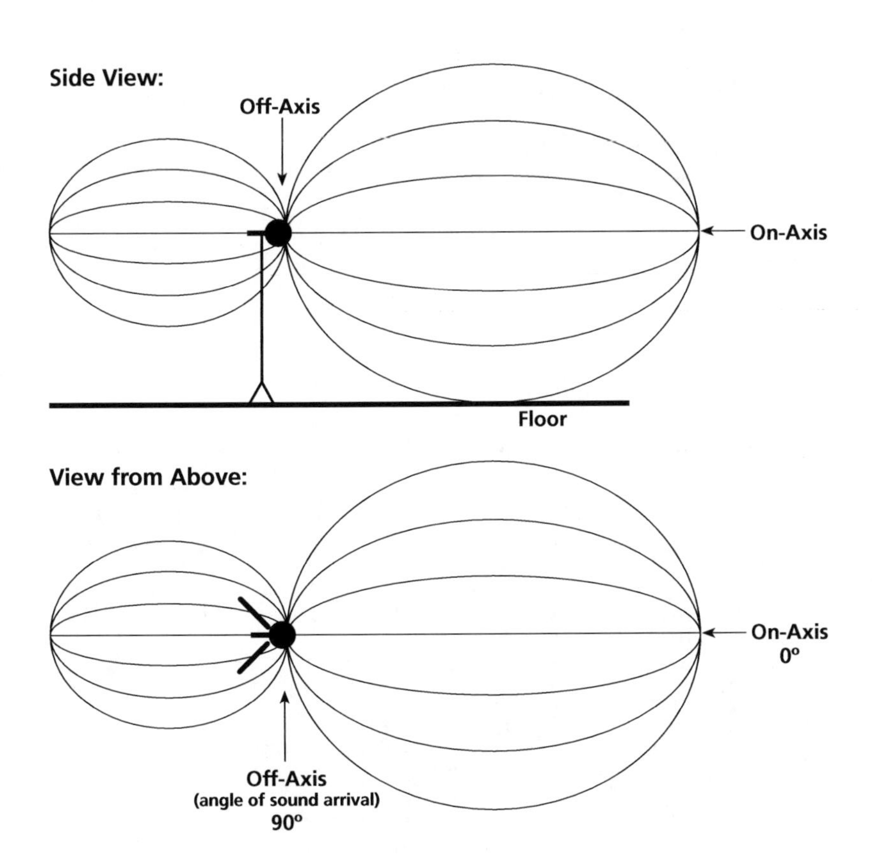


Figure 15.2
Polar pattern spheres and the microphone axis.

arriving from angles, but calculates this sensitivity at only a few frequencies (Figure 15.3).

Sounds directly in front of the microphone diaphragm are considered to be *on-axis*. Sounds deviating from this 0° point on the polar curve are considered *off-axis* and are plotted in relation to the on-axis reference level. The frequency response of most microphones will vary markedly to sounds at different angles. Even microphones that show no pronounced frequency areas of accentuation or attenuation (flat frequency response) on-axis will show an altered frequency response at the sides and the back of the polar pattern (Figures 15.3 and 15.4).

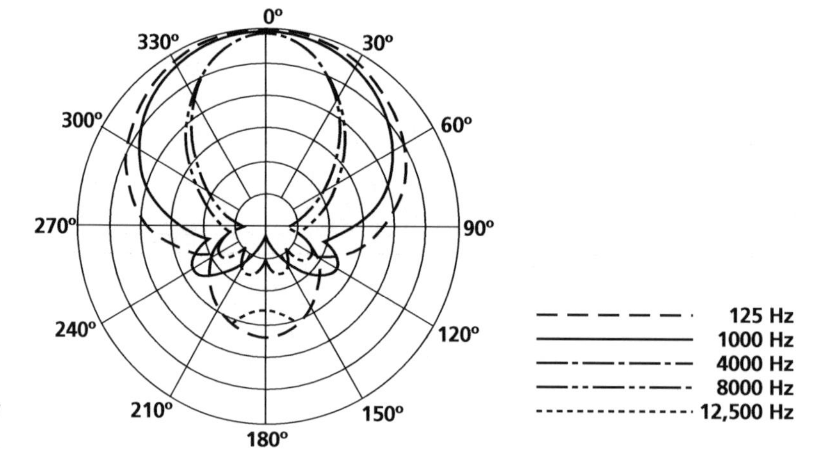


**Figure 15.3**
Polar pattern sensitivity at various frequencies.

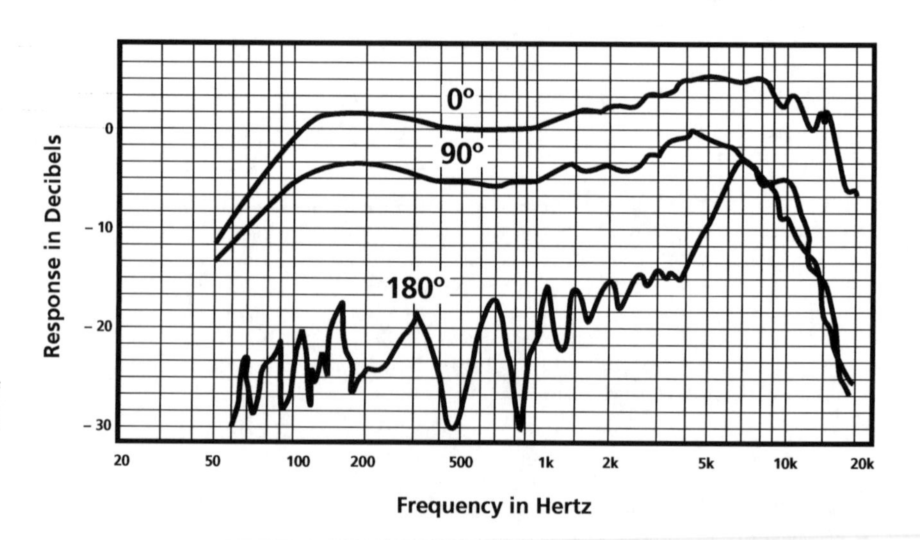


**Figure 15.4**
Frequency response of a hypothetical cardioid microphone on-axis (0°) and off-axis (180° and 90°).

These variations in frequency response at different angles are commonly called *off-axis coloration*. This coloration alters sound-source spectrum, just as on-axis frequency response accentuates and attenuates specific frequency bands. These changes are more pronounced at or toward the attenuated angles of the microphone patterns. In the intermediate angles between directly on-axis and the dead areas of directional patterns, a slight change in the angle/direction of the microphone can make a substantial difference in the frequency response of the captured sound source. Frequencies above 4 kHz are usually most dramatically altered by slight angle changes.

The amount of off-axis coloration is an important measure of the microphone's suitability to a variety of situations. This is especially pronounced in stereo microphone-array recording techniques. Instruments at the edges of the array's pick-up pattern and the reverberant sound of the hall will arrive at the array mostly from angles that are off-axis. The sound qualities of those instruments and of the reverberant energy may be altered significantly by the off-axis coloration. Off-axis coloration has the potential to have a profound impact on the sound qualities of the recording.

### *Transient Response*

Microphones do not immediately track the waveform of a sound source. A certain amount of time is required before the applied acoustic energy is transferred into accurate movement of the microphone's diaphragm. Also, a certain amount of acoustic energy must be applied to initiate this movement, and some of this energy (which is the sound wave of the sound source) can be dissipated in the action of getting the diaphragm moving. Diaphragm mass inhibits transient response, making condenser and ribbon microphones more accurate than even the highest-quality dynamic microphones.

Thus, microphones have different response times before they will begin to accurately track the waveform of the sound source. This *transient response* time distorts the initial, transient portions of the sound's timbre. Slow transient response is most noticeable when the microphone is applied to a sound source that has a fast initial attack time (in the dynamic envelope) and a significant amount of spectral energy during the onset.

Figure 15.5 is a comparison of the transient response of typical condenser and dynamic microphones. It shows that the condenser microphone traces the initial transient of the sound more accurately than the dynamic microphone. The rest of the signal is reproduced at about the same accuracy by either microphone, and is lower in frequency and amplitude. In this way, transient response is often related to frequency response.

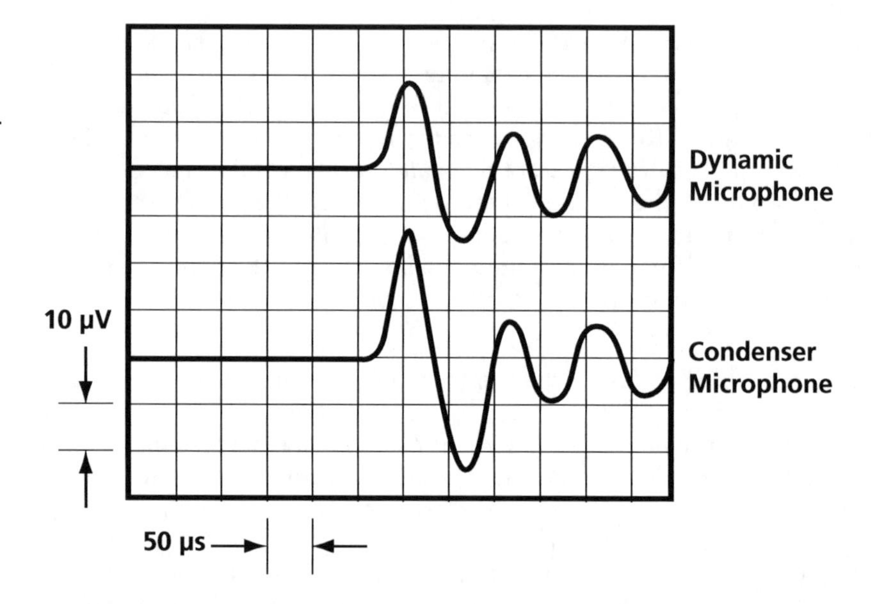


**Figure 15.5**
Transient response of typical condenser and dynamic microphones.

Transient response is a microphone performance characteristic that is not included in the manufacturer specification information that accompanies promotional literature and owner's manuals. Further, it is not calculated by a standard of measurement. It exists as an alteration of the original waveform's initial attack time in both the dynamic envelope and the spectral envelope.

The developed hearing of the recordist can only be used to evaluate this microphone characteristic. The recordist must become aware of differences in timbre that are present in the early time field of the captured (recorded) sound source, as compared with the live sound source. These differences are identified through the critical-listening process and are vitally important to the recordist. Recognizing and understanding how the transient response is altering the source's sound quality will allow the recordist to be in control of capturing and shaping the sound as desired.

Two microphones with identical frequency-response curves may have very different sound characteristics caused by different transient response times.

### *Distance Sensitivity*

All microphones will respond differently to the same sound source, at the same distance and angle. The ability of each microphone to capture the detail of a source's timbre will be different. Microphones will have different sensitivities in relation to distance. The *distance sensitivity* of a microphone is often influenced by the polar response of the microphone and/or its transduction principle (condenser, moving coil, ribbon, etc.).

Directional patterns often are able to capture timbral detail of a sound source at a greater distance than an omnidirectional microphone. This is primarily the result of the ratio of direct to indirect sound, and a masking of timbre detail, but it may also be attributed to the transduction principle, depending on the particular circumstances. Similarly, condenser microphones will have a tendency toward greater distance sensitivity than dynamic microphones, due to their more exacting transfer of energy. Many times a microphone with a small-sized diaphragm will have greater distance sensitivity than a microphone with a larger diaphragm, all other factors being equal.

The concept of the distance sensitivity (sometimes called "reach") of a microphone is an important one. Recordists must also judge this microphone characteristic through acute listening and experience. One must become aware of differences in timbre detail that are present between the miked sound source and the live sound source. Distance sensitivity is a microphone performance characteristic that is not included in the manufacturer-specification information, although it might be measured scientifically if an appropriate scale were devised. Distance sensitivity is a characteristic that must be learned from experience and must be anticipated for the individual environment and recording conditions.

## Microphone Placement

Many variables must be considered during the process of selecting and placing microphones. The primary variables for microphone selection were presented above. The variables of *microphone placement* will directly influence the selection of a microphone, even after an initial selection has been made. The placement of the microphone in relation to the sound source and the performance environment will greatly influence the sound quality of the recording. At times, this influence may be as great as the selection of the microphone itself.

The recordist must consider the following when deciding on placing the microphone in relation to the sound source:

- How sound is produced by and how it radiates from the sound source
- Distance relationships between the microphone, the sound source, and the reflective surfaces of the recording environment (performance space)
- Distance of the microphone from the sound source to be recorded and other sound sources in the performance space
- Height and lateral position of the microphone in relation to the sound source

- Angle of the microphone's axis to the sound source
- Performance characteristics of the microphone selected, as altered by the above four considerations

Microphone placement and microphone performance characteristics interact in complementary ways. The distance of the microphone will be largely determined by the distance sensitivity of the microphone. The angle of the microphone will be largely determined by the frequency response of the microphone in relation to its polar pattern, the characteristics of how the sound source projects its sound, and the characteristics of the environment. The height of the microphone is also a result of the directional characteristics of the sound source (as instruments radiate different spectral information in different directions), the environment, and the microphone's frequency response and distance sensitivity.

**Listen**

to tracks 39–41

for differences in the sound quality of the three recordings of the same performance; these differences are all due to microphone selection (performance) and placement considerations.

***Controlling the Sound of the Performance Space***

The sound characteristics of the studio or performance space can be captured simultaneously with recording the sound source. The recordist may or may not wish to include the sound of this space as part of the sound quality of the sound source. In either instance, the recordist must be in control of the indirect sound of the environment arriving at the microphone from reflective surfaces.

The recordist may seek to record the sound of the sound source within its performance environment. For this to happen, there must be a control of the balance of direct and indirect sound. Through the selection and placement of a microphone with a suitable polar pattern and distance sensitivity, the desired characteristics of the environment may be captured, in the desired amount, along with the sound of the sound source. The distance cues of the initial reflections in the early time field and the amount of timbre detail will be evaluated and weighed against the ratio of direct to indirect sound. Distance and angle of the microphone will be adjusted to achieve the desired sound. When this technique is used, individual sources must be well separated or recorded separately in order to be isolated.

The recordist's objective may be to capture the sound source without the cues of the environment. The sound source may be physically isolated (with portable baffles—gobos—or in isolation booths) from other, unwanted sounds from the environment and other sources, or the leakage of unwanted sounds to the recording microphone might be minimized with microphone pattern selection and microphone placement. This is accomplished through close-miking techniques and direct inputs. It allows the flexibility of being in complete control of the sound source, with environmental characteristics later applied to the sound through signal processing.

Capturing the sound of the environment will be controlled by the relationships of the microphone to the sound source and any other sound sources that may be occurring simultaneously, including the sound of the environment itself.

### *Reflective Surfaces*

The *reflective surfaces* of the recording space (or of any object in that environment) can cause the sound at the microphone to be unusable. Interference problems may be created when the sound from a reflective surface and the direct sound reach the microphone at comparable amplitudes. The slight time delay between the two sounds will cause certain frequencies to be out-of-phase (with cancellation of those frequencies) and certain frequencies to be in-phase (with reinforcement of those frequencies).

The frequencies that will be accentuated and attenuated can be determined by the difference of the distance between the reflective surface and the microphone ($D_1$), and the distance between the sound source and the microphone ($D_2$), in relationship to air velocity. The amount of reinforcement and cancellation of certain frequencies that will occur when the two signals are combined will be determined by their amplitudes. Constructive and destructive interference are most pronounced when the two signals are of equal amplitudes. As the difference in amplitude values between the two signals becomes larger, the effect becomes less noticeable.

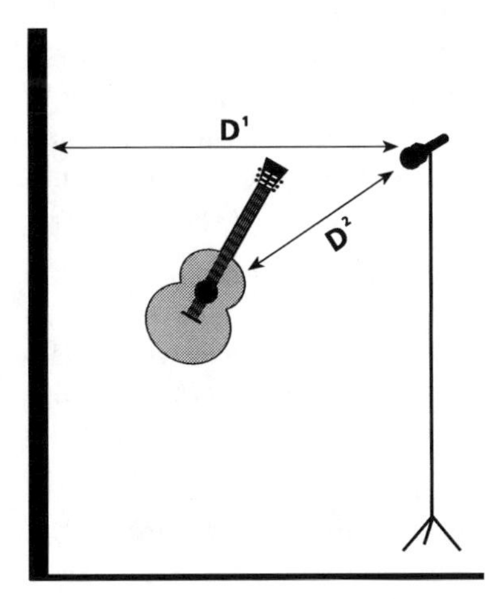


**Figure 15.6**
Distances between microphone, sound source, and reflective surfaces.

The constructive and destructive interference of the combined signals results in a frequency response with emphasized and attenuated frequencies (peaks and dips). These peaks and dips of the frequency response curve of the sound's spectrum have been compared, in analogy, to the tines of a comb. Thus, the term *comb filter* has been applied when a signal is combined with itself, with a slight time delay between the two signals.

### *Microphone Positioning*

The distance of the microphone to the sound source will impact the sound quality of the recorded sound source. This sound quality will also be altered by the distance of the microphone to other sound sources in the environment, along with the height, lateral position, and angle of the microphone. These are variables that may be used as creative elements in shaping sound quality (and the musical material), in the same way as microphone selection. These positioning variables will most significantly influence the following aspects of sound:

- Environmental characteristics (and secondary distance) cue—ratio of direct to reflected sound
- Environmental characteristics cue—early time-field information created by initial reflections
- Distance cue (definition or amount of timbral detail)
- Capturing of the blend and shaping the overall quality of the source's timbre

The distance from the microphone to the sound source is the primary factor that controls the ratio of direct to reflected (indirect) sound. As the microphone is placed closer to the sound source, the proportion of direct sound increases in relation to reflected sound.

Reflective surfaces that are located near the sound source will cause inconsistencies with this general rule. If at all possible, sound sources should be placed at distances from any reflective surface at least three times the distance from the microphone to the sound source. The height and angle of the microphone in relation to the sound source, coupled with its polar pattern, may allow the microphone to keep from picking up the reflected sound off surfaces close to the sound source.

Distance cues can be established by microphone placement when an audible amount of sound from the recording environment is present in the source's sound quality. While this is not usually the case with close-miked sound sources, many sound sources in multitrack projects are recorded from a distance that will capture certain information from the recording environment.

All distant-miking techniques will capture a significant amount of the performance environment's sound.

Recordings made with a moderate distance between the microphone and the sound source will likely have prominent reflection information in the early time field. The floor and objects immediately around the sound source will create reflections that arrive at the microphone at a similar time to the direct sound. While only a few reflections may be present in the final sound, and the reflections may be of significantly lower amplitude than the direct sound, they will impart important environment information. These reflections can provide environmental cues that could distort a reverb program applied to the sound.

Microphone placement location and performance characteristics will play vital roles in establishing distance cues, through defining the amount of timbre detail present in the recorded sound source. This is important to remember, especially if placing a source in "proximity" is planned. This timbral detail must be captured in the recording process, as it cannot be created later.

Generally, the recorded sound source will have greater timbral detail the closer the microphone to the sound source, and/or the more sensitive the microphone in terms of distance sensitivity and transient response. This provides a distance cue that may place the listener near the reproduced image of the recording, sometimes unnaturally near the source. This level of timbral detail directly establishes the level of intimacy of the recording. The depth of sound stage will also be greatly determined through these cues.

Distance cues are often contradictory when a high degree of timbre detail and a large amount of reverberant energy are present simultaneously. This unrealistic sound may be desired for artistic purposes, and has been used to great effect.

The sound quality of the sound source may potentially be altered by close microphone placements. As above, these alterations can be used to creative advantage, or they can interfere with obtaining the desired sound quality. The sound quality may exhibit the following alterations when the source is close miked:

- An increase of definition exceeding known, naturally occurring degrees of timbre detail
- Altered spectral content of the sound source
- An unnatural blend of the source's timbre, caused when the source has not had enough physical space to develop into its characteristic sound quality, or microphone placement does not capture portions of the sound radiated from the instrument
- Noises from the performer or from the instrument

A cardioid or bidirectional microphone placed within two feet of a sound source may have an altered frequency response, and thereby alter the

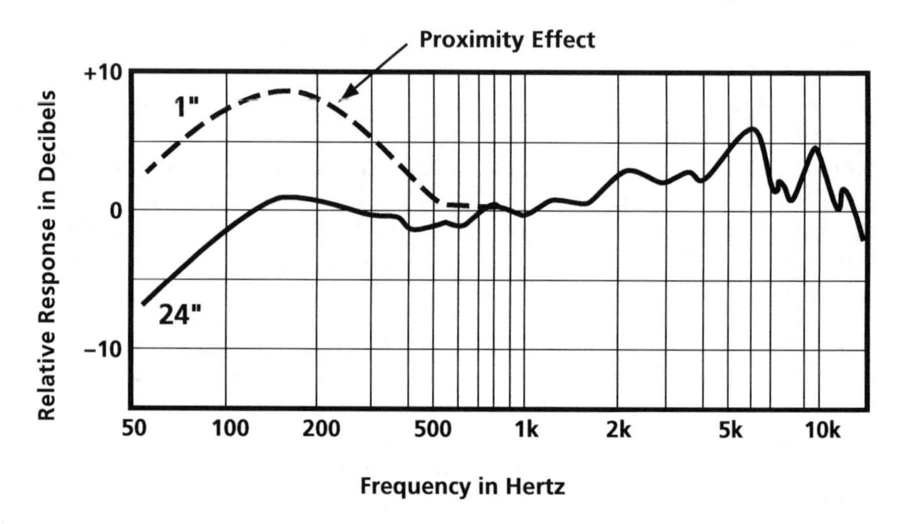


**Figure 15.7**
Altered frequency response of a sound source at 1 inch from microphone, caused by the proximity effect.

source's spectral content. This altered frequency response of certain microphones is the *proximity effect.* The response in the low-frequency range rises relative to response in higher frequencies. Figure 15.7 demonstrates the low-frequency boost of a sound source at 1 inch (dotted line) compared to the source at 24 inches.

Performers and their instruments can create noises—noises from lips, fingers, feet, frets, chairs, picks, sticks, keys, pedals, valves, and so many other potential sources. With a more distant microphone placement certain noise sounds may become acceptable parts of the sound. With increasingly closer placement, instrument or performer noise becomes more pronounced, and the location or the angle of a microphone can make it impossible to obtain a desirable sound.

### *Blend*

The sound source and the way it radiates sound must be considered in relation to the sound's environment. All sound sources radiate sound in a unique way. A polar pattern similar to a microphone's is created as sources radiate different frequency-response curves in different directions.

Sound sources require physical space for the sound quality to develop and coalesce. The sound quality of instruments and voices is a combination of all of the sounds they produce. When a microphone is placed in a physical location near a source, it can be within this critical distance necessary for the sound to develop into its characteristic single sound wave. The sound source might not have the opportunity to *blend* into its unique, overall sound quality. The microphone might capture only a portion of the sound. This resulting sound quality can be very different from how the sound source exists in acoustic environments. It is important that the recordist be aware of this space and in control of recording the desired blend of the source's sound.

Generally sounds need several feet of space for their sound quality to form. This distance is too far for a microphone placement, which would isolate the source from other sources performing at the same time. Close microphone techniques change the sound quality of sources, sometimes markedly. This can be used for great creative advantage, or to the detriment of the project. Often several close microphones are used on a single instrument to "manufacture" the blend of an instrument, and to maintain the isolation of sound sources. The recordist blends the various parts of the source's sound instead of relying on the sound source's natural blend. Pianos are commonly recorded in this way. This concept can be applied to many sound sources, such as the cello and guitar tracks on the companion website.

## Stereo and Surround Microphone Techniques

Stereo and surround microphone arrays are composed of two or more microphones (or diaphragm assemblies) in a systematic arrangement. They are designed to record sound in such a way that upon playback (through two channels or surround) a certain sense of the spatial relationships of the sound sources present during the recorded performance is reproduced. These techniques have the potential to accurately capture the sound qualities of the live performance and of the hall, the performed balance of the instruments of the ensemble, and the spatial relationships of the ensemble (location and distance cues), with minimal alteration but with subtle and unique qualities.

Many microphone techniques have been developed. A few of the most commonly used stereo techniques are:

- *X-Y* coincident techniques
- Middle–side technique (M–S)
- Blumlein *X-Y* (crossed figure-eights)
- Near-coincident techniques (NOS and ORTF)
- Spaced omnidirectional microphones
- Spaced bidirectional microphones
- Binaural system (artificial head)
- Sound field and other specialized microphones and systems.

Among currently used surround microphone techniques are:

- Sound field microphone (coincident array)
- Double MS: four coincident cardioids (coincident array)
- Decca Tree: spaced omnidirectional microphones

- Fukada Tree: five cardioid microphones and two additional omnis
- Hamasaki Square: four figure-eight microphones
- IRT Cross: four spaced cardioids, surround ambiance
- Frontal arrays, plus ambiance microphones.

Stereo and surround microphone techniques are often used in recording large ensembles, in large acoustic spaces, and from a rather distant placement. The techniques are very powerful in their accuracy and flexibility, and may be applied to either a single sound source or to any sized ensemble. They may be used from a rather distant placement (perhaps 15 or more meters, depending on the pertinent variables), to within about a meter of the sound source. Close placements of stereo arrays are commonly applied to drum sets, for example, sometimes supplemented with accent microphones, sometimes not.

Stereo and surround microphone techniques can be thought of as a preprocessing of the recorded sound. Preprocessing alters the source's sound quality before it reaches the routing and mixing stages of the recording chain. Microphone techniques will add and capture spatial information to the recording.

**Listen**

to tracks 50 and 51

for unaltered stereo microphone technique recordings.

Microphone arrays can significantly alter the sound source in a number of ways. All stereo and surround microphone techniques have their own unique characteristics and their own inherent strengths and weaknesses. The inherent sound qualities of the microphone arrays can be used to great advantage if the recordist understands and is in control of their sound qualities.

The following artistic elements are created or captured, with varying characters and realism, by the stereo and surround techniques listed above. By evaluating the sound qualities of each of the various arrays, their unique qualities can be learned.

- Perceived listener-to-sound-stage distance
- Amount and sound quality of the environmental characteristics of the performance space
- Perceived depth of the ensemble, sound source, and/or sound stage
- Perceived width of the sound source or the sound stage
- Definition and stability of phantom images and distance imaging
- Musical balance of the sound sources in the ensemble
- Sound qualities of the entire ensemble or of specific sources within the ensemble.

The microphone placement variables are also factors in microphone techniques, as they help determine where the array is to be placed.

Placement of the array will considerably impact the quality of sound, as will the microphones selected for the microphone array, as they impart their own unique sound characteristics, as described above. Matching specific microphones (type of transducer or model/manufacturer) to the stereo or surround microphone array is also important in ensuring the effectiveness of the array's capturing of the desired sound qualities of the performance.

The aesthetic approach to the project will guide how the microphone technique is selected and used. Techniques can provide a clear documentation of a performance, can enhance a performance, and can impart significant and unique characteristics to a recording.

The recordist will often envision an *ideal seat* for the performance to assist in determining the placement of a microphone array. This type of placement of the array will seek the sound qualities that are most desirable for the particular ensemble, performing a specific piece of music in the performance space. The recordist will seek to balance the hall sound with that of the ensemble, capture an appropriate amount of timbre definition from the ensemble, retain all performed dynamic relationships, and establish desirable and stable spatial relationships in the sound stage. The balance of the total energy of the direct sound and that of the reverberant sound will be carefully considered in determining the placement of the microphone array. The point where the two are equal has been identified as the *critical distance*. The recordist will focus on this ratio of the reflected to direct sound to identify a desired balance.

*Accent microphones* are often used to supplement stereo and surround microphone techniques. These are microphones that are dedicated to capturing a single sound source, or a small group of sound sources, within the total ensemble being recorded by the array. The accent microphones are placed much closer to the sound sources than the array and may cause the recordist to consider some of the close miking variables discussed above.

Accent microphones are most often used to complement the array. They assist the overall array by bringing more dynamic presence and timbre definition to certain sound sources in the ensemble, and they allow the recordist some control over the musical balance of the ensemble. Accent microphones also create noticeable time differences between the arrival of the sound source(s) at the stereo array, and the arrival of the sound source(s) at the accent microphone.

**Listen**

to tracks 50 and 37

for an accent microphone added to a stereo microphone technique: track 50 is an ORTF recording of drum set; track 37 adds an accent microphone to the ORTF recording of track 50.

Adding accent microphones will diminish the realistic sound qualities of the microphone technique. The sound relationships of the performance will be altered by the dynamics, sound quality, and distance cues added by the accent microphone(s). The microphone array's ability to accurately capture the performance will be diminished by using accent microphones to alter the sound present in the performance hall.

## Equipment Selection: Application of Inherent Sound Quality

All recording/reproduction devices impart a unique sonic imprint on the sound. Just as was discussed with microphones above, individual recorders, mixing consoles, signal processors, analog-to-digital converters, and all other devices in the signal chain each have unique sound qualities, which are the result of their performance characteristics. This extends to all software that acts upon sound, such as functions within DAWs, plug-ins, etc. They all modify the original signal; some do this almost imperceptibly, nearly transparently, and others quite noticeably, or profoundly. Any individual device will be evaluated for how it transforms the frequency, amplitude, and time components of the original signal. The modifications of the original signal caused by the basic performance characteristics of a device (real or virtual) create its *inherent sound quality*. Further, different technologies have distinct sound characteristics, or the potential to display or produce certain characteristics.

In defining the materials of the project, the specific pieces of recording equipment need to be determined. These will be to realize the desired sound qualities of the individual sound sources and the project's overall sound. The recordist can approach this problem directly by evaluating and understanding the inherent sound characteristics of the available individual devices, and the inherent sound characteristics of the technologies of those devices, then comparing them against the unique needs of the individual project. A compatible match will be sought for the device, technology, and sound source to arrive at a desired sound quality.

Just as with evaluating microphone characteristics, the recordist's focus must shift between critical and analytical listening purposes at a variety of perspectives in order to evaluate the sound qualities of devices. Careful evaluation out of context of a recording will allow traits to be evaluated most easily. This process can be applied to evaluating the sonic differences between any technologies, devices, or software available for the project.

## Technology Selection: Analog versus Digital

The debate over the superiority of analog and digital devices and their inherent sound qualities, and other characteristics, continues to occupy many recordists and listeners alike. Decades of debate and technological advancements have shown both formats can produce stunning recordings, with some unique qualities and with many similar qualities, both "positive and negative." "Positive and negative" are terms best applied to realizing the specific goals of the project, rather than the personal preferences of the individual.

Digital recording, processing, and editing equipment is not necessarily "better" than analog equipment, nor is the opposite true. Both have great potential for artistic expression, and both can generate recordings of impeccable quality. No technology is inherently better suited than another for generating, capturing, shaping, mixing, processing, combining, or recording sound. Some devices may have functional features that are more attractive to an individual recordist, and some people develop personal sound preferences. These are, however, matters of taste. Technologies are simply different in how they sound, how they retain the original signal characteristics, and how they alter the sound source.

Analog technology has certain inherent sound characteristics. Digital technology has certain inherent sound characteristics. The characteristics of one technology may or may not be appropriate for a particular project. Inherent sound qualities are inherent sound deficiencies if they work against the sound quality the recordist is trying to obtain. Inherent sound qualities are desirable if they produce the sound quality the recordist is looking for.

It is difficult to make generalizations as to the characteristics of analog versus digital technology. The sound qualities of both technologies vary widely depending on the particular unit and the integrity of the audio signal within the particular devices. An 8-bit digital system is significantly less accurate and flexible than a 32-bit system. A consumer-model analog system is significantly noisier and less accurate than a professional recorder with high-end noise reduction.

Differences often exist between the two technologies in:

- Accurately tracking the shape of the waveform (especially the initial transients of the sound wave, in both technologies)
- Processing all frequencies equally well (especially frequency-response linearity in analog or quantization issues in digital)
- Storing the waveform without distortion from the medium (especially tape noise floor or A/D and D/A conversion accuracy)

- Altering the waveform in precise increments and precisely repeating these functions (a measure of signal processors)
- Performing repeated playing, successive generations of copying, and long-term storage with minimal signal degradation (a measure of recording formats)
- Noise and distortion added to the signal.

Many other, subtler, differences exist, especially between specific devices of each technology.

### Selecting Devices and Models

No recording device is inherently "better" than another similar device. While certain devices might be more flexible than others, and certain devices are certainly of higher technical quality than others, the primary artistic concern for selecting a piece of equipment is its suitability to the particular needs of the project, at a given point in time.

The advantages of any particular device in one application may be a disadvantage in another. The sound quality of one device may be appropriate to one musical context, and not to another. The measure of the device will be in how its inherent sound qualities can be used to obtain the desired sound qualities of the recording. Pieces of recording equipment should be evaluated for their sound qualities, and their potential usefulness in communicating the artistic message in the piece of music. This evaluation is performed through a critical-listening process similar to that used to evaluate microphone performance characteristics.

Recording equipment (including computers) and software are tools. The tools may be applied to any task, with consistent results. The recordist needs to decide if the particular tool (device or software) is the appropriate one to craft the sound quality that is required of the particular project.

Musicians often carry with them a number of musical instruments. They will use a different model of the same instrument (perhaps made by a different manufacturer) to obtain a different sound quality of their performance, depending on what is required by the musical material. The recordist should recognize this is similar to their situation.

In selecting recording equipment, the recordist is, in essence, selecting a musical instrument. The sound quality of sound sources, or of the entire recording, may be markedly transformed by the piece of equipment, all while under the control of the recordist. This is the way a traditional musical instrument is applied by a traditional performer.

Recordists will develop sound-quality preferences and working preferences for particular pieces of equipment, and perhaps for a particular technology. Developing such preferences may or may not be artistically healthy. The recordist may become inclined to consider a certain technology to be "better" than another simply because it is the one they are most familiar with, not because it is the one that is most appropriate for the project. Personal preferences (or personal experiences) might become confused with the actual quality or usefulness of a device or a technology.

In contrast, one's own "sound" can result by developing one's own production preferences and sound-quality preferences. Equipment selection will contribute much to this when a recordist has strong preferences to use certain devices; how the recordist shapes the artistic elements will also play a significant role. A person's own sound is developed over time, and after considerable experience. How a recordist handles the mix process and the final dimensions of the recording ultimately define their "sound."

In summary, audio devices have inherent sound qualities that are determined by technology and the device's unique performance characteristics. The recordist will be using these devices to shape the sound of the music recording. Learning the inherent sound qualities of as many devices as possible will increase the creative tools of the recordist, and provide them with more options in obtaining the sound they want. These devices are the musical instruments that are used to craft the mix, the recording.

## Monitoring: The Sound Quality of Playback

During these preliminary stages, the recordist must decide on how the recording process will be monitored. Monitoring most often takes place in a recording control room—whether a commercial facility, a home studio, a closet at a remote location, or something else. All of the sounds and relationships of the project are presented to the recordist through the playback system.

The monitor system is much more than a pair of loudspeakers. In terms of hardware, it includes power amplifier(s), loudspeaker drivers, crossover networks, loudspeaker enclosures, decoupling platforms and pads, and even connector cables. The monitor system also encompasses the listening room itself; the placement of the loudspeakers in the room, and the interactions of the room and the loudspeakers become part of the sound quality of the monitor system.

The monitor system has the potential to transform all sound qualities and relationships in the recording. The system must not considerably alter the original signal, or at the very least the recordist must know how the sound is being altered. The monitor system needs to accurately reproduce the spatial qualities and the frequency, amplitude, and time information of the recording.

If recordists are to be in control of the recording process, they must be able to evaluate the recording itself, not the recording modified by the monitor system. The recording will only have the same qualities in another (neutral) listening environment if the sound quality is not originally altered through the recordist's system.

The most desirable monitoring system is transparent. The playback system reproduces sound while interacting with the control room acoustics without altering the quality of sound. In actual practice this is impossible. Still, sound alterations caused by monitor systems can be minimized. High-quality audio systems can be assembled and appropriately located in suitable environments to provide accurate and nearly transparent sound reproduction.

The following considerations must be factored into establishing an accurate monitor system:

- High-quality playback system (loudspeakers and compatible amplifiers)
- Loudspeaker and control room interaction
- Effective listening zone
- Sound fields: direct/near field versus room monitoring
- Monitoring levels.

### High-Quality Playback System

The playback system for accurate monitoring will be of high quality and carefully engineered. It is designed to provide unaltered, detailed sound, while working within the acoustic characteristics of a control room or similar environment. It is inherently different from systems intended for home use (even high-end systems intended for the audiophile market).

Most consumer playback systems seek to blend sound in pleasing ways and alter the sound with characteristics consumers (or certain segments of the general public) might enjoy; they often seek to smooth out imperfections in recordings and remove timbral detail that may be startling to listeners in their target market. The purpose and function of a pro audio playback system is very different. It seeks to provide the recordist with extreme clarity and detail of sound, minimizing any added fusion of sound elements. While it is common for some recordists to use one or several sets of consumer-type loudspeakers (sometimes in and often outside the studio) to obtain an idea of how their project might sound over home-quality playback systems, listening during the recording process requires great attention to subtleties of sound that are often only apparent over speakers designed to deliver great sonic detail.

All components in a high-quality playback system must perform at a high level and have qualities that complement the other components of the system. From the input of the power amplifier(s) to the output of the loudspeaker drivers, the signal should undergo minimal alteration, distortion, and added noise.

Loudspeaker systems are comprised of multiple drivers—at a minimum, a low-frequency woofer and a high-frequency tweeter, and often an additional mid-range driver(s) and/or a subwoofer. These drivers are housed in a loudspeaker enclosure that will impart sound qualities onto the reproduced sound, as well as impact the efficiency of the system. The subwoofer is the exception, as it will have its own enclosure and will reproduce the lowest frequencies of all channels. The drivers are fed a signal in the frequency range of their optimal performance by a crossover network.

A crossover network is a series of filters that produce the required frequency-limited signals for each driver. The network may be inserted into the signal chain after the power amplifier; this is a passive network that filters the input signal from the power amplifier into appropriate frequency bands at its outputs. Active crossover networks are inserted into the signal chain before the power amplifier; since the total frequency range is divided before the amplifier, a separate amplifier channel is required for each driver. A biamplified system contains a two-driver loudspeaker, two power amplifiers (or a two-channel amplifier), and an active crossover network.

A topic of some passionate debate, the qualities of the wires (cables) that connect devices may alter properties of the signals they carry. This seems especially noticeable with the cables between the power amplifier and the loudspeaker.

Detailed information on performance specifications of all of these devices, and the delicate science of matching components, is well outside the scope of this writing. The reader is encouraged to explore this material, and also to listen carefully to many different playback systems to learn their various characteristics—preferably when using the same or similar recordings, and placing the system in the same location(s) in the same room.

A high-quality playback system should meet the following criteria:

- Efficient transfer of power between components
- Exhibit ±1.5 dB deviation in all frequencies from 40 Hz up to 17 kHz
- Even lateral loudspeaker dispersion at all frequencies to cover the effective listening zone equally with the same frequency response
- Meet quality performance levels in dynamic range, power supply

and output noise, frequency response, slew rate, damping factor, total harmonic distortion (THD), intermodulation distortion (IMD), transient intermodulation distortion (TIM), time coherence

- Noise floor of the system in relation to the acoustic noise floor of the listening room

The system should produce sound with a high level of clarity and detail, strong time/phase coherence throughout the listening range (especially around the crossover frequencies), accurate tracking of dynamic changes (especially for high-pitched sounds with fast attacks), and a flat frequency response (especially in the lower octaves and above 10 kHz). The monitors should supply stable imaging and not draw the listener's attention to the loudspeaker locations. It is important for the recordist to be able to identify if alterations to sound quality are a product of the components of the playback system, or if they are being created by the interaction of the loudspeakers and the listening room.

### Loudspeaker and Control Room Interaction

The control room itself can alter the sound that is heard. The acoustics of the control room can cause radical changes in the frequency response and time information of the sound. Ideally, any listening room would have a constant acoustical absorption over the operating range of the loudspeakers, and would appropriately diffuse the sound from the loudspeakers to create a desirable blend of direct and reflected sound at the mixing/listening position. Such conditions rarely exist.

The influence of the control room on the sound coming from the loudspeakers can be minimized with: nonparallel walls; nonparallel floor and ceiling; acoustical treatments to absorb, reflect, and diffuse sound where needed; careful selection, placement, and installation of loudspeakers; and sufficient volume (dimensions of the room).

The room should absorb and reflect all frequencies equally well: should produce very short decay times that are at substantially lower levels than the direct sound from the loudspeakers. The room should not produce resonance frequencies and should not produce reflections that arrive at the listening position at a similar amplitude as, or within a small time window (2–5 ms) of, the direct sound.

As needed, rooms may be tuned for uniform amplitudes of all frequencies by room equalizers, and tuned for the control of reflections (time) with diffusers, traps, and sound-absorption materials. The ideal control room would include very specific acoustical treatments in such a way that minimal room equalization is needed.

Studio designers have widely divergent opinions on the most desirable acoustical properties of control rooms. Conflicting information and

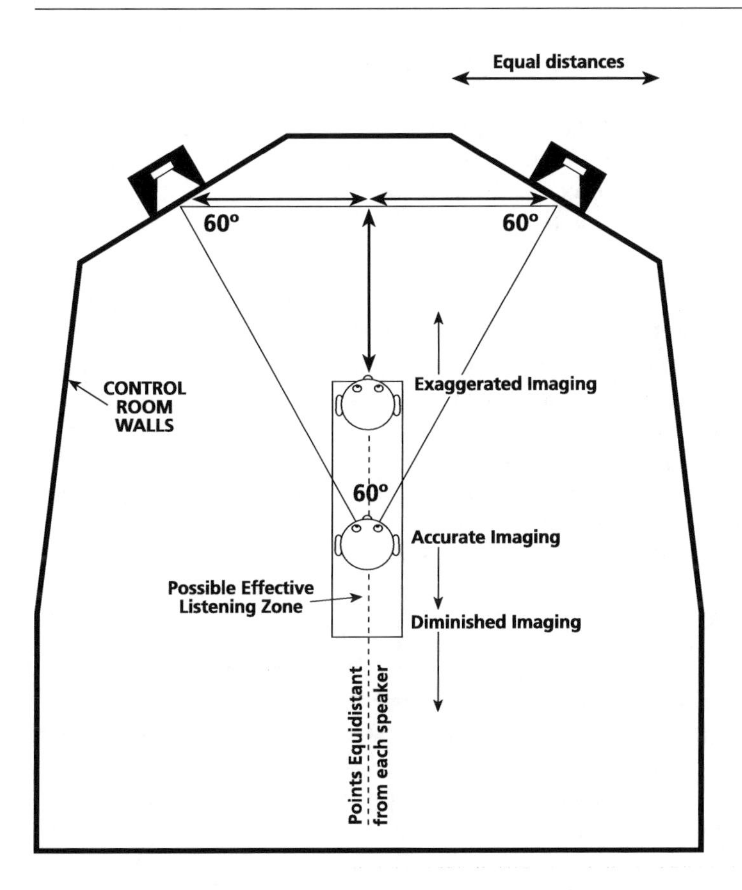


**Figure 15.8**
Loudspeakers as part of the control room and the effective listening zone.

opinions are common. The objective of all designers is very similar, however. It is to establish a listening environment where sound can be accurately reproduced.

Loudspeaker placement and performance specifications are factored into the design of control rooms. One common design approach dictates that the loudspeaker should be a part of the wall, mounted within the wall itself so the front of the loudspeaker is flush with the face of the wall. This negates the usual boost of low frequencies that results when a loudspeaker is placed near walls, ceilings, and floors (and especially in corners).

Loudspeakers can also be freestanding within the room in a direct field to the listener. Direct-field loudspeaker placement away from side walls by 4 feet, and from the front wall by at least 3 feet,will minimize the boost of low frequencies (that occurs when the omnidirectional low frequencies reflect off the wall surfaces and combine with the direct sound from the speaker).

The loudspeakers should be aligned on the same vertical plane, with the high-frequency driver at ear height for a seated person or slightly higher. It is important that the meter bridge of the console, or any other object such as a computer monitor, not be in the path between the loudspeakers and the mix position. Strong reflections off the console, tabletop, etc., also must be minimized. The loudspeakers should be aligned symmetrically with the side walls.

The *effective listening zone* is an area in the control room where the reproduced sound can be accurately perceived. In a control room, the mix position and/or the producer's seat are located in the effective listening zone. In most control rooms the size of the effective listening zone is quite small. It is an area that is equidistant from the two loudspeakers in stereo and all loudspeakers in surround. The area is located at roughly the same distance from each speaker as the speakers are from each other. Angling loudspeakers correctly provides optimal imaging in the effective listening zone, given complementary room acoustics. The reader should review stereo and surround loudspeaker positioning described earlier.

For accurate spatial perception, it is necessary for the effective listening zone to be carefully evaluated. The listener must be seated in the proper location, and the volume level must be the same at each speaker when identical signals are applied to each channel. Moving the listening location closer to the speaker array exaggerates imaging, and moving away from the loudspeakers diminishes imaging relationships. Rooms are built and/or speakers are placed so that few strong reflections arrive at this area. The control room should be virtually transparent (add or subtract no characteristics to the sound) in the listening zone.

### Near-Field, Direct-Field, and Room Monitoring

Loudspeakers can be incorporated into the structure of the control room, as described above. These room monitors utilize the acoustics of the room to their advantage, to complement their performance. Loudspeakers are also found freestanding and on desktops in control rooms. *Near-field* monitoring seeks to eliminate the influence of the control room on the performance of the loudspeakers, and *direct-field* monitoring seeks to minimize this influence. Near-field monitors are typically two-way speakers with dome-shaped tweeters and 6-inch woofers, and are designed to be placed 3 to 4 feet from the listener. Direct-field monitors are typically a bit larger (8-inch woofers are common) and are usually placed 4 to 5 feet from the mix seat (monitoring location), but can be somewhat further away, depending on the listening room. Near-field monitoring and the use of freestanding loudspeakers at close distances (direct-field monitoring) have become common as small project studios dominate the market. Quality, accurate, and reasonably priced monitor systems have brought detailed and accurate reproduction systems to be widely accessible for homes and small studios.

For direct- and near-field monitoring, speakers can be located on stands in front of a DAW's computer screen, seated on a meter bridge, or placed elsewhere as appropriate. The speakers will be 3 to 5 feet apart, and the listener will be at the same distance from the speakers (Figure 15.9). Sound is heard near the speakers, before the acoustics of the room can alter it. The goal is for the reflected sound in the control room to have little or no impact on what is heard in direct- and near-field monitoring because of the listener's close position to the loudspeakers. A room with highly reverberant characteristics, such as highly reflective parallel walls and other surfaces, may still alter sound in direct- and near-field monitoring.

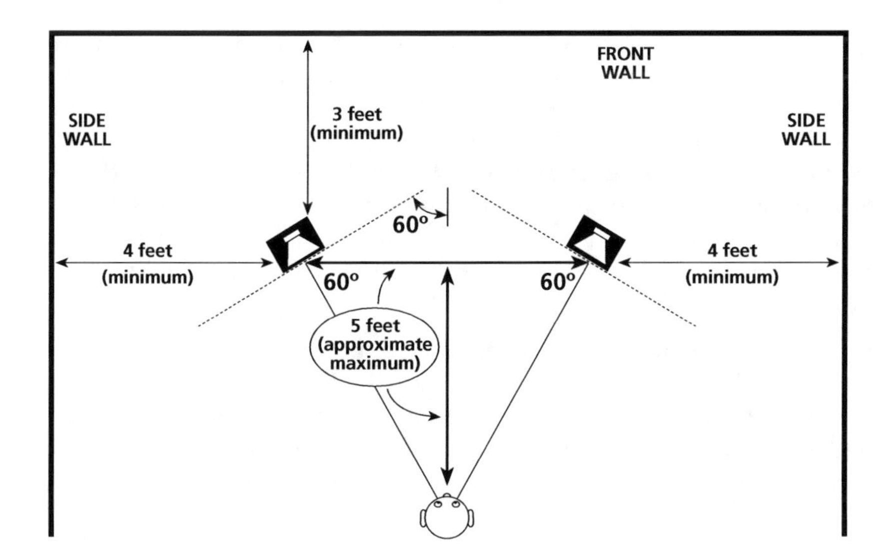


**Figure 15.9** Freestanding direct-field and near-field loudspeaker relationships to the listening environment and the listener location.

Since the control room has very little influence on the sound quality of near- and direct-field monitoring, this approach is often the exclusive monitoring system of control rooms that have poor room acoustics. In this case, high-quality monitors specifically designed for professional direct/near-field monitoring are used. These speakers are preferred over the large studio monitors by some recordists, especially during long sessions when the more intense energy of (some) large speakers can fatigue the ear.

Often problems with recordings can be more easily identified using high-quality near-field monitors rather than monitors that interact with the room. This is especially true of rooms with decay times above 0.25 seconds. Near-field monitors can be used for quality control to more readily identify subtle issues and noises in recordings and to facilitate editing, because subtle details are often perceived as clearer when room sounds are not perceived. The ambiance of the room does not pull detail from the sound and the listener is able to more easily focus attention on a nearer source.

The size of the effective listening zone for direct-field and near-field monitoring is very small. The listener must be precisely centered between the two loudspeakers, and at a distance that places them at the apex of an equilateral triangle relative to the two speakers. The listener should be located at or very near the same distance from each loudspeaker as the two loudspeakers are from each other. Should the listener be located out of this narrow effective listening zone, or "sweet spot," the effects of the room will immediately come into play.

It is not unusual for recordists to periodically switch back and forth between room and near-field monitors. This allows the recordist to evaluate the sound qualities and relationships of the recording from different listening locations and with loudspeakers that have different sound qualities. The recordist will be attempting to create a consistency between the sounds of the two monitor systems inasmuch as this is possible. The speakers are usually switchable at the mixing position, with each pair having a dedicated amplifier(s), set to match monitoring level while switching speakers.

Bookshelf-type, consumer loudspeakers are sometimes used in the studio in a near-field setting. This provides a reference to a consumer playback environment, and can represent the sound of typical, moderately priced speaker systems. They often tend to have a narrow frequency response, de-emphasizing high and low frequencies areas, and diminished detail.

### Monitor Levels

The sound pressure level (SPL) at which the recordist listens will influence the sound of the project. Humans do not hear frequency equally well at all amplitudes, as discussed in Chapter 1. A recording created at a *monitor level* of 100 dB (SPL) will sound considerably different when it is played back at the common home-listening monitoring level of 75 dB or lower. The bass line that was present during the mixdown session will not be as prominent in playback at the lower level. Similarly, the tracks that were recorded at 105 dB during a tracking session (because someone in the control room wanted to "feel" the sound) will have quite different spectral content when they are heard the next day at 85 dB, during the mixdown session.

The recordist will need to develop consistency of listening levels. Ideally, monitoring levels should be reasonably consistent throughout a project—tracking, mixing, editing, and mastering. Calibrating your monitor system for 85 dB SPL at the monitoring location to match 0 VU reference will assist this greatly.

The most desirable range for monitoring is 85 to 90 dB SPL; a nominal listening level of around 85 dB with peaks reaching to 90 dB (or very slightly beyond) is a fairly loud home-listening level, but it is a level that can be sustained throughout a work day. Recordings made while monitored in this range will exhibit minimal changes in frequency responses (≤ 5 dB at the extremes of the hearing range) when played back as low as 60 dB SPL. Monitoring at this level will also do much to minimize listening fatigue during prolonged listening periods (recording and mixing sessions).

The possibility of hearing damage is a very real job hazard for recordists. If the recordist primarily listens in the 85–90 dB range, they will have accurate hearing much longer than if they consistently monitor at a level 10–15 dB higher—they may very likely also have a longer career.

Loudness is our perception of the physical sensation of amplitude. It is possible to become sensitive to amplitude, and therefore loudness, by attention to the physical sensation of sound-pressure levels. It is unreasonable to believe the listener might develop an ability to accurately identify a specific SPL or a specific increment of change of level. It is, however, possible to learn how a narrow range of SPL "feels," and thereby learn to recognize a safe and effective listening range. The reader is strongly encouraged to work through the Loudness Perception Exercise (Exercise 15.1) at the end of this chapter, and to purchase an inexpensive sound-level meter and check listening levels frequently.

## Headphones

*Headphones* are sometimes used in recording processes for monitoring. They are required for creating and listening to binaural recordings, which minimize reliance on acoustic interaural signals. Headphones are often used for remote recording and for projects that have the performers in the same space as the recordist. In these instances, headphones may be the only feasible monitoring option. Other than these situations, monitoring is most accurately accomplished over loudspeakers, with few exceptions.

Headphones will distort spatial information, especially stereo imaging. The listener will perceive sound within their head, instead of as occurring in front of their location. This distorts depth of imaging. The interaural information of the recording is not accurately perceived during headphone monitoring, causing potentially pronounced image size and stereo-location distortions. Further, the frequency response of headphones will not match that of loudspeakers and is inconsistent. Frequency response of headphones will vary with the pressure of the headphones against the head and the resulting nearness of the transducer to the ear. In listening to a recording over headphones, the recordist must also imagine what the recording will sound like when played back over loudspeakers. Some recordists have developed this skill very accurately.

High-quality headphones paired with quality headphone amplifiers can prove useful. They can provide low-distortion, broadband frequency response, and excellent transient response, and deliver exceptional timbral clarity and detail. Headphone monitoring with such equipment can be very useful for editing, especially when listening for subtle technical details. At times the ambiance of a control room can make processing artifacts, timing errors, faint clicks, ticks, and edit point issues very difficult to hear; such sounds are often easier to perceive over headphones, where the speaker is nearer the ear. Sonic detail and accuracy of spatial information remain compromised, however.

The sound evaluations that are covered herein can only be accurately performed over quality loudspeakers, in a transparent (or complementary) listening environment. Rarely is it preferable to monitor over headphones if both options are available.

## Summary

Before sounds are recorded, the recordist makes many decisions that shape the recording project. Many of these choices will limit how the music might be shaped later in the recording process and must be approached with knowledge and sensitivity.

Selecting and defining sound-source timbres, capturing those sought qualities with effective microphone selection and placement, crafting new timbres with synthesis, selecting recording devices throughout the signal chain, and selecting monitoring systems and conditions are all decisions that can profoundly impact any project.

When the entire project is considered at this planning stage, decisions in many areas can be made that will not limit the recordist as the project progresses. A clear sense of artistic direction will be established that will allow the music's potential to be realized and a high-quality recording to be made.

These preliminary stages prepare for successful projects of good quality, that make efficient and effective use of studio time. They also allow recordists to be in control of their craft and shape the artistic qualities of music.

# Exercises

## Exercise 15.1

*Loudness Perception Exercise*

Humans experience a physical sensation from the amplitude of a sound wave that is transferred into loudness. Excessive amplitude can cause pain in humans. Becoming sensitive to the physical sensation of listening at an appropriate loudness level is equally possible, with practice, attention, and diligence.

1. Purchase an inexpensive sound-level meter, and keep it in front of your listening location for your monitor system. Set your monitoring level so that the average SPL registered is between 80 and 85 dB SPL. It is acceptable for peaks to hit 90 dB or slightly above and for soft passages to dip below 80 dB by several dB.
2. Begin your practice by listening to recordings you know well or are studying at this loudness level, checking the meter frequently to verify listening levels. In the beginning, keep the meter on whenever you are listening. Become aware of the physical sensation of your hearing created by listening at that level. You will begin to notice energy impacting the hearing mechanism (inner ear) with your focused attention, over time. You will begin to develop an increased sensitivity to the physical sensation of loudness level.
3. Be consistent in listening at this level when working on your own projects. Check your meter regularly. Over a few weeks (or a bit less or longer) of being aware of the physical sensation of this loudness-level region, you will develop a memory of the sensation.
4. Listen to a recording you know well on your monitor system by starting with the monitor level off and without the meter on. Bring up the level gradually until you believe you have reached this average level of between 80 and 85 dB SPL. Check yourself with your meter after you believe you have established this level. Repeat this exercise regularly (several times per day if possible, with a minimum of once per day) until you begin to have success in recognizing this level.
5. Turn next to listening in other environments you do not know as well, perhaps your automobile, a home system, or a monitor system in another location. You will notice you are not as accurate at first, but this accuracy will increase as you become more aware of focusing on energy impacting your hearing mechanism. In time this skill will carry over from listening environment to listening environment, if you continue to try to establish a correct listening level in a new location, and check your accuracy with your meter.

Just as with pitch reference, this skill will take time and effort over an extended period, and the skill may not become completely reliable. It will, however, improve your listening skill considerably and serve as an important point of reference. In time you will gain an awareness of when you are listening at an average level of between 80 and 85 dB SPL. It is also possible that you will come to prefer an average level toward either the high or low side of this range. Arriving at a reliable sense of average loudness level will greatly assist one in maintaining a stable sound quality of monitoring, as well as an accurate listening level.

Most importantly, you will become very aware of being in an environment where the SPL extends well above 90 dB SPL. That higher level brings increased pressure on the hearing mechanism that will be very apparent and uncomfortable. This will trigger your awareness that hearing fatigue will happen quickly, and a sense of urgency that if this level extends far above 90 dB, you are in danger of damaging your hearing.

### Exercise 15.2

*Identifying and Comparing Microphone Characteristics*

The purpose of this exercise is to learn the special way a microphone will transform sound. This is most easily accomplished by comparing several different microphones located in a very similar location, while recording a single performance.

1. Set up to record a single performance of a sound source you know well, and that is easy for you to record accurately, given your system and recording space. Identify your preferred location to capture the sound of that instrument or voice.
2. Place the 2, 3, or 4 microphones (on appropriate stands) as near the same location as possible.
3. Assign each microphone its own track on a multitrack recorder or DAW, and obtain suitable record levels.
4. Record the sound source performing pitches in the extremes of its range, performing loudly and softly, short sounds and sustained sounds, sounds with a fast attack and with slower attacks (of course this is not possible on some instruments), and other material you might find helpful.
5. Establish playback levels where each track can be compared to all others at the same loudness.
6. Listen carefully to the sound quality of each microphone to identify its unique qualities. Listen to how the microphone captured the spectrum and the dynamic envelope of the source, and how quickly

it responded to changes. Listen to different "blends" of the source's timbre captured by the individual microphones, and how quickly each microphone responded to changes to the source.

### Exercise 15.3

*Comparing Microphone Placements*

The purpose of this exercise is to learn how microphones have different sound qualities when placed at different distances. This can be accomplished by creating a recording of a single performance of the same microphone at a number of locations. You will need several (2, 3, or 4) of the same microphone make and model for this.

1. Set up to record a single performance of a sound source you know well, and that is easy for you to record accurately, given your system and recording space. Identify your preferred locations to capture the sound of that instrument or voice.
2. Place the 2, 3, or 4 microphones (on appropriate stands) at those locations. Take the time to measure the distances and angles of the microphones to the source, and write them down.
3. Assign each microphone its own track on a multitrack recorder or DAW, and obtain suitable record levels.
4. Record the sound source performing pitches in the extremes of its range, performing loudly and softly, short sounds and sustained sounds, sounds with a fast attack and with slower attacks (of course this is not possible on some instruments), and other material you might find helpful.
5. Establish playback levels where each track can be compared to all others at the same loudness.
6. Listen carefully to the sound quality of each microphone to identify its unique qualities. Listen to how the microphone captured the spectrum and the dynamic envelope of the source, and how quickly it responded to changes. Listen to different "blends" of the source's timbre captured by the individual microphones, and how quickly each microphone responded to changes to the source. Finally, note the "reach" of the microphone at these various locations, and bring your attention to the accuracy of timbral detail and the ratio of direct to reverberant sound. If you can, listen for reflections in the room that are causing portions of the source's timbre to be emphasized and attenuated.

### Exercises 15.4

*Additional Exercises for Microphone Techniques*

Stereo and surround microphone techniques and their placements can also be compared and evaluated by substituting arrays for the individual microphones found in the above two exercises. In these evaluations, bring your attention to the distance relationship of the source to the listener location. Listen for the degree of blend of the instrument sound (or the ensemble sound). Next bring your attention to recognizing how the selection and placement of the stereo or surround technique is impacting the width of the sound stage, and its depth; move the arrays to notice shifts in these dimensions, and seek to find a placement of the array that provides a sound that is the same as the third row of a hall (or some other location that you might wish). Next observe the arrays as they influence the spaciousness of environmental cues of the hall, and the balance of direct to reflected sound, and how they influence the sound stage.

# 16 Recording and Tracking: Capturing and Shaping the Performance

It is time to record. Some final decisions on approach, then the recording begins. We will explore the capturing and shaping of the performance, in real time and out of time.

Here we will explore approaches to recording and tracking, and shifting our learned listening skills to the live experience and to interacting with others. Recordings will be made, some all at once and some in pieces and layers. And after the sessions, more tasks to prepare for mixing. These and more are covered in this chapter, making final preparations for the mix.

## Approaching the Recording: Tracking and Recording Sessions

Every project will be unique in the ways and formats its sounds are recorded. The differences may be slight or great, but they are based on how the performance is recorded and the amount of mixing that takes place during the recording. Projects can be recorded with musicians performing individually, or in subgroups (such as a rhythm section), or with all parts of the music performed at once (such as a live band or an orchestra). The project will be unique in how the recording process is used to keep instruments separated, or to the extent instruments are combined during the recording process, and the resulting format of their initial storage—individual instruments and voices as completely isolated tracks in a multitrack (DAW or tape) versus all parts mixed to mono as extremes. It is this difference that brings us to the terms "tracking" and "recording" to underscore this distinction.

"Recording" and "recording sessions," as used here, are recordings of all the parts being played at once; the entire musical texture is recorded as a single sound, and in a single performance. "Live to Two" recordings are made directly to two-track stereo; all sounds are recorded, processed, and mixed to stereo in real time with the live performance. The recordist is in effect part of the performance. This approach can be used in recording to mono and to surround as well.

Tracking sessions are the recording of the individual instruments or voices (sound sources), or small groups of instruments or voices, into a multitrack format (DAW, analog multitrack recorder). This is done in such a way that the sounds can be mixed, processed, edited, or otherwise altered at some future time, and without altering other sound sources. For this control of the sound source to be possible, it is necessary that sound sources be isolated and recorded without sound from other sound sources intruding on the track. The performances can happen simultaneously and still be isolated under the right conditions. Performances can also happen one at a time, individually as overdubs or electronically as MIDI synths or otherwise. As such, these performances will take place at a different time—perhaps immediately after one another, or that day, in a few weeks, or some other time or place. In fact, tracking and mixing can, and often do, take place in different studios.

In actual practice these two extremes are not so exclusive. Recording a live band to make a multitrack recording project is common, and combines some of these concepts. Typically the band members are separated in the studio to have some sound isolation; but they will still be able to communicate visually or musically while performing. Their sounds will be fed to separate tracks for later use. Certain instruments might share the same space and have a bit of bleed between them and are mixed subtly during tracking. Some instruments are recorded using more than one microphone, and may receive some mixing at this stage—this is very common for drum sets and for piano. All of the parts of the live performance are not usually meant to be final performance. *Scratch* performances, intended to be replaced later by overdubs, are common, and are meant to be a guide for the band. For instance it is expected that overdubs for the vocal will be redone later, and the final lead vocal will likely be compiled from many takes of overdubs. Including it in the live performance is vital to the feel, timing, and structure of the song. Other parts will be planned for addition or replacement, such as a solo added by the guitarist, and their rhythm track redone completely.

## In Session: Shifting Focus and Perspective

Recording and tracking sessions require the recordist to continually shift focus and perspective, while listening to both the live musicians and the recorded sound. Further, they will move between analytical- and critical-listening processes. The musical qualities of the performance will be constantly evaluated at the same time as the perceived qualities of the captured sound. The recordist is responsible for making certain both aspects are of the highest quality and exist accurately and without distortion state in the storage medium.

In Parts Two and Three listening skills were developed for individual aspects of sound. These skills are used here, during production, in actual practice. The great difference is that earlier the reader was allowed to focus on a single aspect of sound and to remain at a single level of perspective. This is very helpful for developing one's listening abilities. In actual recording practice, those many skills will be used in rapid succession, as the recordist switches between elements, levels of detail, and types of information very quickly and deliberately. Used in practice now, this listening process is done immediately, in real time, requiring immediate recognition and understanding of many pieces of sound information—all while musicians are expecting the recordist to be capturing their performance impeccably and quickly.

Focus will have the listener's attention moving freely but deliberately between all of the artistic elements of sound and all of the perceived parameters of sound. The recordist will need to have the skill to quickly shift focus between artistic elements (perhaps shifting attention between dynamic levels, pitch information, or spatial cues), then immediately shift to a perspective that evaluates program dynamic contour (of the overall sound). The recordist is required to continually scan the sound materials to determine the appropriateness of the sound (and its artistic elements) to the creative objectives of the project, and to determine the technical quality of the perceived characteristics of the sound.

Within recording and tracking sessions, the recordist is concerned about the aesthetics of the sound quality that is going to "tape," and they are concerned about the technical quality of the signal. Depending on their function in the particular project, they may or may not be in the position to make decisions related to performance quality. They should none-theless be aware of the performance and be ready to provide their evaluations when asked, or to anticipate the next activity (repeat a take, or move on to the next section, as examples). It is imperative that the recordist always be aware of the technical quality of the recording. It is their responsibility for high-quality sound to be recorded.

### Analytical-Listening Concerns

There are many performance-quality aspects of music that will engage the recordist in analytical listening at this stage. Most of these aspects are related to:

- Intonation
- Control of dynamics
- Accuracy of rhythm
- Tempo
- Expression and intensity
- Performance technique.

The pitch reference, time judgment, and musical memory skills gained in Part Two will now be invaluable. Staying in tune at a specific reference (such as A440) is necessary for "Take 1" to be useable against, say, "Take 617." Tempo also must not change unintentionally, and dynamics and performance intensity must remain consistent or deliberately adjusted throughout the session(s). Dynamics can be altered by performance intensity, and these changes in expression should be under control and factored into dynamic-level considerations. How a musical idea is played is often just as important as the idea itself; the expressiveness of the performance and its suitability to the project are important aspects of great recordings (such as the ones we seek to make). While these areas may not always be under the purview of the recordist, flawed material needs to be identified at all times. Just how and when issues get pointed out to the artists, producer, or others varies with the project, roles, and personalities involved.

The performance technique of musicians is critical in recording. The performer's sound cannot be covered up by the other players or assisted by the acoustics of a performance environment, and will be apparent in the final sound. The ways performers produce sound, or the instruments themselves, can create the desired musical impact, or they may not. The person responsible for the project will need to make evaluations and to offer necessary alternatives. The recordist should know any performance-technique problems and the natural acoustic sound properties of all of the instruments (while well outside the scope of this writing).

At this stage the recordist should be getting a clear idea of the intended overall qualities of the final recording, and how each track will contribute to those overall qualities. Tracking will capture or shape sound significantly, and will allow the recordist to achieve the desired sound, or it will fall short. Careful attention must be given to the sound quality of the sound source for timbral detail, performance intensity, dynamic contour, spatial concerns, sound quality (for pitch density), and more.

Depending upon what source is being recorded at the time and the final intended sound quality, some artistic elements might be more important than others; still, all contribute important information and must be considered. All of the analytical-listening skills gained in Parts Two and Three will be used in this process.

## Critical-Listening Concerns

The technical quality of the recording will be reflected in the integrity of the signal. The recording process must not be allowed to alter the sound, unless there is a particular artistic purpose or technical function for the alterations. The perceived pitch, amplitude, time elements, timbre, and spatial qualities of the recorded sound sources should be accurately captured in the recording, and should be the only sounds present in the reproduced recording. Any extra sounds are noise, and any unwanted alterations to the sound are distortions. The recordist must know the recording process and devices instinctively well for this to be accomplished fluently. The goal is for impeccable technical sound quality to be achieved effortlessly, so that attention can be mostly given to the artistic qualities of the sound and the performance.

Accurate recording and reproduction of the audio signal must be consistently present for professional-quality music recordings. Skills in evaluating sound quality, time information, and dynamic contours at all levels of perspective are important here. Sound will be evaluated for flaws and added sound/noises by scanning all parameters of sound at all levels of perspective.

Each device in the signal chain has potential to add noise to the signal, and to alter aspects of the signal in undesirable ways. Myriad noises (such as clicks, hiss, hum, digital artifacts, quantization noise, and countless others) can be added to the signal by any device. They might also degrade the quality of the signal by introducing clipping, total harmonic distortion, phase shifting, drop out, intermodulation distortion, latency, and more. These can all be caused by misuse or malfunctions of the devices (i.e., a microphone overdriving a microphone preamp, or a dirty potentiometer on a mixing console).

Interconnections of devices must also be correct to preserve the quality of signals throughout the signal chain. Gain staging, connection points, analog-to-digital and digital-to-analog conversions, conversions of digital formats, and many other issues also have the potential to diminish the integrity at any point throughout the signal chain. It is the recordist's responsibility to recognize all these events, their sources, and more, to ensure a high-quality recording.

Other critical-listening applications during recording and tracking sessions relate to microphone issues discussed in Chapter 15:

- Isolating sound sources during tracking
- Achieving a desirable sound quality through matching an appropriate microphone with the sound source
- Identifying an appropriate placement angle and distance for the microphone
- Eliminating unwanted sounds created by performers or instruments.

Sounds will have a certain degree of isolation from one another. If the sounds are to be altered individually in the mixing stage, they must be isolated. When a group of sounds function as a single unit, it may not be desirable to isolate the sounds from one another, as they are often blended by the performance space. Problems arise when unwanted sounds leak onto tracks that contain sound sources that were intended to be isolated from all other sound sources. This leakage can be the cause of many problems later on during the mixdown process.

Sound quality should be carefully evaluated as the microphone captures it. The amount of timbral detail, blend, etc., of the sound sources will be determined now. These are major decisions that will have a decisive impact on the sound of the final recording, and are largely determined by microphone selection and placement. Evaluation of the sound out of the context of the music is helpful in discovering subtle qualities that are often missed when one is listening analytically.

Many recordists rely on equalization and other processing to obtain a suitable sound quality during this initial recording. This use of processing alters the natural sound quality of the sound source and does not alter the timbre equally throughout an instrument's range: for example, equalization will add a new formant region to the track/sound source. It can be used effectively when the processed sound is desired over the source's natural sound, or when practical considerations limit microphone selection and placement options. Often it can compromise the recording and should be approached carefully. Processing—especially equalization—is often used at this stage to compensate for poor microphone selection or placement.

Related to performance technique, the ways performers produce sound on their instruments, or the instruments themselves, can create unwanted sound qualities that must be negated during the tracking process. Often these aspects of sound quality are very subtle and go unnoticed until the mixing stage (when it is too late to correct them).

Instruments are capable of making unwanted sounds, as well as musical ones. The sound of a guitarist's left hand moving on the fingerboard, the breath sounds of a vocalist or wind player, and the release of a keyboard pedal are but a few of the possible nonmusical and (normally) unwanted noises that may be produced by instruments during the initial recording and tracking process.

These sounds are easily eliminated during the initial recording and tracking process through altering microphone placement, through slight modifications in performance technique, or through minor repairs to the instrument. The tracks should be as free of all unwanted live-performer sounds and sound alterations as possible. These sounds will be much more difficult to remove later in the recording process. They may be comprised of certain performance peculiarities that (depending on the situation) can only be alleviated by signal processing (such as the use of a de-esser on a vocalist).

## Anticipating the Mix during Tracking

The mixdown process can be anticipated while compiling basic tracks. There are many advantages to doing this. First, determine how much control in combining sounds is needed or desired during the mixing process. Tracking, submixing, and isolation decisions can then be made accordingly.

Any mixing of microphones during the tracking process will diminish the amount of independent control the recordist will have over the individual sound sources during the final mixdown process. Some mixing will often occur during the tracking stage, as submixes, to consolidate instruments and open tracks, or to blend performers who must interact with one another for musical reasons.

*Submixes* will be carefully planned at the beginning of a session, and must be executed with a clear idea of how the sounds will be present in the final mix. Drums are often condensed into submixes (either mixed live, or through overdubbing and bouncing). Other mixes that will occur during the tracking process include the combining of several microphones (and/or a direct box) on the same instrument(s). The recordist must be planning ahead to how these sounds will appear in the anticipated final mix, especially for musical balance, sound qualities, distance cues (definition of timbral detail), and pitch density.

*Preprocessing* (such as adding compression while recording) alters the sound before it reaches the mixing stage. Preprocessing also diminishes the amount of control available over the sound during the mixing process. At times, it is desirable to preprocess signals; often it is not.

Desirable preprocessing might include stereo microphone techniques used on sound sources, effects that are integral parts of the sound quality of an instrument (such as distorted guitar), or processors that are used to provide a specific sound quality (a compressed bass). At times preprocessing is used to eliminate unwanted sounds during tracking (such as noise-gated drums). Once a source has been preprocessed, the alterations to the sound source cannot be undone.

The recordist should be confident that they want the processed sound before recording it. Often processing is most effectively or appropriately applied during the mix itself.

Some initial planning of the mixdown sessions will typically begin during the tracking process. Certain events that will need to take place during the mixdown session will become apparent as the tracking process unfolds. Keeping a tally of these observations will save considerable time later on and may help other tracking decisions—keeping this documentation clear and detailed is imperative if there is any possibility that the project will be mixed by another recordist.

Examples of items that should be noted for the mixing process are:

- Changes in the mix that may be required because of the content of the tracks
- Certain processing techniques that are planned
- Any spatial relationships or environmental characteristics that may be desired for certain tracks
- Track noises or poor performances of certain sections that will need to be eliminated (muted) during the mix.

These are just a few examples of the many factors that may become apparent during the tracking process.

## An Overview of Two Production Sequences

Remembering that every music recording project is unique, some general observations about production techniques and sequences will be made. The details of how these are accomplished will create differences between projects. The sequence of events in the individual recording production and the use of the recording chain will be adapted to suit the needs of the individual music recording.

Some projects will require more session preplanning than others. One project may require more mixdown planning and preparation than another. Other projects might have very different requirements from more conventional projects. The order of events will be mostly consistent with the outlines below; some overlapping between the events will be common, as well as some alterations to the orders of the events—or portions thereof.

## Multitrack Recording/Tracking Sequence

A complete sequence of events for a multitrack recording might be:

1. Session preplanning: conceptualize project and pieces of music to be recorded (writing the music if necessary), rehearse musicians, define sound sources and their timbres, select microphones, plan track assignments, determine recording order of the tracks, plan the recording's sound stage
2. Tracking session: record reference tracks (vocals and accompaniment, etc.) and basic tracks (primarily the rhythm tracks)
3. Editing of basic tracks for out-takes, and to create the basic structure and length of the piece; reorganize tracks
4. Overdub sessions: replacing reference tracks with final performances; adding solo parts and secondary ideas to the basic tracks, refining the musical material; composing and recording any additional parts to fill newly discovered requirements of the piece; comping the lead vocal and adding backing vocals
5. Processing and mixdown preparation sessions: finalize the sound qualities of sound sources; edit the source tape/tracks for mixdown (reorganize tracks, remove unwanted sounds)
6. "Compose" the mix: defining the artistic elements of dynamic levels, spatial properties, and sound quality for each sound source, and by considering the interrelationships of the mix and the musical materials of the piece; rehearse the mixdown sequence with people who will assist in the session; remember, different sections can have very different mixes
7. Mixdown session: performing the mix(es), mixing the multitrack down to two tracks or surround (often occurring during the same session as Step 6)
8. Compiling the song: assembling and processing a final version of the song by combining the section mixes and by matching levels and applying any global signal processing.

## Direct-to-Master Recordings

These are live-to-two-track recordings, or live to surround or any other format. Such an approach is common in music recordings for film, television, and many other applications. These recordings are mixed to final relationships (master) during the recording session—and the mix/recording processes will often impose minimal alterations to the ensemble's sound.

This approach is common in recording classical or art music (such as orchestral, choral, or chamber music) and much jazz. It is also suitable for concert recordings, and for folk, ethnic musics, popular, rock, or any other music when the musicians (or the conductor) want to be in control of the musical relationships within their performance, or when the function of the recording is best served by having all of the musical parts performed at once (often the case for film scoring or archival recordings, as examples).

The process of making direct-to-master recordings is strikingly different from the multitrack recording process discussed previously.

The act of defining the sound quality, as previously presented, is shifted from the sound source to the perspectives of the overall ensemble, of groups of instruments within the ensemble, or to the perspective of a limited number of individual soloists.

A complete sequence of events for a direct-to-master recording might follow the outline below. These events will be discussed in detail and in the form of a commonly occurring sequence in the following paragraphs. This sequence and the details that follow are guidelines to be altered for the individual project—sometimes markedly.

1. Session preplanning
2. Creating the sound quality of the recording
3. Consultations with the conductor (musicians)
4. Recording session
5. Selection of takes
6. Editing to compile a final version of the recording and music.

**1.** Session preplanning always begins this production sequence. Once the music to be recorded is known, the recordist will need to know the skill level of the musicians (performers) and the location of the recording session (if it will not take place in the studio). This will allow suitable microphones to be selected, an appropriate stereo or surround microphone technique to be identified (if desired), and microphone placements to be planned. The acoustics of the recording environment will also need to be evaluated when the recording is to take place in a space unknown to the recordist.

The recordist and the conductor of the ensemble, or the primary performers, will next determine how the music can be effectively divided into sections, or the recordist might make these decisions alone. Problems of editing sections together into the master recording of the work, and issues in stopping and starting the ensemble, will be considered when making these divisions. The order in which the sections will be recorded may then be determined. The recordist's and conductor's scores, and the musicians' parts, should be marked to identify these sections, to make starting and stopping the ensemble during the recording session clear and efficient. A discussion between the recordist and the conductor (or performers) should also clarify the recordist's artistic role in the project.

Further discussion, and perhaps some recorded rehearsals in the recording space, should clearly define the sound qualities that will be sought for the recording project.

**2.** Crafting the recording's actual sound quality can be accomplished by monitoring a final rehearsal of the ensemble within the performance space in which the recording will take place. Alterations to microphone selection and placement will be made to achieve the desired sound quality of the recording, previously discussed. The microphone selection and placement will largely determine the spatial properties, dynamic-level relationships, and the sound quality within the recording. Balancing of microphones and signal processing will do the final shaping of the sound quality of the recording. Any necessary signal processing (environmental characteristics, time delay, EQ, and dynamic processing being most common) for accent microphones and the stereo array (or arrays) will be added and tuned at this stage of the production sequence.

After the final sound quality has been established, portions of the rehearsal are recorded for later reference and discussion. Any changes in the mix that may be required for the recording session are determined. In the best of situations, these changes will be thoroughly rehearsed during this rehearsal of the ensemble.

**3.** The recordist and conductor (or musicians) will listen to the reference recording that was made during the dress rehearsal. Often their discussion will be solely on the subject of sound quality; all musical considerations may be determined between the conductor and the musicians, or amongst the musicians themselves. Any alterations that must be made to the recording's sound quality are determined during this discussion; the recordist needs to obtain a clear idea of the sound qualities required for the project.

All requested changes to the sound quality of the recording are worked into the recording process. The microphone and recording equipment set-up for the recording session will reflect these changes.

At the recording session itself, sound quality is rechecked during the musicians' warm-up period, before the session begins. The recordist and conductor (musicians) might now make final evaluation of the changes that were made to the sound quality and confirm that the sound quality is correct, or the recordist may be responsible for making this judgment alone. The recordist, conductor, and musicians will briefly clarify the logistics of the recording session (how stopping, starting, slating, etc., will be handled).

**4.** The recording session follows. During the recording session, the sections of the piece are performed in the prearranged order. Many takes may bc performed of each section of the work until two suitable takes (of each section) are recorded. Each take of each section is monitored by the recordist, with a focus on consistency of loudness levels, tempo, intonation, performance quality, and the expressive qualities of the performance. An assistant engineer or second engineer may be used to assist in sound evaluation. This person would typically focus their attention on the technical and critical-listening aspects of the sound of the recording. If possible, another assistant will be used to maintain a record of the content of the session takes, making notes on the recordist's observations of each take, and of the observations of the musicians' spokesperson (usually the conductor, if one is present).

Any changes in the mix that are required in the recording were choreographed during the rehearsal session(s). These changes in the mix are performed, in real time, during the musicians' performance in the recording session. The recordist may coordinate the activities of one or more assistant engineers, who would physically perform the actual changes in the mix. The recordist would remain focused on the accuracy level of all of these changes, as they are made, as well as their relationships to the performance.

A multitrack recording of the session may be made simultaneously with the reduction mix to make a safety recording of the session. This will allow the recordist to perform a remix of the session at some future time, should this be necessary. All microphones will usually be sent from the console directly to a DAW or multitrack recorder, often routed from the console's direct or patch outputs. No balancing of dynamic levels or extra signal processing would normally be performed on these tracks. This would defeat the purpose of the multitrack backup.

**5.** The conductor (musicians) will often listen to the session tapes with the recordist. Usually they will listen to only a few takes of each section of the piece. These takes will have been preselected by the recordist, using the conductor's observations (that were written down by the production assistant) during the recording session as a guide for the selection of takes. Takes with technical problems will also be identified. The takes that will be used in the final recording are determined during this conference between the recordist and the conductor. Both parties will discuss their perception of the sound qualities of the takes, and the recordist may or may not be asked to evaluate the musicality or accuracy of the performances of each take. Any specific aspects of the sound quality that are undesirable will be identified. The recordist might determine signal-processing alterations to attempt to solve (or minimize) any problems and may play some of the possible alterations for the conductor (performers) during this session. A remix from the safety multitrack might be considered at this point as a last resort in the event of poor session results.

**6.** Any signal-processing or mixing alterations that were determined in the review of takes are performed by the recordist in the mastering session. These changes may be performed before or after the master recording has been compiled, depending on the type of alterations that need to be made. The master recording is created by editing or "splicing" together the selected takes—at the correct locations and in the correct order.

Any global signal processing will be applied to the overall program after the edited version has been assembled. In this case, a master recording will be made by playing the edited version through any signal-processing device(s) or plug-in(s), to record the actual master of the work. The recordist will arrange for the conductor (musicians) to hear the master recording, for final observations and approval. Any final alterations to sound quality (etc.) requested by the conductor will then be performed by the recordist and will complete the project.

## Editing: Rearranging and Suspending Time

With analog tape, the recordist can physically hold time in their hands and move it around. Audio recording transfers sound, which can only occur over time, into a storage medium where the sound is physically located, suspended out of time. This is very significant, and this concept is getting lost as more people have never experienced or worked with analog tape. The sound can then be changed and reordered by physically altering the storage medium itself (as in cutting analog tape), or by altering the way the storage medium reproduces the sound (i.e., replaying a portion of a digital recording/sound file). The sound may be altered at any time, present or future, and may be replayed forwards, backwards, at any speed (even at uneven speeds).

In *editing* sound, the recordist is able to precisely shape material out of real time. Editing typically combines or joins several different time segments, each time segment being composed of a group of any number of sounds. It can also remove time segments and the sounds they contain. The time segments may exist as pieces of analog tape or as computer data.

In joining the sound segments, the recordist can significantly alter the piece of music and its artistic message. These alterations may serve many functions. They must be accomplished in an artistically sensitive manner and must be inaudible in all areas of technical quality.

It may be impossible to perform technically inaudible or artistically sensitive edits under some circumstances and in some locations in the piece of music. The recordist will identify potential edit points to carefully calculate an edit before it is made. In analog, this may even involve rehearsing the edit on a copy of the master tape.

Editing is often used to compile a master of a recording session, as above. In this process, a few or a good many separate sections of the piece of music are joined into a single performance. The most appropriate material, or the most accurate and/or pleasing performances, will be selected for the master. A single performance is compiled from the many takes of segments of a piece or of a direct-to-master recording, or a single performance is compiled from joining the few individual mixes of multitrack recording.

It is possible to reorder sounds through editing techniques. The major sections of a piece of music may be rearranged. Entire measures may be exchanged, or sounds within a measure may be reordered.

In analog recording, the editing process will alter all sounds present. A reordering of sounds cannot occur unless they are isolated. It will not be possible to reorder the sounds of instruments in a drum fill without also moving the sounds that occur simultaneously with the drum sounds. Likewise, it is impossible to cut a sound source into numerous time segments and reorder the sound, unless that sound is isolated from other sounds. Digital audio workstations make all of these and much more possible, depending on how tracking was accomplished.

### Identifying Edit Points

Edit points (also called splice locations) are calculated by anticipating the sound that will be created when the two segments are joined. A critical-listening process of evaluating sound quality is used. Each segment to be joined will be evaluated for its sound qualities to determine the most appropriate location of the splice. Beginners often find the edit points in the music through trial and error. With developed skill, the listener will more readily identify these locations by listening carefully to the sound qualities of the two segments, remembering what was heard, and comparing the two sound events. How the edit impacts the musical materials will also be considered by the recordist.

Audible edits are nearly always unacceptable and may be created by many factors. Both the critical-listening concerns of audio quality and the analytical-listening concerns of the musical materials must be considered in determining suitable edit points. The sound must be evaluated for any changes that might be caused by the edit process itself, and for any noises that may have been added. In calculating the edit, the recordist will scan all artistic elements, or perceived parameters of sound, at all perspectives, to determine a usable edit point.

It is not possible to perform an inaudible edit when large differences exist in any of the elements of sound, between the two time segments. Such a splice would result in a sudden alteration of a component of sound at the point where the two segments meet; the sudden change would be audible and unacceptable. As soon as an edit has been made, it will be checked for accuracy and to be certain it is inaudible, and that no noises were added in making the edit.

Under unique circumstances, sudden changes between segments may be desired, as in creating a master recording where the splice actually joins very different musical ideas. In these instances, the recordist must make certain that the editing process does not create noise at the edit point, and that the sudden changes are presented as a part of the musical materials (have significance and are handled artistically).

Edits are most easily made at points where loud attacks are performed by prominent instruments or the entire ensemble, or immediately before or after (not during) areas of silence.

Sound sources that are sustaining over the edit point or that are present in each time segment make the edits more difficult. Changes in the sound source will make the edit point audible.

Among the most common of inconsistencies that are present between two time segments are differences in loudness levels. Even subtle changes can be quite audible. Calculating the loudness levels between various takes of an entire ensemble can be quite difficult, but will be developed through learning to focus on program dynamic contour. Beginning recordists will often only notice problems in this element after the edit has been made.

Tape noise is part of analog recording. The amount of noise on the tape may or may not be consistent throughout a recording. Changes in noise floor at edit points are very noticeable.

Differences in sound quality of individual instruments and of the overall ensemble are easily overlooked. The potential exists for sound sources and an ensemble to undergo significant changes in sound quality from the beginning of a recording session to the end. Performer fatigue, performance intensity, artistic expression, even a change in temperature or humidity in the performance space may cause these changes in sound quality. The subtlest of changes of sound quality can have a marked impact on the technical quality and musicality of the recording.

Changes in pitch between the two segments can be the most noticeable of all changes. The recordist must be well aware of any inconsistencies in this element. Inconsistencies may occur within a particular sound source, or there may be a change of the reference pitch level (tuning) of the cnsemble. Care must be taken to monitor the tuning of the ensemble and the intonation of the performers.

No changes in spatial properties should occur at the edit point, unless they are planned. It is common for spatial properties to be considerably different between time segments, when they represent different mixes of a multitrack master. Sudden shifts of distance locations are common between sections, and typically they represent few potential technical problems. Changes of lateral locations of images are different. Although sudden shifts of surround or stereo location are equally common between time segments as distance changes, both musical and technical problems can be created. Among the most common of these are unstable images, noises caused by phase differences between similar sounds at the edit point, and jarring movements or shifts of sounds.

Sound sources or environments that have a lengthy decay may need to be carried over across the edit point. This may or may not be possible, depending on the musical context, how the sounds were recorded, and the nature of the sounds themselves. Edits at these points are sometimes possible when other sources mask portions of the sound, but must be carefully handled to avoid audible changes of sound quality. These edits may need to be planned before the recording session, with suitable alterations made to the performances at the session, for instance starting the performers a bar before a planned edit point to have reverberation present across the edit.

The musical material must remain in rhythm: It is possible to add or subtract time in making an edit. Rhythm changes are very noticeable in their effect on the performance, and measures will appear to be extended or shortened by fractions of a beat.

Tempo changes or inconsistencies can occur between takes. The tempo of the performances will be carefully monitored during the recording process, but like all of the above it must be reevaluated during editing. Any tempo differences that are present between segments will make the edit point very noticeable, and will also make significant changes to the music. An entire take may be unusable, solely because of tempo inconsistencies.

Unwanted noises and technical issues can also be created at edit points, and be very audible or almost indistinguishable. The possibilities of things that can go wrong can seem limitless.

## Editing Techniques and Technologies

Analog and digital recording systems have some different characteristics specifically related to their technology. The inherent qualities of each format create advantages or disadvantages, depending on the application of the recording and the specific nature of the recording session. Either analog or digital recording may be the most appropriate choice, depending on the individual recording project. Sound is edited very differently in the two technologies.

### *Analog Tape Editing*

In an *analog recording*, a physical image of the sound is present as oriented magnetic particles on tape, and the physical characteristics of the image are directly proportional to the sound wave. In editing an analog tape, the tape itself is physically cut with a razor blade. Two cut ends of magnetic tape are joined (usually at 45°) with an adhesive tape.

Splice locations are found by slowly moving the tape across the playback head of the recorder. By rocking the tape across the head, the recordist is able to identify the edit point. The edit point is physically located on the tape at the playback head. The tape is marked, removed from the recorder's tape path, placed in an editing block, and is cut.

Once an analog tape has been spliced, it is difficult to redo an edit. Splices are difficult to separate without causing damage to the magnetic tape (which contains the sound—music). If the recordist is successful in undoing the splice without damaging the tape, it is difficult to cut thin time segments (pieces of tape) off the end of a magnetic tape (should the original splice be just a bit too far to the left of the desired edit point). It is almost impossible to add a small piece of magnetic tape onto the beginning of a tape segment (should the original splice be a bit too far to the right of the desired edit point). Identifying analog edit points, and the actual cutting and taping activities required of analog editing, all require significant skill gained through practice and experience.

Difficult edits are sometimes rehearsed. Copies of portions of the session tapes are made, and the copies are edited. The recordist gains confidence, or finds the precise edit points that are usable, on the copies of the tape, thereby allowing most errors to be made on tape that will not be used in the final version of the project. Obviously, this is a time-consuming process.

### *Digital Sound Editing*

*Digital recording* formats are quite different from analog. Sound exists as digital information, stored as data files. Specialized computers or specialized software for personal computers are used to edit the waveform. The digitized waveform can be altered by modifying and/or rearranging its digital information, but this need not be so. In many systems, edits are simply "play lists" of select portions of select files at precisely defined starting and stopping points.

A disadvantage of digital editing is that the sound cannot be held in the recordist's hand. There is no physical location of the recording and its component sounds. All editing is accomplished on a computer and must be conceptualized more abstractly than analog editing practices. Decisions might rely on the eyes, rather than on listening.

Conversely, the primary advantage of digital editing is that the sound is not physically present in the recordist's hands. The sound exists as computer information and may be acted upon in ways that are not restricted by physical limitations. The following items are the most commonly used among the many functions of most digital editing systems:

- Precise edit points may be identified and saved for future use, with great time resolution
- An edit may be heard, changed, reheard, and evaluated by the recordist; in many systems an edit might never need to be permanent
- Edits can be undone, quickly and easily, as the original data remains unchanged
- The edit does not alter the original material; the original recording is not edited; a copy of the original recording is edited, as a computer file (with no generation loss)
- Overall dynamic levels of the time segments on either side of the edit may be controlled to match at the edit point
- Edits may be made by cross fading from one segment to the other, or by suddenly switching from one take to another (called a butt edit)
- Some systems allow the signal to be heard as the recordist moves the cursor point (simulating the rocking of an analog tape across the playback head)
- Time, dynamics, and frequency processing are usually available to address specific types of inconsistent sound quality and relationships between the two time segments
- Special effects, such as looping and reversing sounds, are common

It should be evident that digital formats allow more flexibility within and control over the editing process than was available in analog editing.

## Signal Processing: Refining Sounds and Music

Whether incorporated into the tracking/recording process or within mixing, signal processing can play a large role in shaping the sound qualities of a project. Specific devices are chosen because their individual, inherent sound qualities lend themselves to the particular project—its desired overall sound and/or the qualities of its sound sources.

Signal processors each primarily control or alter one of the three basic properties of the waveform: frequency, amplitude, or time. The results of processing will reveal themselves in all of the artistic elements—including spatial dimensions and timbre.

The three generic types of signal processors each function on a particular dimension of the waveform. As learned in Part One, an alteration in one of the physical dimensions of sound will cause a change in the other dimensions. Furthermore, alterations of the physical dimensions will cause changes in timbre (sound quality). The three types of processors do not only cause audible changes in the three characteristics of the waveform, but they may also alter the timbre of the sound source. If considered according to how processors alter sound, signal processing can be simplified and approached with clarity.

- Frequency processors
- Amplitude processors
- Time processors

Frequency processors include equalizers and filters. Compressors, limiters, expanders, noise gates, and de-essers are essentially amplitude processors. Time processors are primarily delay and reverberation units. Effects devices are hybrids of one, two, or all of these three primary categories. Some examples of these specialized signal processors include flanges, chorusing devices, distortion, fuzz, and pitch shifters; and there are more. Many devices also include spatial qualities of location and environmental cues.

### Uses of Processing

Signal processing is used to shape sound qualities. It is applied to the sound source to complete the process of carefully crafting sounds. This is done for the character of the source's sound quality, and to shape sounds to complement the functions and meanings of the musical materials and creative ideas.

In the recording chain, signal processing can occur at a number of times. It may be incorporated in the tracking as preprocessing, and is most often used to bridge the tracking and mixdown processes. It can also be added subtly in mastering. Individual instruments or voices can be directed through any number of signal processors. Similarly, groups of instruments can receive the same processing, either in the same way or in differing amounts (such as any number of instruments, each sending a different amount of signal to a buss feeding a reverb unit). The entire recording might also be processed, as is common in the mastering process.

Signal processing often occurs separately between the tracking sessions and in preparation for mix session(s)—usually without performing actual re-recordings of the basic tracks. Tracks are evaluated and signal processors (hardware devices or the software equivalent plug-ins) are applied to the tracks to determine final sound qualities. Processor settings are often noted in session documentation for incorporation into the final mix, and the recordist performs signal processing in real time during the mixdown. While it is not common to change processor settings throughout the course of the mix, the ratio of processed signal to unprocessed signal might be adjusted, especially for reverb.

Signal processing sessions on a DAW are simpler. When the desired settings in one or more plug-ins have been established, they are saved as part of the project file. Of course it is also possible to send signals out of a DAW for signal processing, and this is a reasonably common practice, especially when one might want a specific device (reverb unit or compressor, for instance) instead of a plug-in.

### Listening and Processing

Signal-processing decisions involve critical and analytical listening. Critical-listening decisions will be made to establish qualities of the sound, for its own sake. Decisions will also be made concerning how the sound relates to other sounds and to the entire program. The recordist will focus on the smallest changes in the source's timbre, and on any number of higher levels of perspective. Careful attention to detail is needed, as well as attention to how small changes in one element or level of perspective impact the sound in other ways. Many of these changes are difficult to detect at first, and misperceiving results is common, especially when under pressure to make a quick assessment, as happens during a project with a client present.

The recordist will use the skills of Parts Two and Three to focus on the component parts of the sound qualities of the sound sources being processed. Small, precise changes in sound quality are possible with signal processing. This requires the recordist to listen at the lowest levels of perspective, and to continually shift focus between the various artistic elements (or perceived parameters) being altered. These changes are often subtle, and can be unnoticeable to untrained listeners. Often, beginning recordists are not able to detect low levels of processing. This is a skill that must be developed.

Most signal processing involves critical listening. The sound source is considered for its timbral qualities out of context and as a separate entity. In this way, the sound can be shaped to the precise sound qualities desired by the recordist, without the distractions of context.

Attention to and knowledge of the physical dimensions of the sound, and of human sound perception, are great aids for successful signal processing.

After the sound has been reshaped, the listener will use analytical listening to evaluate the sound. The altered characteristics of the sound source and the overall sound quality of the source will be evaluated as they relate to other sounds and to their function in the musical context. They will ask, "Are these changes appropriate for the music, or do they achieve the desired sound for the musical instrument?" The sound-quality shifts of processing will be evaluated according to their appropriateness to the musical idea.

## Preparing for the Mix

Mixing may not happen immediately following recording or tracking. Even if delayed for only a few weeks, the intervening time will bury the details of the project in the fog of memory. Documentation from the session is critical for saving time and ensuring quality. Make documentation a part of concluding sessions, and preparing for the next step of the production process.

Much time and attention is often needed in getting tracks ready for mixing. The flurry of activities in tracking or sessions opens many possibilities for things getting recorded that shouldn't be, or that aren't as they should be. This leaves work to do before the project is ready for mixing. This step is an often overlooked or under-recognized one; one that can add hours to the mixing process if not accomplished before starting the mix. Preparing for the mix ensures the artistry of mixing is not diminished by mundane and thought-occupying tasks that detract from one's attention on the music.

For a recording to be ready for the mixing process, the song (or album) should no longer need editing, arranging, track replacements, vocal tuning, etc.

All of the tracks should be recorded completely. If there are unresolved matters of arranging, such as "the band might still want to lay an harmonica track in the middle eight," the project is not ready for mixing. If the guitarist is not accepting of their solo, and continues to talk about replacing it, the project is not ready for mixing. If there is the possibility of a need to drop in a drum fill, or anything else on any other track, the project is not ready for mixing. This list could go on—indefinitely—but you are getting the idea. The arrangement must be done, and completely recorded.

Recordings nearly always need some editing to prepare them for mixing. These will be related to noise issues or the performance issues. One should not try to mix a project that is unmixable. Noisy tracks are unmixable. They have inherent problems that will need to be worked around. Foot noises, amplifier buzzes, bumped stands, squeaky foot pedals, door creaks, mouth sounds, fret screeches, count downs, double breaths or unnatural sounds in vocals, pop sounds, coughs—whether loud and noticeable, soft and barely perceptible—will all suddenly appear within a mix, if not dealt with here. Take a few minutes now and go back and listen to the early versions of "Let It Be" for noises, as encouraged during Chapter 14. These things can get through to the consumer, and can diminish the quality of the recording; and we cannot expect to be so fortunate to have our mistakes disappear because the music more than makes up for it.

Making certain the tracks are ready for the mix session also is about getting the performance tracks finalized and polished. Edit the performances as needed to get them into their final form. Re-arrange phrases, drop lines, make timing and tuning adjustments (sensitively, as needed), and more. Nothing that will detract from complete focus on mixing the musicality of the performance should pass through. Even *comped vocals* are most effective when reduced to a single track; correct timing and tuning as needed (desired), remembering slightly out of tune vocals can be ignored or can even add to the emotional impact of a great performance. This is where we learn a great deal about the successes and shortcomings of our tracking sessions, as we hear the results in a new day and find out what was really being recorded. You can learn a great deal by carefully listening to your results, and evaluating what other options you might wish to pursue next time.

Signal processing is often best delayed to take place during the mix, where its effects can be evaluated against other sounds as the mix progresses. In some situations, adding signal processing may be appropriate for some sounds at this stage of track preparation. Be certain of all consequences before making permanent adjustments. Some general ideas with documentation of potential processing can be productive during this process of finalizing the tracks, however.

If you will be handing off the project to another mixer, make contact with them regarding their preferences for preparing the tracks and other workflow issues. Discuss logistical issues like processing sounds during the mix versus preparing the tracks before delivery, how they would like tracks organized and labeled, documentation of processing or plug-in settings, mix suggestions and perhaps DAW automation of certain lines, and more. This conversation is about how to best serve the client and the project; making a smooth transition between recordists and doing all you can to make the next person's work more productive will ensure this, so do as much as is in your control to make this happen.

## Exercises

### Exercises 16.1

*Signal-Processing Exercises*

The reader will benefit greatly from methodically exploring the following topics. Each topic should become the focus of an exercise that would have the reader explore:

- How does the device/plug-in or process transform the waveform?
- What changes to the waveform can I recognize?
- In what ways do these transformations impact the audio signal and the sound source?

The companion website provides tracks that can be used as sound sources for many of these exercises, whether it be the sound of track 24, or 25's cymbal, one of the solo cello or solo guitar tracks, the piano pitch of tracks 10 through 13, or any others that might be of interest or appropriate. These tracks could be downloaded and fed into your signal chain as appropriate. Other source material available to the reader can certainly be substituted.

In all exercises, the reader should begin work on exaggerated settings. These will likely be far from artistically pleasing, but will cause the changes to be readily apparent and easier to perceive in first encounters where subtle alterations to the sound or changes in effect level might well go undetected. Upon repetitions, make changes to the source material smaller and smaller, subtler and subtler, to refine your ability to identify the changes and to observe the alterations created to the qualities of the sound source(s). Any exercise will benefit from repetitions with different sound sources.

Do not think about the musical result of these exercises at this point. Allow yourself the opportunity to simply learn sound qualities and relationships. Learn these devices and processes, as they are your tools. When you have control of your tools, you can start bringing your attention to making artistic decisions.

Spend time with the manuals (or Help files) of the devices or plug-ins to learn about and experience their potential. Again, start with extreme settings to identify the perspective and artistic element that must be the center of your focus, and then move to more subtle changes.

**Exercise 16.1.a**

*EQ*

1. Process to a single drum or cymbal sound by emphasizing the primary pitch area; seek to identify the area before starting this process, then boost that area to verify your accuracy; tune in on the primary pitch area by adjusting EQ settings; listen carefully to the qualities of the changes created.
2. Next, process an entire acoustic guitar track; listen for an EQ setting that will alter the attack characteristics of the instrument; notice the content of those changes, and try to control the changes throughout the instrument's range; the EQ will act differently on different pitch levels by "adding formants."

**Exercise 16.1.b**

*Noise Gate*

1. Route a bass drum sound through a noise gate; make adjustments to create extreme changes in the sound; gradually change one setting of the device to minimize its affect on the sound, and repeat this with each other setting; bring the device to the point of barely operating, and notice the subtle changes.
2. Repeat this process with a single cymbal sound.

**Exercise 16.1.c**

*Compression*

1. Apply compression to an entire cello track (such as track 39); make adjustments to the threshold and compression ratio to create extreme changes in the sound; gradually change the threshold of the device toward barely affecting the sound; return to original setting and modify ratio, bring the device to the point of barely operating, and notice the subtle changes.
2. Repeat this process on a full drum track (such as track 38).

### Exercise 16.1.d

*Delay*

1. Route a snare drum sound through a delay plug-in; set the drum to repeat at a moderate speed; adjust the delay time to create an eight-note delay, then a sixteenth-note delay.
2. Now change the delay time to 2 ms, then to 8 ms; note the difference between the two; if a difference is not easily heard, compare 2 and 20 ms delays with the delayed signal at a higher amplitude than the direct sound.
3. Repeat this process to recognize different delay times while using a single guitar pitch.
4. Experiment with creating rhythmic patterns with the delay, and pan the delayed signal separately from the source signal.
5. Finally, work with an entire guitar track (such as track 42 on the website) to identify and learn the effects of adding delay to individual sounds and to an entire track.

### Exercise 16.1.e

*Reverb*

1. Route a single bass drum sound through a reverb device/plug-in.
2. Examine the variables of the device and create a long reverb time, and set it at a pronounced level.
3. Experiment with the decay time and the dynamic envelop of the decay (as possible); learn these sound experiences.
4. With a setting for a long decay at a pronounced level, make a pronounced change in the low-pitch-area frequency response of the decay; listen to learn this change, then notice the impact of the reverb on the overall character of the sound.
5. Return the frequency response to normal, then make another adjustment to the area around 2 kHz; listen to learn this change, then notice the impact of the reverb on the overall character of the sound.
7. Now delay the onset of the reverb by 2, 4, 8, 12, 16, then 20 ms; notice these qualities, and try to learn their sounds.
6. Repeat these processes on the guitar, track 42, to learn how they impact the guitar sound; learn the effects of adding various reverb settings to individual sounds and to an entire track.

### Exercise 16.1.f

*Other*

Examine your DAW or rack for other signal-processing options. Explore their potentials systematically by following the same processes as above. Experiment and listen and learn the sound qualities of these sound modifications as well as equipment functions.

### Exercises 16.2

*Tracking Exercises*

There is no substitute for working with a performer in front of a microphone, while being responsible for capturing the sound of the performance. Create opportunities to experience this, no matter your situation. Quality and types of performer, equipment or studio/performance space may vary widely and yet still offer the opportunity to learn and experience, and to develop your skills and knowledge.

For the following, secure a single performer who is willing to play for you, and perhaps learn with you. As you progress in understanding, add more performers to the experience. In these beginning stages, it is important that you are not seeking to make a recording that will have value or use. Allow yourself the opportunity to fail in safety, knowing that in doing so, you will learn and develop your skill and ultimately acquire your craft.

### Exercise 16.2.a

*Microphone Selection and Placement*

1. Set up a single microphone in front of your performer; ask them to perform a simple passage slowly and repeatedly, holding the last note until it dies and leaving a moment of silence between repeats. Listen carefully to the microphone for timbral detail and related frequency response and distance cues. Cycle through these dimensions, listening carefully for changes and characteristics.
2. Try another microphone placement and repeat.
3. Repeat this process with a different microphone; then another.
4. Expand your listening for blend of source's sound, pitch area, and sound quality.

5. Continue to substitute microphones; alternate between several placements of the same microphone.

---

**Exercise 16.2.b**

*Listening to Record Levels and Qualities*

1. Alternate listening to input of console/DAW and playback of recorded signal (output of sound card, I/O interface, or record deck).
2. Seek to identify alterations to the source signal caused by the recording medium and signal chain, and listen for added noise of any type.
3. Trace any changes to their source, and learn to identify the sounds of these alterations.

---

**Exercise 16.2.c**

*Performance-Related Issues*

Have your performer lay an overdub over an existing track. Work on listening carefully during tracking and in evaluating recorded tracks for:

- Tuning of instruments, tuning from one take to another
- Tempo remaining consistent throughout a take, and changes of tempo between takes
- Loudness levels of instruments and a recording consistent within and between takes
- Musical expression and performance intensity: Do they match the qualities you (or the performers) are trying to pull out of the music? Do they remain consistent between takes?

---

**Exercise 16.2.d**

*Advanced Performance Issues*

1. Bring in two performers willing to patiently assist you, and repeat 16.2.a and c.
2. Next work with three performers. Track them performing simultaneously to separate tracks. Engage the issues of 16.2.c.

Remember to focus on sound qualities and performance matters exclusively. Allow yourself the opportunity to attend to only one issue at a time.

---

# 17 Crafting the Mix, and Finalizing the Production

The successful and effective mix is a vehicle for the music and its story. The qualities of the mix complement the qualities of the song, in all of its dimensions. The mix creates and establishes the context for the song—a world of sound qualities and relationships unique to the piece of music, to the recording.

Each mix will be unique, just as each song is unique—each mix composed as a unique set of ideas and relationships, some tried and true, and some newly devised. The mix does not need to be revolutionary to be unique; it just needs to effectively present what is unique about the piece of music. And yes, some mixes are revolutionary and astonishingly different from all others that have preceded it, but creating such a mix, that is also effective, will necessitate a song equally revolutionary in some manner; this opportunity rarely presents itself.

This chapter will allow us to explore how to craft the mix to support and most effectively present the arrangement of the song, how to use the dimensions of the mix to add richness to the song's instruments and voices. We will also consider the song as story, and its meaning, and bring focus to the real purpose of the mix: to most effectively communicate the core concepts and all of the elaborate details of the song's text, as primary to the song's musical materials.

The acts of composing and performing the mix will follow. The mix is crafted as a creative act, with much in common to composing and orchestrating, but with a different palette of sound dimensions and sonic relationships. Realizing or performing the mix follows next in this chapter. Though some of the actual act of realizing the mix might occur simultaneously with creating the mix, performing the mix is a separate act, one that brings focus on shaping the sounds of the mix in real time.

Mastering follows mixing, and is the final process before the recording is available to the consumer. This section explores that process, and how it fits into the project.

The consumer's interaction with the recording is the final topic of the chapter.

## Mixing to Support the Music and the Text

The mix is an integral part of the recorded song. It adds substantial dimensions to the song's arrangement, enhances its musical materials, and brings enrichment to the song's text. The mix can assist in delineating and connecting sections of songs, and can bring coherence and contrast to the song's evolving story.

This section will bring together the sections typically found within modern songs, and how they typically function and appear, song lyrics, and the stories and meanings they may present. These will be examined in relation to the mix, and the mix will be examined in relation to song sections and song lyrics.

### Sections of Songs and Song Structure

Sections of songs have materials and character that cause them to be distinct from others. Sections are delineated by their text, by musical materials (melodies and harmonic progressions, among others), and by their instrumentation (different instruments or uses of instruments in different sections). They have functions within the form and structure of the song, and each type of section serves an important role.

We recognize standard ordering of sections, and deviations from known orderings and characters. These allow us to understand the shape of the song and its structure. Most importantly they allow us to understand what makes a song unique, different from all others. Labels are useful points of departure to organize our thoughts; in defining the labels of song sections we understand that as points of departure these are norms to which there will be exceptions. It is with the exceptions that the norms are defined, and through which the art is enriched.

The most readily apparent sections of songs are their verses and chorus. Most readers will be able to recognize these when they hear them, even identify where they start and stop, but may not be able to articulate what makes them one and not the other.

The *verse* tells the story of the song. Verses tend to be personal, typically written in the first person. They speak directly to the listener and are the most intimate sections of a song. Verses present different text each time they appear in the song, as the storyline develops. Sometimes a first verse will reappear at the end of the song.

The *chorus* contrasts with the verse, and is a complete section typically of similar length to the verse. The chorus is repeated regularly throughout the song and retains the same lyrics each time it appears. It is often in the same structural relationship with the repeating verses, and their new textual materials. The chorus is more detached from the listener; its message is typically more objective and wide reaching than the verse. Very frequently it has a character that is larger than life, somehow more substantial and grand than the verse of ordinary life's concerns and situations. It is universal in nature and very rarely written in the first person. The chorus brings the listener to recognize that more people than the storyteller of the verse share the message of the chorus—its message is perhaps even a universal truth. The chorus offers a solution of sorts to the verse's problems, both musically (in its contrasting, answering materials) and textually (in its more objective message). Its text may be simple or more complex; it may even be one line of text that is immediately repeated once or more times during the section. Background group vocals are common; they give name to the chorus, and enhance the notion that many share the message of the section. The chorus will most often contain a line that emphasizes or highlights the title of the song.

The chorus of "Let It Be" is a clear example. The song title appears prominently in the text, and is repeated immediately and numerous times. The group of background vocals enters shortly after the chorus begins, though does not share in the text. The chorus section is repeated in a predictable way in relation to the recurring verse (with its new texts). The chorus may also incorporate some nonsense syllables to elaborate a text as the "yeah, yeah, yeah" in "She Loves You."

The *refrain* is similar to the chorus. Refrains typically replace a chorus, though some songs contain both sections. A refrain is one line long, and returns only after other material is presented. In this way it usually appears as a summarizing line at the end of a verse; it summarizes the ideas that have come before it, or presents an overarching idea that is more detached (just as a chorus). Its line of text is nearly always the title of the song, and its melody might be the hook of the song. Sometimes a refrain will not contain a text, but rather have nonsense syllables sung in its place—to create the same type of literary detachment.

"I Saw Her Standing There" is an example of a song with a refrain at the end of each verse. In "She Loves You" the refrain is stretched into two phrases ("because she loves you, and you know that can't be bad . . .") and becomes a section of its own; it resembles a chorus, until the more substantial chorus arrives after the second iteration of the refrain. A refrain may not contain a text at all, with nonsense vocalizations or syllables replacing a text, such as the lengthy final section of "Hey Jude."

When the verse and the chorus have materials that are significantly contrasting, a *prechorus* sometimes provides a way to make a smoother movement between them. While this section is transitional in character, it provides substance of both musical materials and the storyline of the text.

The *bridge*, as its name implies, connects two sections. It serves to take the music from one major section to another section (as in verse to chorus), while contributing motion and direction. It will have contrasting materials and character to either of the sections it connects; its materials typically have the greatest contrast to the verses' material. Rock songs tend to have instrumental bridges. When the bridge appears in instrumental form, it contains musical materials of little substance in comparison to the main ideas in the verses and chorus, as in the long bridge in "A Day in the Life."

A section with vocals and text may also be a bridge. This bridge will typically only appear once, and at a dramatic point in the work, though it may appear twice and will be nearly identical on its second appearance. It is a contrasting section imbedded within the song. The text typically presents another overarching impression of the subject of the song, and is often more universal in perspective than the chorus (when a chorus also appears). The text will typically not contain the song title, but rather be more dramatic or urgent, more introspective or desperate; it will sometimes present an alternative view of the song's main idea.

This bridge with text is sometimes called *middle eight*, because it is often eight measures long. "Ticket to Ride" contains a middle eight that appears twice, with both appearances identical. This middle eight is actually nine measures in length, as the eighth measure is extended through the ninth. The bridge of "I am the Walrus" momentarily suspends the song's driving motion and provides a fresh perspective ("Sitting in an English garden. . .") and a sonic oasis.

The instrumental *solo* or *break* can create even greater contrast. The length of this section is often similar or the same as a verse or a chorus. At times a lead instrument such as guitar, piano, or saxophone will play a solo melody time. In "Every Little Thing" a half-verse appears in instrumental form between the second and third choruses (see Figure 3.3) as the guitar presents the vocal line. "I've Just Seen A Face" contains a full instrumental verse performed by acoustic guitar mirroring the lead vocal, with alterations only in the final phrase. The break or solo may also present new and unrelated musical materials; consider the materials in "Day Tripper," with the verse's accompaniment gradually breaking down into new "instrumental-vocal" parts without text.

Many songs begin with an *introduction*. The introduction is any instrumental music that occurs before the entrance of the voice. Introductions most often foreshadow material from the verse or chorus, but on rare occasions borrow from another section. Their length is often the same as either the verse or the chorus, or a significant phrase of either, and sets the stage for their entry. Notice how the introduction of "Come Together" prepares the listener for the entry of the verse; with the same instrumentation, rhythmic materials, and the inclusion of vocals acting as an accompanimental instrument, its phrase length matches the first mid-level phrasing of the first verse.

Great variation of length is possible, as an introduction might be as simple and short as the single sustained chord in "A Hard Day's Night." "Love You To," on the other hand, is 39 seconds in length and foreshadows a melodic idea for the song only near the end, with the remainder observing North Indian Khyal conventions of Indian classical music materials and structure, featuring Harrison's sitar passages and improvisation.

On occasion introductions can be abbreviated versions of sections; most often this will be the chorus or the refrain and will include the song title. "Help!" is a great example of this, as it begins with a shortened version and a different arrangement of the later chorus. "Paperback Writer" begins with its signature multi-part vocal phrase of the song title before falling into a more typical introduction of rhythm section establishing the foundation for the entry of the verse.

The *coda* is an ending for the song. It typically contains material from either the verse or the chorus, sometimes both simultaneously, to bring all of the ideas of the song together to form a substantial ending. It rarely is contrasting, and "Strawberry Fields Forever" is an example of this exception, with its large coda that follows a false ending. Often the coda is a repetition of the chorus or part of the chorus with a fade, such as found in "Lucy in the Sky with Diamonds." A change of mix might be incorporated into the repetition and fade, and perhaps a subtle new line added, as appears in the last repetition of "Lucy in the Sky with Diamonds."

A multitude of potential endings have been used with material originating in the verses or choruses. From the simple and straightforward, as in the last vocal line being repeated instrumentally in "Yesterday" and the more unusual ending of the multiple piano chord of "A Day In the Life," to the unusual and more complex materials and coda fade-in of "Rain," or the new material and faster tempo of "Ticket to Ride," or the psychedelic results of the aleatoric processes that established the unusual effects of the entrances and treatments of voices, radio broadcast snippets, and excerpted *King Lear* performance superimposed on the irregular phrasing of instruments during the prolonged coda of "I am the Walrus," there are many more types of coda structures and materials that permeate The Beatles' catalog and are found in use today.

These definitions are not exhaustive, and this list of sections is not complete. Yet this is a reasonable summary for our purposes, and for this writing. As you, the emerging recordist, learn your craft, you will find greater richness in the sections of songs as you observe more artists and their creative ideas. Great richness also exists in how these sections combine to create the unique structures of songs, and their overall shapes of form.

Song structure follows some established practices. Exploring these practices is valuable for the recordist, but regrettably is well beyond the scope of this writing.

### The Song as Story and Message

The song is much more than music. The song communicates a message, and often tells a story. It is a story on a musical journey: a special message elevated by its musical setting. The mix adds dimension to the character of the story and the message, and enhances how the story is told, how the message is delivered. How to shape the mix to best support the story and the storyteller (the music performance) is at the heart of the mix engineer's craft.

The ideas the song communicates create drama, and progress from a beginning to a conclusion, typically through a series of verses. Along the way many songs will often have a recurring section to ground the story in a bigger context; this is typically either a refrain or a chorus. Further commentary and another perspective may be presented in a middle eight, or bridge. The structure of the song supports the presentation of the text, and reinforces its shape and how its parts come together.

Song lyrics are poems of sorts, but different from literary poetry. Literary poems do not commonly have the traits that make for quality successful song lyrics. They are two separate art forms: There are certainly some similarities; there are also many differences. Some songwriters have crafted lyrics that rival the intricacies and abstract nature of literary poetry, but these are rare; these also rarely contain the chorus stanza or refrain lines that ground the tradition of song lyrics.

### *Content: Meaning and Ideas of the Text*

The song lyrics will contain primary, or central, concepts and ideas. Nearly always these will be reflected in the title of the song. This central concept is what the song is "about," and forms the basis for its story and its message or meaning.

The primary concept also provides a common thread to all other ideas and subplots within the text. These other ideas, images, activities, etc., in the text are supporting materials or sub-ideas/concepts. These are ideas that either provide added substance and dimension to the story or its message, or supply ornamentation, detail, and richness to it. The ideas and concepts of the text establish layers of activities and meanings that can bring great complexities to some songs; these ideas can even stretch outside of the song itself and bring in associations to other works of art, other poems, other events, or ideas unrelated to the poem itself.

This all establishes the literary meaning of the text. There may be a deeper meaning than the surface text, or not. There may also be associations to outside works or influences, or the song may be self-contained. Still, the lyrics will communicate a story of sorts—sometimes quite literally, sometimes more abstractly—in communicating a message without the drama of action.

The lyrics will often contain drama: an unfolding of the story in individual lines, combining to larger sections as stanzas, gradually revealing to the listener action, meaning, emotion, etc. The story may take place entirely in one time period, such as the present, or the immediate past. It is reasonably common for a song to shift time periods, with (as an example) the verses to be occurring in the present and the chorus (in all its appearances) to be in a past or a future tense. The shift of tense can allow the chorus to present a more profound message, or assume a more substantial character. This juxtaposes action and commentary, immediacy and detachment of communication with the audience, perspectives of the personal and the universal, the present tense of today and the knowledge of the past or projection of a possible future, and more.

### *Structure: Organization of the Text*

The structure of the lyrics is usually reflected in the structure of the music. Major sections of the text are organized into stanzas, and the stanzas typically set to music in separate sections of the song. Texts are structured, or organized, by:

- Number of stanzas
- Recurring stanzas (for the chorus, if present)
- Number of lines per stanza
- Recurring lines (for the refrain, if present)
- Number of syllables per line.

The number of lines per stanza is typically the same from one verse to another, and with each appearance of the chorus. It is common for the verses and chorus to each have the same number of lines, though it is quite common also for them to have a different number of lines.

The number of syllables per line will typically create a pattern within the stanza. The pattern will create consistency and variety to bring some order and unity to the text. While numerous patterns are possible, let's examine one. For example, a four-line stanza might have these syllables per line:

12 syllables
9 syllables
9 syllables
12 syllables

The pattern 12, 9, 9, 12 provides order and variety at the perspective level of each line. The repetition of 12 at the beginning and at the end provides consistency and order, as does the repetition of 9 in the inner lines. The overall shape of the outer two lines being the same, and the inner two being the same, provides the unity of a recognizable pattern. If the inner two lines instead had 9 and 8 syllables per line respectively, the stanza would have more variety at the expense of reduced consistency; this might be desirable or may not be, depending on the stanza and its context.

This number of syllables per line creates a rhythm to the text. Patterns of syllables create rhythmic patterns, as the rhythm of words and their sounds combine with the rhythm of the melodies that present them. These are sometimes one and the same, and sometimes quite different. The mix can provide opportunities to draw attention to these rhythms and the relationships of lines.

Consider the non-poetic, fabricated stanza of Figure 17.1 for syllables per line and rhythms within the text.

Notice the internal rhythms of the inner two lines, patterning of syllable rhythms and word rhythms: "syl-la-bles long" and "comes to the song." Notice the first phrase of the first and last lines for similar rhythm and pacing.

This idea of number of syllables can also be applied to rhyme schemes. Returning to the above lines, note the end of the lines. Lines two and three rhyme, and lines one and four end in different word sounds.

Now return to the list from above: an abstract idea of a stanza with 12, 9, 9, 12 syllables per line respectively. Consider that the first and last lines might rhyme. The middle two lines may also rhyme between themselves or they may each end in a different word sound. Other

First line of the stan-za, in-tro-du-ces sub-ject (12 syllables)

The next line is nine syl-la-bles long (9)

Or-gan-i-za-tion comes to the song (9)

Stan-za's last line con-cludes, sub-ject ends with no rhyme (12)

**Figure 17.1**
Sample stanza structure of four lines, with syllables per line of 12, 9, 9, 12 respectively, and inner lines rhyming.

rhyming schemes could exist: Lines one and three could rhyme, and two and four could rhyme, even though they have different relationships of syllable content; lines one, two, and three might rhyme, and line four end with a different sound. This sense of word-sound rhyme brings an added layer and sense of rhythm to the text.

Building on this, repeated words, repeated word sounds, and word rhythms can also be used to organize a text, or to add ornamentation to the text. The difference is the level of significance to the word sounds or word rhythms in determining the substance of the text's structure and shape, or a state of providing embellishment to the text's substance. Notice the repetition of words in the first and fourth lines in the above stanza; the lines both use "stanza" and "subject," placed in different locations within the line and with different rhythmic emphasis.

Rhyme schemes do not need to be employed directly for word sounds to bind a text. Alliteration can be used, beginning important words with the same sound quality. Texts may make use of words of similar but not precisely the same sound qualities, and treat them as the same or as rhymes. There are countless additional approaches to organizing text by word sound. All serve to bring structural coherence to the text, to enhance its message and meaning, and to allow the song structure and the text to find synergy in creating stronger connections and contrasts.

## Shaping Song Structure in the Mix

The mix of the song and the song's arrangement are fundamentally linked. The mix is how the arrangement appears in the recording. This is how the mix engineer acts as an orchestrator, spreading the timbral details of the instruments and voices across the dimensions of the mix.

Just as the arrangement supports the structure of the song, so does the mix. Just as the arrangement adds to the unfolding, presentation, drama, and effectiveness of the song, so does the mix. The mix brings the arrangement into the recording, gives it added dimensions, and brings it to life. The mix delivers and supports the story of the song, and its meaning.

The mix shapes sound structure by supporting the arrangement, which presents the musical materials. The mix pulls together all of the instruments and voices, and provides them with appropriate sound qualities and sound relationships, to best support the musical materials and the story or message of the song. The mix shapes the experience of the song: it is the vehicle for the journey, the voice of the message of the song.

### *Mixing Song Sections*

Separate sections of the song will have a mix with sound characteristics that are different from other sections. Choruses will typically have different sound relationships than verses, and bridges will have mix dimensions that are different from either the verses or the chorus.

Introductions and codas will typically have some strong connection to the verse or chorus, and will be mixed similarly so a connection and sonic continuity are evident. Other sections such as bridges, solos, and prechoruses will tend to have mixes that have dimensions in contrast with the verses and choruses; often the contrast is significant.

There will typically be some aspects of the mix (instruments and voices) that retain the same or very similar sound dimensions from one section to another. These will establish continuity of the mix throughout the song, or at least through the majority of the various sections of the song.

Importantly, there will be some sound sources that will change between sections, perhaps markedly. Vocal parts in particular exhibit change; this is especially noticed between verses and the chorus. Notice the change in George Harrison's vocal as the song changes from the verse to the prechorus in "Something"; the image of the lead vocal contains further width changes and distance shifts to additional locations as the song unfolds. Some instruments have different levels of importance from section to section, as their roles and their musical materials change; how these instruments sit in the mix will reflect this, in order for the mix to fully support the musical materials and the song's character. Listening again to "Something," notice the change in Harrison's lead guitar lines throughout the song; the guitar emerges from the fabric of the background instrumentation in various forms at various times. The guitar is sometimes left, sometimes more centered and wider; its loudness shifts gradually louder and sometimes suddenly louder, with performance-intensity cues often not aligned with loudness.

The mix may incorporate abrupt and/or gradual changes to sound dimensions of instruments or voices in the mix. Abrupt changes in the mix between sections are common; most songs have them. Abrupt changes within sections are not as common, but are also not rare.

The overall character of a section is supported by and reflected in the mix's overall sound qualities. The mix will allow the nature of the section's music and ideas to be evident if it supports them sonically. As one example, the way the mix blends or delineates sounds relative to one another brings their musical materials to do the same; the recordist must recognize which is appropriate and realize that sound in the mix. This holds for all of the other dimensions of the mix, as will be described later.

Not only will the overall character of the section be reflected in the mix, the overall qualities of the music and its story are also reflected in the mix. Successful mixes do this convincingly; less successful mixes might do this in a more general or rather cursory manner. We will explore this next.

#### *The Mix's Overall Shape*

As we know, the mix has sound qualities and relationships that exist at the highest level of perspective. In total, these qualities will provide shape and drama, movement, build and release, context and points of reference, and more. The overall qualities of the mix will support the overall qualities of the music and its journey through the text, and all that the text represents.

The mix aligns with the structure of the song. Changes in the mix typically occur at new sections (or the return to previous sections); these changes are typically abruptly changed as the section changes. Gradual changes in the mix typically happen within sections that connect one section to another (such as bridges or introductions or the prechorus); they help provide motion and direction.

Overall motion and direction, shape, drama, and movement are created by the relationships of sections, more than by changes within sections. This more or less holds for the mix, the arrangement, the musical materials, and the text.

As an example of a very common approach, while a song progresses, its repeated sections receive slight changes in musical materials, text, arrangement, and mix; sometimes these are marked changes. These changes build so that each following repetition of the chorus is slightly more urgent than the one before; each new verse brings more angst or drama in the unfolding story. This continues, sometimes with other sections inserted, until the peak tension of the song is reached; after the sustain of that tension (whether momentary or prolonged), the sections become less urgent, intense, complex, and the tension of the song relaxes until it ends.

Of course there are many variations to this concept, and each song navigates this tension and relaxation, build and release, increase and diminish, in its own way. These changes are sometimes quick and substantial or fast and small; they sometimes happen slowly and are extreme, or gradual and subtle. The changes can happen in any aspect of the mix, of the musical materials, and within the arrangement.

There are many possible overall shapes for the drama and tension of the text, of the music, and of the mix. This is just one form in which it regularly appears. Further complicating our understanding, but enriching our experience of the song, the overall shapes of the text, the music, and the mix are not necessarily parallel in how they proceed—in fact, they rarely are.

An example of changes in levels and states of the mix occurring within a section can be found in "Here Comes the Sun." By reexamining Figure 8.3 we can readily identify the rising dynamic levels within the middle section. As each phrase of the section is repeated, the characteristic dynamic contour appears at a louder level. The overall loudness level (program dynamic contour) is shaped by the mix process, as each repetition gets louder and pushes the tension of the middle section to be greater and greater until the peak is reached.

Building the complexity and intensity by making changes between sections is quite common. We can see this in the accumulating instrumentation of "Let It Be" from the album *Let It Be*. Table 17.1 shows us the building instrumentation of the song by section until the bridge. The unfolding of the arrangement, adding instrumentation gradually by section, supports the building tension and expanding nature of the song. By comparing the mixes of the single release and the album release (and perhaps the re-mastered versions and the *Let It Be . . .Naked* version), we can see stark contrasts in how the mix can reflect and enhance this unfolding arrangement, while being significantly different. We can ask ourselves about effectiveness of how well the mixes support the musical ideas and the message and story of the song. How have the overall qualities of the song and its musical materials and message/story been altered or enhanced by each of the different versions?

**Table 17.1** Instrumentation building by section in "Let It Be" from *Past Masters, Volume Two*

| *Introduction* | *Piano* |
|---|---|
| Verse 1 | Lead vocal, piano |
| Chorus 1 | Lead vocal, piano, background vocals |
| Verse 2 | Lead vocal, piano, high hat, organ, bass |
| Chorus 2 | Lead vocal, piano, background vocals, complete drum set, organ, bass, brass |
| Bridge | Electric piano entrance |

## The Mix: Composing and Performing the Recording

It is often helpful to consider the mix process in two stages: one artistic, composing (crafting) the mix, and one technical, performing (executing) the mix. While the two often happen almost simultaneously, they require different skills and thought processes. It is a duality of function that challenges many people, as the creative flow is diverted by the technical, with great complexity in both realms.

The process of planning and shaping the mix is very similar to composing. Sounds are put together in particular ways to best suit the music. The mix is crafted through shaping the sound stage, through combining sound sources at certain dynamic levels, through structuring pitch density, and much more. How these many aspects come together provides the overall characteristics of the recording, as well as all its sonic details. Consistently shaping the "sound" of recordings in certain ways leads some recordists to develop their own personal, audible styles over time.

The process of executing the mix is very similar to performing. Mixing often involves controlling sound in real time. Controlling the loudness levels of tracks, changing panning, muting tracks, altering processing, and more, are routinely performed during mixdown—especially in systems that do not have automation. The sense of performance is visceral and real.

Many technical decisions also occur during mixing; many are out of real time. Recordists will develop their own approach to sequencing those activities and decisions, and ultimately create their own working methods.

Often the acts of composing the mix and performing the mix happen within the same time frame. The piece is shaped as parts come together, and creative ideas are refined as new ideas emerge from hearing portions of the work as they are being completed. Sometimes new ideas take the project in new directions entirely; sometimes sounds or relationships of parts are "discovered" during the mixing process. Creative ideas get refined as a mix unfolds.

Keeping the technical process and all of the devices and controls from absorbing the creative energy and disrupting the flow of the project is necessary, and is not an easy thing to accomplish. It is necessary to learn to fluently operate all devices and software of the recording chain so one is in complete control of the recording process. Only with the technical processes under control and in the foundational background of the recordist's mind can attention truly be directed to crafting the artistic dimensions of the recording.

During the mixdown sessions, the separate tracks that were recorded during tracking and synthesis will be combined. Many of the recording's sound relationships are crafted in the process. To compose the mix, the recordist shapes the artistic elements—we have discussed this in detail previously. It can be helpful to conceptualize the elements into three groups, or broad areas of focus. Approaching each separately in planning and executing the mix can simplify matters, and help the beginning recordist clarify their ideas and approaches to a mix. As we already know, each of these significantly shapes the recording, and each will also impact the other two.

These three groups are listed below, with the elements and some other concerns they contain:

1. Musical balance and loudness
   a. Loudness relationships
   b. Prominence versus loudness
   c. Dynamic contours by loudness levels and by intensities
2. Spatial qualities and relationships
   a. Dimensions of the sound stage
   b. Lateral placement and size of sources on the sound stage
   c. Distance placement of sound sources, and listener-to-sound-stage distances
   d. Environments of sources and depth of sound stage
3. Pitch and sound quality
   a. Pitch density and timbral balance
   b. Performance intensity and sound quality
   c. Environments of sources and sound quality
   d. Dynamic contour by density and/or register versus loudness

Successful mixes find a unique balance of sounds and sound qualities to best serve the message and musicality of the song.

Before we explore these three groups of elements individually, we will examine how the mix interacts with the music's message and its musical materials.

**Listen**

to tracks 48–53

for these musical balance, pitch and sound quality, and spatial quality groups and the elements they contain, found in these six different mixes of the same drum performance. Note that all of the materials discussed in this section play directly into these different sound qualities. Observe these various production aesthetics and the sound stages and sound qualities that are created. Use any graphs from Part Three that might be helpful.

## Composing the Mix: Presenting, Shaping, and Enhancing Musical Materials

The mix is crafted to bring musical ideas and their sound qualities together. The song is built during the mixing process by combining the musical ideas, focusing on shaping the sound qualities of the recording. The recording is the song, and it is the performance of the piece of music.

A successful mix will be constructed with a returning focus on the materials of the song, and the message of the music. The musical ideas that were captured in tracking are now presented in the mix in ways that best deliver the story of the text and the character of the music. Remember: The song is about more than the music.

Referring back to the hierarchy of musical materials in Part One, we remember that the structure of a song, or piece of music, is created by primary and secondary musical elements. These take place at different levels of importance to the overall musical structure, and at different levels of perspective for the listener. How the musical ideas are crafted and combined is in the area of music theory, and, while not covered here, this information is of great value to the recordist. The ways the recording process will combine the sound sources' materials will present their musical ideas in new relationships, and can profoundly shape and enhance them. The mix will greatly influence the musical style of the piece of music. This influence can provide great support for the musical ideas and elevate the music, and the opposite is possible with unskilled applications of these concepts and techniques.

While it is important to establish a clear idea of the final sound qualities that are desired for the song/recording, just where to start will vary between individuals. Some people work from small ideas, adding and building to create large sections. Some people need to have an idea of where they want to arrive, before they begin crafting the smaller details. Often individuals will use different approaches for different songs. *How* one arrives at a vision of what the song needs to sound like is not important. What *is* important is having a strong sense of the desired overall sound qualities of song, and a strong sense of how some (or most, or all) of the details of the music will bring this to reality.

A general starting point for this process may prove helpful. Consider the three groups of elements (from above), reduced to the actions of balance, panning, and processing. You may wish to think of these as signal flow, or input module layout: fader; pan pot; EQ, compressor/noise gate, FX send/return. Within these three production steps the mix is realized—composed and performed.

Remember you are shaping the overall sound qualities, as explored previously. As you work, establish some conceptual distance from the mix periodically, and consider:

- The song's sense of intimacy with the listener, created by the perceived listener to nearest sound-source distance
- The energy level, intensity, and expressive qualities of the song that create a reference dynamic level
- The overall dynamic shape of the song (program dynamic contour)
- How the song uses the pitch registers to create a timbral balance or spectrum of the overall recording
- The width and depth of the sound stage (stereo or surround)
- The impression of the song's performance space, or perceived performance environment
- All of these and the musical structure coming together into a global shape and concept of the song, its form.

Some simple and direct questions can sometimes aid in determining or clarifying these areas:

- How does the story of the text unfold?
- How does the music support this?
- How can the mix support this?
- What relationship do I want the listener to have with the music? (Observing from afar or intimately close? Maintaining a comfortable distance? In a large performance space or small? Vocal dominating the texture or blending into a grouping with the band? And many more.)
- Are there special qualities that are needed to most effectively present what the song is trying to portray?
- What must be done to achieve these?

Certainly many other questions are equally valid and potentially important.

The clearer the vision of what is sought in the areas, the smoother and more effectively the mix will progress.

As we have learned, mixes create relationships of individual sources or small groups of sources when they are assembled. These sources (and the musical materials they are playing) are all given their own individual or a shared "place" in the mix. This "place" might be a "space" or "location," a "level" or "area," a "character" or a "set of characteristics" in each artistic element, such as lateral location, register, dynamic level, etc. In this way, mixing is the act of putting everything in its place, while giving any final shapes to the sounds.

Just where to place a sound source is a matter of what will best suit the song, and what will deliver the desired sound qualities. Every song is unique in some way; your mix will be unique in a complementary way.

Among important questions to ask are:

- How can this instrument/voice, presenting this musical idea, be placed in the musical balance to contribute most effectively?
- What spatial qualities and relationships will most effectively present this instrument/voice and its musical material?
- What sound qualities are best suited to this instrument/voice presenting this musical idea?
- How can I best present or enhance this sound source and musical idea?
- Should this idea and instrument (bass, for instance) be emphasized, or should it be blended with others (bass, keyboard, and bass drum)?
- What special qualities do I want to bring to this instrument to enhance the song?
- How can these instruments be combined to provide the sound qualities and sound stage that is desired?
- What musical materials/instruments contribute most to defining the energy or the character or the message or the sound quality or the expression of the piece—and how should they be treated?
- What musical materials are supporting the primary ideas, and how can they be made to do this most effectively?

The answers to these questions (and others that are similar and others that are more detailed) will lead the recordist with direction and purpose to crafting a mix that supports the music and tells its story in the most appropriate way.

### Crafting Musical Balance

The entire mixdown process is often envisioned as the process of determining the dynamic-level relationships of the sound sources. As we now know, the mixdown process is actually much more. It combines many complex sound relationships, of which dynamics is only one. But dynamics may well be the most important one.

Working with musical balance is often the most effective place to begin constructing a mix. In a way, getting at least a rough musical-balance mix is a necessary first step to crafting the mix, as sounds are not present until they have a loudness level. Refining the musical balance will often take place throughout the mix process, though it will be the focus at one time or another.

The musical balance crafted here is the relationship of the dynamic levels of each instrument to one another and to the overall musical texture. The individual sound sources are combined into a single musical texture, each source at its own loudness level. Small changes in level of a single sound source can make a large impact in the mix. It is critical to obtain the ability to recognize and control loudness changes and relationships. A return to Exercise 8.3 could prove valuable now.

Every skilled mix engineer will have their own preferred way of mixing loudness levels. Some individuals will be more flexible in their approach than others, and some may change approaches to best suit the client or project. Among commonly discussed approaches to starting a mix balance are:

- Adding one instrument at a time, starting with the most important sources
- Bringing up the vocal, then adding instruments one or a few at a time as seems appropriate
- Adding one instrument at a time, starting with drums (there are widely varied approaches to which drums or microphones to begin with)
- Adding groups of instruments, starting with background instruments
- Bringing up all of the faders at once, then start adjusting.

Establishing an appropriate balance of loudness levels is often a painstaking process. Too much level or too little level skews the importance of musical lines, disrupts the blend or clarity of sound qualities, timbral balance, and so much more. It can take a mix from good to breathtaking—or to boring or utterly forgettable.

Some things are important enough to bear repeating. First, perception of actual loudness level is often mistaken for other things. It is often easy to confuse the prominence of a musical idea or instrument with loudness. A sound may be prominent because of some special quality (register placement, for instance), while actually being at a lower loudness level than other sounds. A sound can be the most prominent in the listener's consciousness, while being at a lower dynamic level than other sounds. The other artistic elements have equal potential to provide outstanding qualities to the sound, and to cause the sound to stand out of the musical texture.

Second, loudness is often confused with or distorted by the listener's attention to certain aspects of a song, by unexpected events in the music, and by the meaning of a text or the music. The listener may be drawn to the text of a song, and the singer might be perceived as the loudest musical part; while this can be the case, it often is not. This understanding can sometimes allow the recordist to lower the loudness level of the vocal without moving the attention of the listener.

The musical balance graph of Part Three can be used to plan musical-balance relationships, or to take notes during production. While a complete graph will never be created in production practice, beginning recordists might find writing out a few actual relationships of important instruments and vocals helpful in understanding and planning their early mixes. Most importantly, the graph can help one to learn to bring focus to loudness alone, and not to confuse other perceptions as being loudness related.

### Creating a Performance Space for the Music and the Recording

Panning is where we begin to give sound sources spatial qualities in the mix. These qualities provide an illusion of a performance space for the recording—an imaginary place where the music was performed, with dimensions and sound qualities. The mix establishes these dimensions and qualities.

As we have learned, individual sources are shaped in terms of placement on the sound stage and environmental characteristics. In placing sounds on the sound stage, they are distributed in space. Sounds are allowed their own place, size, distance, dimension, space—or they are blended with others in one way or another.

Spatial properties for the overall program are also crafted in the mix. The spatial properties created in the mix provide an illusion of a space within which the performance takes place. As sounds are placed at locations within this perceived performance environment, the dimensions of the sound stage are defined.

All of the recording's spatial qualities are crafted during mixing, utilizing and relaying on any spatial qualities that were captured when the sounds were recorded. Again, the dimensions are:

- Dimensions of the sound stage
- Placement and size of sources on the sound stage (lateral and distance)
- Listener-to-sound-stage distance
- Environments of sources and depth of sound stage
- Perceived performance environment

During the mix, sounds will be placed at specific lateral locations, as phantom images or at speaker locations. These images will be given a width anywhere from the breadth of the entire available sound stage to a narrow point in space. This lateral location can have great impact on separating sounds in the mix, or blending them. The size and location of sounds can provide prominence or importance for a sound that would otherwise be less noticeable.

It is helpful to remember humans do not localize sounds equally well at all frequencies. Further, we primarily use interaural time information to localize sounds below 2 kHz, and interaural amplitude cues above 4 kHz to localize sounds. This requires the recordist to consider the pitch area of the sound source, and to work with amplitude, time/phase, and spectrum appropriately, to accurately create stable images. In this way, time processing joins the pan pot in determining image location and size.

Taken as a sum, the lateral placements of all of the sound sources provide the listener with a sense of width of the sound stage. The listener develops a sense of "where" the sounds are and the size and location of the stage where the song is being performed. This shapes an overall quality of the recording that is very important to the context of the music.

The location and size of the sound stage has become increasingly significant, as the industry has engaged surround-sound production practice. This is a major factor that separates different approaches to surround-sound production, and revolves around how the rear channels are used for different types of program materials (especially musical materials and reverberant sound). When sound sources are placed behind the listener, for instance, they might perceive themselves as sitting within the sound stage. The rear channels might also be used to pull the sides of the sound stage wider than is possible with two-channel stereo playback. Of course it is also possible (and not uncommon) for the sound stage to remain in front of the listener, and for the lateral imaging to be very similar to a traditional stereo recording, with the addition of natural or manufactured ambient sound appearing behind the listener.

The other dimension of the sound stage is distance. Sounds are placed at a distance from the listener, and provide the illusion of depth to the recording. As a group, the distances of all of the sound sources provide the front-to-back dimension of the sound stage.

Distance placement can provide a special quality to a sound source. It can allow a sound to be clearly apparent in a musical texture or to blend more with other sounds. The importance of distance cues is often underestimated. These cues bring musical materials and instruments/voices into a physical relationship to the listener that can be profoundly effective in helping the musical message or expressive nature of a line. Distance can bring a musical idea to have a strong and close physical connection to the listener, or can provide a sense of separation from the performance and the musical idea; this has the potential to greatly shape the listener's sense of the musical idea. A very different sound stage exists when all sources are at approximately the same distance, as they appear to be performing in a similar area and have a sense of connection in space, than when all sound sources are at even slightly different distance relationships, as they appear to extend the sound stage and bring sources to have differing relationships to the listener. In recordings where sources are extended from close proximity into the far areas, with many sounds in between, the sound stage can achieve vast proportions and bring a substantial new dimension to a mix.

Distance information is captured in recording; it cannot be created or added to the sound. Distance can be altered by all functions: balance, panning, and processing. Balance and panning can allow the timbral detail present in the sound to be audible in the mix, and control the extent to which this detail is present or is blended away. Processing will only increase distance—pushing the sound further from the listener by diminishing timbral detail.

It is important to remember that distance cues are primarily the result of timbral detail. A high degree of low-level spectral information brings the timbral detail that must be present for a sound to appear very close. Sounds become more distant as this detail is removed. Placing a sound in the mix can alter its timbral detail, as it blends with or is masked by other sounds. Mixing in a way that is sensitive to pitch density and lateral location is especially helpful in preserving a sound's timbral detail.

Reverberation can alter distance. This can be because of the ratio of direct to reverberant sound, but is most often because the reverberant sound masks the timbral detail of the direct sound. The reverberation's arrival time gap and the reflections of the early sound field will also provide subtle cues that can shape the imagined distance of the sound source. When adding reverb to a signal, it is important to bring attention to how it is impacting distance.

The listening perspective of the recordist will alternate between locating individual sound sources on the sound stage, comparing locations of sources to one another, observing how all sources create an overall sound stage (and observing how balanced the stage might be), and recognizing how the sound qualities of sources might be transformed by placing the sounds on the sound stage. These observations will be made for both distance and lateral location cues.

Sound stages can be planned and evaluated using the diagrams of Figures 17.2 and 17.3. These will prove helpful in balancing the sound stage and in creating variety and interest—as desired—of image locations and distances. These diagrams are snapshots of time and may represent any time segment, from a moment to a complete song. They will prove useful to sketch sound stages of mixes, whether an entire symphony or individual sections of a song (such as verses and choruses).

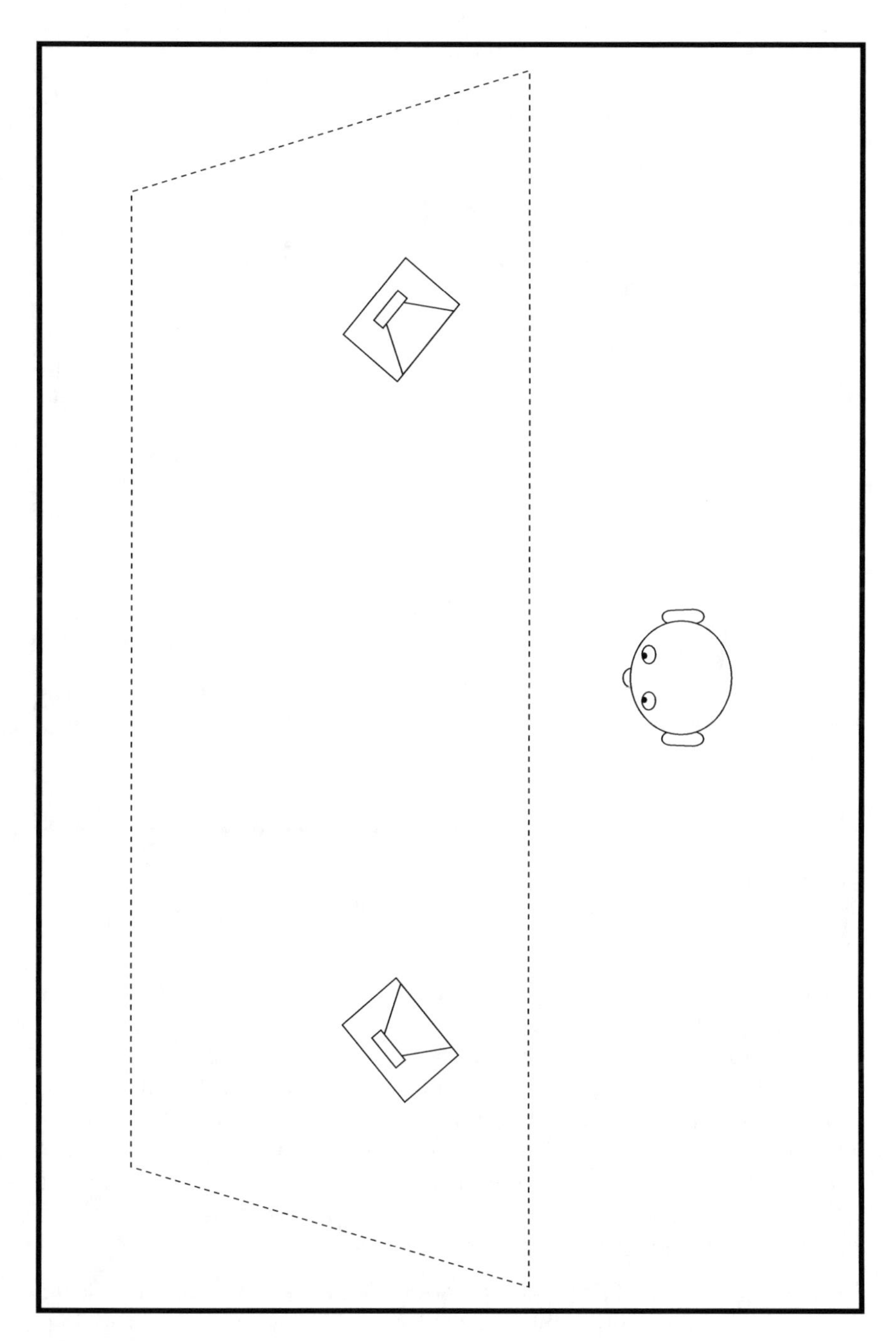

**Figure 17.2**
Sound-stage diagram for two-channel recordings.

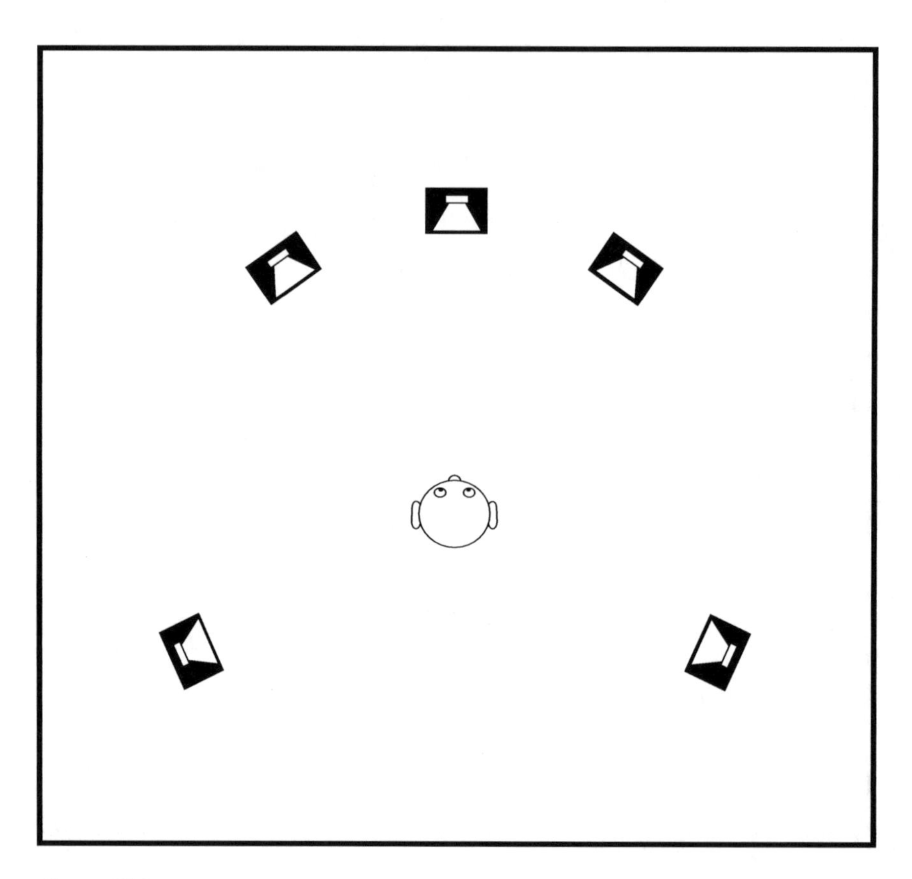

**Figure 17.3**
Sound-stage diagram for surround-sound recordings.

These diagrams allow imaging to be recognized and understood. The dimensions of the sound stage can be drawn around the loudspeakers in each figure, and the listener location will be determined in two-channel recordings. The diagrams should show the front edge of the sound stage, giving the recordist a reminder of the sense of placing the listener at a specific conceptual distance from the recording, and the depth of the sound stage, providing important environment size and distance information. A significant set of sound relationships can be planned, crafted, and evaluated with these diagrams.

Typically a product of processing, each individual sound source is placed in an environment during the process of making the recording. Qualities of the recording environment may have been captured during tracking, and these may provide all of the environmental characteristics that are desired. Very often environmental characteristics are added to a sound source or a group of sound sources during the mix. If a sound is added into the mix without environmental characteristics (such as a direct-in electric bass), the listener will imagine an environment (often a very small one).

Environments have timbre and time properties. When a source is placed in an environment, it acquires the spectrum and time/reflection components of the environment. The environment and sound-source timbres are fused into a single and new sound quality.

Environmental characteristics can add important dimensions to a sound source, and shape their sound quality in significant ways. They also can add to the dimensions of the sound stage. These characteristics are added during this panning stage if part of the original tracks, or during processing if they are added by a reverb device or plug-in.

Crafting spatial properties is as important to the mix as any other element. Approaching panning as one of the three primary phases of constructing the mix will bring the beginning mixer to embrace this directly.

### Pitch and Sound Quality Concerns in Combining Sound Sources

The mix also combines the pitch- and frequency-related sound qualities of instruments and voices. When sounds are combined in the mixing process, the timbres of the instruments/voices are blended. This blending of sound-source timbres can bring sounds to fuse together into a group, or if handled differently the sound sources can retain all or some of their unique characters, even if they occupy a very similar pitch area. Carefully employing the other elements of sound (such as stereo location, loudness, etc.) can keep similar timbres from fusing in the mix.

The sound quality of the sound source plays a significant role in the successful presentation of the musical idea. The instrument or voice's timbre is shaped to most effectively present the musical material, and the listener will ultimately come to identify the musical idea by the instrument (or singer) that delivers the musical idea. This process of shaping the sound quality of the sources began in tracking, with instrument selection, microphone selection and placement, performance intensity and expressive qualities of the performance, and perhaps signal processing. Sound sources will now receive final shaping during mixdown by signal processing (whether hardware or plug-ins). All changes, subtle or significant, are provided to sounds to achieve their final characters and their final placement in the mix.

This final shaping can be used to enhance the character of the sound, or to help a sound combine more effectively in the mix. Time, spectrum, and amplitude processing may all be employed for these purposes.

The sound quality of sound sources also contains the dimension of environmental characteristics. We recall that the sound source and its host environment fuse into a single impression. In this way, the sound qualities of the environment become a part of the sound qualities of the sound source. Shaping of environmental characteristics must therefore be viewed from the perspective of how they impact the sound quality of the source. This can have a significant impact on the source's pitch area.

Signal processing will impact the mix's dynamic balance and its sound stage. One must continually return to considering these aspects while sounds and the mix undergo signal processing. Adjustments to balance or sound stage may be needed after processing, but it is usually more effective to have processing performed carefully, so as not to disrupt the sound stage and balance of sounds previously crafted with care—and completed. The frequency/pitch content of sounds that will be approached next can alter these greatly.

When instruments and voices get placed in the mix, they are evaluated for their frequency content. This is composed of the pitch(es) they perform plus the spectrum of their timbre (including environmental characteristics). In this way, all sounds occupy an "area" in the frequency range of our hearing. This is a bandwidth, of sorts, where the instrument's sound and musical material combine and occupy. This is the pitch density of the musical idea. The pitch densities of all of the sound sources combine to create the timbral balance, or the "spectrum" of the overall texture.

It is common for styles of music to emphasize certain pitch registers over others. For example, the rhythm section of the typical rock band will have the bulk of its pitch-plus-timbre information (or pitch area) in the "low," "low-mid," and "mid" ranges. The timbral balance is weighted in these low-frequency ranges, and instruments and voices performing above these registers occupy a very different pitch area and are easily perceived even when performing at lower dynamic levels or placed in the same locations on the sound stage. Thus, when sounds are combined, a source's pitch area can be exploited to bring the source to blend with others or to be more readily perceived.

As this song progresses, the rhythm section might thin out at times or perhaps stop. This causes a shift in the timbral balance. Such shifts are common in music. Shifts of timbral balance often happen with changes in the mix, at climactic points in the music, with the entry and exiting of instruments/voices, between a verse and chorus, and more. Shifts are common; some are very subtle and some are striking. This emphasis of some pitch registers over others provides an overall sound quality to the song that comprises its timbral balance. Individual songs within a style will bring further refinement to this approach.

Timbral balance can contribute to shaping the program dynamic contour of the work. The density of pitch areas and register placement can increase or decrease the overall loudness of the program. In this way, the dynamic level of the overall program is shaped by the number of sound sources present and the registers in which they sound, as much as the loudness levels of the sources. This is an important consideration when adding instruments and voices to the mix, or pulling sounds out.

Pitch density may be used in innumerable ways to assist in shaping the timbral balance and dynamic contour of the music and in defining the relationships of the individual sound sources. The timbral-balance graph of Chapter 12 can be used (with or without a timeline) to plan the mix and to keep track of sound-source registers. It can be a useful tool in composing the mix or in evaluating the recordings of others.

It is helpful to remember here, crafting sound qualities and creating pitch densities and a timbral balance are new forms of orchestration, and the mix and the arrangement have a mingled relationship in presenting the musical materials of the song.

## Performing the Mix

In performing the mix, the sound qualities and relationships of the recording are realized. Just how everything comes together depends on technologies and work methods. Generally, the more computer-based the production system, the less real-time oriented the mix performance.

The process of executing the mix is similar to performing, in that sound is being controlled, often in real time. Many technical decisions also occur during mixing; many are out of real time. These activities are planned, just as sound qualities and relationships were planned while "composing" the mix. Details pertaining to specific recording equipment, technologies, and techniques will directly impact how the mix is planned.

Work methods will vary by individuals, but they will all center on establishing a logical, efficient, and effective sequence of activities. The goal is to best use the selected equipment and technologies to craft the mix as planned, and to establish and maintain a high-quality signal. Equipment and technologies vary greatly, but the process remains the same; the basic signal chain and the events of the production sequence remain constant and should remain clearly in mind. Once the recordist has obtained knowledge of technologies and mastery of software and equipment usage, they can develop their own approach to sequencing production activities and decisions, and might ultimately create their own working methods. As these methods are technology and equipment dependent, they are outside the scope of this writing.

Recording aesthetics and the planned qualities of the recording will shape the production process. Many possibilities for approaching the mix exist, and the correct approach for one project is not necessarily correct for another. For example, multitrack mixes are often "performed" a track or two at a time, but sometimes any number of tracks are crafted simultaneously; in direct-to-two recordings all tracks are performed simultaneously. In another example, sounds might be placed in the mix before sound qualities receive their final shaping (with final EQ or reverberation added later) in one project, but frequently sound qualities are finalized well before sources are placed in the mix and combined with other sounds.

Skill in performing the mix relies on knowledge of the technologies being used in the signal chain, on mastery of equipment and software usage and the interface points of the signal chain, and on dexterity with the techniques of creatively using these devices and technologies to craft the mix.

## Restarting the Mix

A flow gets learned, developed by experience. The discussions that brought you here will guide you in developing your most effective and productive flow. They will also bring you to sensitively shape all of the sonic dimensions of the mix to best suit the music and its message. A work flow for creatively crafting your mix and then executing its performance will gradually become known. This flow will change with new technologies, but most will be considerably refined with practice. The flow is not always effective; it does not always work.

Creating a mix is a creative act that can bring one to a place of dissatisfaction with the results, or at a loss with how to proceed, or just plain frustrated and ineffective. Should you find yourself at a place where you have mixed yourself into a corner, where you are lost or confused and do not know how to salvage your mix, remember you can start over. Be willing to start over. The idea of this is startling at first, when one considers hours have been spent getting to the place of being "stuck." Getting "un-stuck," however, can be more time consuming, frustrating, and less productive than just plain starting over. Working to fix a failed mix can often lead to a slightly more polished failed mix, after even more time, effort, and frustration.

Starting over is not a failure. It embraces your knowledge, and celebrates your discovery of what does not work and what will not serve your project. Pull the faders down, reset the tracks and plug-ins, zero the board and processors; begin again.

Come back to the song with fresh ideas for the mix and a memory of the effective aspects of your previous attempts. Beginning again, this next mix will be better. And there is the continual possibility that there will be another, better mix on the horizon—until you run out of mix time. Your skill levels and creative dexterity will grow with your artistry, practice, and exposure to possibilities of what is possible in the mix.

## Mastering: The Final Artistic Decisions

The final master recording of the piece of music (joined with all other pieces in the project) is the result of a mastering process. The mastering process is the last chance to shape the recording before it goes to the duplicating plant, or otherwise prepared for sale and distribution. This is the last creative step in crafting the individual song, as well as shaping the whole album into one experience. Mastering fills the void between mixing and replication, where sound can be enhanced one last time and any problems repaired before the recording is finalized. Sonic consistency between tracks is achieved in mastering, as well as the dynamic shape of the project as a whole and the dynamic relationships of tracks. Mastering provides the final touch to make the record album sound finished, and also will seek to ensure the recording will retain its sound quality when played on a variety of playback systems and formats.

Mastering can leave the individual piece of music sounding much as it did in the studio, after mixdown and any final processing. It can consist almost solely of ordering the songs of the album and perhaps adjusting loudness levels between tracks (songs). In earlier times, when the primary final format for recordings was LP records, these were the primary tasks of the mastering engineer. Subtle equalization (to compensate for frequency differences between the inner and outer grooves of the disc), gentle compression (to ensure the level remained above the noise floor), and limiting (to protect the cutter head) may have been often used, but the sound quality of the recording made in the studio was not intended to be altered by the mastering process.

This approach might still be found, but is now quite rare. Currently the mastering process is much more involved, and will typically transform a recording, sometimes significantly, from what was made in the studio. While these transformations may often be subtle to untrained ears, they are significant to the success of the recording. These transformations can be used to address problems with the recording and to enhance its sound qualities. As many recordings are now made in project and home studios that often have inaccurate monitoring, and many times are made by people with limited experience, the importance of the mastering engineer and their influence on the final recording has increased substantially. Their carefully calibrated monitor systems are in a quiet and acoustically neutral room in order to be very accurate.

The mastering engineer will shape sound at the highest levels of perspective to enhance or correct dimensions of the overall characteristics of the recording; they will also listen at all other levels of perspective to focus on how their alterations impact all of the other aspects of sound. They engage in both artistic and technical tasks, and use both critical and analytical listening. Where mixing utilized methods for improving the recording by manipulating the sound characteristics of the individual sound sources within it, mastering requires techniques for enhancing and correcting completed mixes. Mastering can turn an ordinary mix into an extraordinary recording, and it is also possible for a stunning mix to be greatly diminished by poor mastering decisions.

The sequence of the mastering process follows. This ordering of events will be largely consistent between projects:

1. Assembling all final versions (mixes) of all of the songs (pieces of music, or movements of compositions) of the project (album, film sound track, etc.) into the song tracks of the final release
2. Establishing the time length of spaces between the songs
3. Editing to remove noises in the recording and to minimize distortion or unwanted sounds
4. Establishing an appropriate timbral balance (also called spectral balance) for the album, and sometimes for an individual song
5. Adjusting the dynamic levels of individual tracks (songs) as needed
6. Establishing an appropriate dynamic level for the album, and leveling the individual dynamic levels of all tracks
7. Coding for replication, or reformatting for digital downloads.

## Assembly

The process begins with assembling the project. The individual tracks are sequenced, or placed in an order. This is not a simple matter. Projects sometimes have an overall concept (an idea that may have originated with *Sgt. Pepper's Lonely Hearts Club Band*), though this is not common. They will, however, very often have some type of a theme or direction throughout or a central song to bind the album into a single experience. Rarely is an album simply a random collection of songs. The songs will be carefully ordered to provide a diverse experience that brings changes of tempo, mood, subjects, materials, intensity, and sound characteristics. The record album will be compiled to create the most rewarding musical experience from beginning to end. This is the idea that the album is a single experience, and that individual songs are enhanced because of their relationships to the other songs on the album.

Important to successfully creating this single experience is the length of the silences, or gaps, between songs (tracks). The lengths of these silences are based on the music, not on the clock. No formula or preset amount will provide the right movement from one song to another. This is, in effect, the rhythm of the record album. A pace is established between songs that is sometimes lengthened (to slow the progression), sometimes shortened (to hasten it), at times eliminated (to have one song lead directly to another); at times the length of silence between tracks might be quite long to heighten anticipation, and songs might even overlap in a cross fade for a length of time to form a segue. Some pieces of music end in silence. This time is used for listener reflection, for a sense of drama, or to allow the music to reach its own sense of conclusion. This silence might even represent the song's reference dynamic level. The lengths of silences between pieces will be carefully calculated to effectively serve the individual pieces of music and the overall project. Identifying this time unit requires an understanding of how the songs relate to one another and to the overall idea of the album.

### Editing

The album and all tracks receive final editing. Noises and unwanted sounds should not have made it into, let alone through, the mix process, but they often do. Any noises that might still exist will be eliminated in as much as this is possible after mixing; these may be performer noises, poor edits, technical issues, and more. Song heads and tails are sometimes altered at this stage; they might be shortened by editing out some material or extended by copying and pasting material; fade-outs might be created. In classical music and concert recordings the sound of the hall or concert venue is typically needed between tracks; room tone, applause, audience sounds, and the like might be added to fill this space. Clicks, thumps, pops, and other noises will be eliminated. Noise-reduction techniques might be employed to reduce or eliminate narrow or broadband hiss, or hum and buzz (harmonics of hum) that can originate in the recording process or from storage media.

### Adjusting Timbral Balance

Establishing an appropriate timbral balance (this is also called spectral balance and tonal balance) for the album follows. One of the most important roles of the mastering engineer is to achieve a desirable timbral balance for the album (and sometimes for each song individually). This greatly shapes the overall sound quality of the recording, and can provide the recording with a unique "sound." The goal of adjusting timbral balance is to bring out the most desirable qualities of the recording and to address any spectral problems that might exist. Often, subtle changes in spectrum will have a great impact on the recording.

Timbral balance is adjusted for a variety of purposes. Individual tracks (songs) might be treated separately, and their spectral issues handled separately. Some songs might deliberately have a unique sound quality that is celebrated and treated differently from other songs on the album; one song might have a very different spectral problem from another, causing each song to receive a very different treatment. Often an album will have an overall sound quality that helps bind it into a single artistic impression; spectrum will play a decisive role in this, and the timbral balance will be crafted carefully to shape an appropriate overall quality. As discussed above, we find that different types of music (pop, blues, rock, classical, jazz, metal, rap, etc.) have different timbral balances that contribute to the character of that type of music; this is especially prominent in how different types of music use the extreme upper and lower octaves of the spectrum.

This adjustment to the overall spectrum of the recording will also take into account the need for the recording to play back most effectively on a wide range of listening systems (home entertainment, automobile, MP3 players) and playback formats/media (from lossy codecs to high-resolution formats), or to be adjusted to be most effective on a single, specific consumer playback medium. The mastering engineer must be able to project how the recording will sound after it has been transformed in these ways.

Equalization is the primary way timbral balance is altered in mastering. It will be applied before dynamics processing, because changing EQ will alter the dynamic level of the signal. Previously set dynamic relationships and dynamic levels will be altered, and previous dynamic processing, such as a limiter's peak protection, could be undone.

Equalization of the mixed program will be a balance of compromises, and is very different from processing an individual instrument in tracking or mixing. Applying EQ to the overall program can cause a change in the prominence of one sound source over another within the mix. Also, changes designed to correct the timbral balance of one aspect of the sound may well cause problems in another. Small adjustments in one frequency band can result in significant changes in others (for example, adjusting the amount of 100 Hz can alter the perception of 5 kHz to a remarkable degree); many times multiple solutions are tried before one is found that will not create more problems than it fixes.

Large sections of the album are routinely auditioned carefully to try to determine the full impact of even the subtlest change in equalization. Focus will shift deliberately between the perspective of the spectra of all of the individual sound sources and the overall perspective of timbral balance. Well-developed listening skills are required to perceive and shape the nuances of spectral information of a complex, overall program. A similar routine and skill levels are required of the mastering engineer in making adjustments to the overall dynamics of the album.

We can compare the original CD of "Here Comes the Sun" from *Abbey Road* to the remastered version. The subtle change of timbral balance makes pronounced changes in the mix. The instruments have greater detail and the imaging is more discrete and exact; McCartney's bass is more prominent in the remastered version, but is not louder. There is also less background noise.

## Adjusting Dynamic Levels and Relationships

The dynamics of the recording can be altered at two levels of perspective in the mastering process. These are (1) the dynamic levels within and dynamic contours of the individual songs, and (2) the dynamic levels and contours that comprise the program dynamic contour of the entire album.

Adjusting the dynamic levels of individual songs might occur between sections of a song. In this case the loudness levels of the sections might be brought closer together to create less dynamic range (as might be necessary to make a wide dynamic range more appropriate for a home environment), or dynamic range might also be increased by lowering or raising the loudness of a section to create more contrast in the song. Dynamic levels and contour changes can also be made within sections, such as a fade out, or perhaps to increase the loudness of the dynamic peak of the song.

Changes in the dynamic level of the program can change other aspects of the recording, sometimes markedly. For example, even subtle compression on a mix can cause sound-stage imaging to change, depending on the program material. The location and width of images, the perceived distance of sources, and the ambiance of the sound stage can all be altered as compression raises the loudness levels of less prominent sound sources in the mix above a certain threshold.

The album as a whole will have a dynamic contour, relating the dynamic contours and levels of the individual tracks to one another. This may be accomplished by identifying an appropriate reference dynamic level (RDL) for the album, and leveling the individual dynamic levels of all tracks accordingly. A reference dynamic level provides a point of reference to which all other loudness levels can be compared. This level might be any level that is appropriate for the particular project, such as the highest point of the loudest track, or perhaps the most poignant moment of the album's title track, as examples. With this reference, the other songs can be given a related loudness level as appropriate to their own unique qualities and how they fit or contribute to the album. In this way, dynamic relationships between tracks are established.

The songs may have slightly different loudness levels, or quite different levels, depending on the project. The entire project will, however, be reasonably consistent in terms of loudness. This will allow for the dynamic relationships of the songs and the album to be correctly transferred to any format. In some applications, dynamic peaks are limited to allow the project to transfer more easily into certain formats (such as media broadcast) or for sound-quality purposes.

The album will also have an overall loudness level that is established in the mastering process. In recent years this level has been pushed higher by a good many engineers, and CDs may vary in average loudness level by as much as 15 dB. Selecting this actual loudness level is important to the technical quality of the recording as well as artistic product. Distortions and alterations in sound quality can occur when pushing for more program loudness (which provides a sound preferred by many consumers) or when lower levels bring diminished resolution of the waveform.

The remastered Beatles' catalog had limiting applied, gently. The desired result was an increase in overall level, without other artifacts of equalization changes or pumping. The loudest part of the song would be limited slightly and the rest would be level corrected. This resulted in increased timbral detail, as well as increased loudness level. Listening to the remastered version of "Strawberry Fields Forever" compared to the original, most sound sources are noticably nearer to the listener; of interest, sources are not affected equally, causing some distance-location relationships to switch. Notice these alterations, and also changes to musical balance and performance intensities—and timbral-balance changes act in tandem with the higher dynamic level. This comparison is an excellent step to hearing how mastering can create substantial shifts in musical results through subtle changes of loudness and timbral balance.

### Coding for Replication, Formatting for Digital Distribution

Concluding the mastering process, the PQ codes necessary for CD replication are encoded into the digital master (which will usually be referred to as a premaster by the replication plant). This code will establish the locations of track numbers, to start the CD at specific points to play individual songs tracks. It will also be used to create an index for the disc and provide other information needed for playback and replication.

A similar process for either 2-track or 5.1 mastering creates the production master for distribution of the DVD and Blu-ray Disc formats.

Formatting for other consumer media may replace the more traditional production master. Masterfiles are created from this formatting, and are used for online, digital distribution. Digital download compression schemes and other digital download options are created for digital distribution outlets.

No matter the final format, this stage of the process is one of translating the recording from whatever its original source media (analog or digital format) to a target consumer media (such as CD, iTunes, Blu-ray, or the next new format).

## The Listener's Alterations to the Recording

The listener may shape the final sound of the recording. This may be through a conscious altering of the original characteristics of the recording, by accident, or as a product of circumstance. It is common for some listeners to intentionally alter the sound qualities of a recording to align with their own personal preferences (such as markedly boosting low frequencies), thereby changing the sound qualities that were crafted by the recordist. Unintentional alterations to the recording will also happen, usually as a product of one of the multitude of life's circumstances.

For instance, the sound-reproduction systems of the final listener to the recording will in all likelihood be significantly different than the system that was used as a reference during the production of the recording. The listening environment and equipment used for home playback are almost never similar to (let alone the same as) those used in determining the final sound of the recording. The differences between studio and home-listening environments, and studio and consumer sound-reproduction systems, cause great changes to be made in the sound qualities of the recording during playback in home listening environments. This situation has become more pronounced with home surround systems that need to be set up carefully and more accurately calibrated in order for the intended sound to be heard.

Further, the listener alters the original sound characteristics of the recording through a number of activities: adjusting playback equalization and loudness level, their selection of playback equipment, the location of the playback system (especially loudspeakers) in their homes (where often visual aesthetics and the logistics of everyday living win out over sound quality), and through the playback of the recording in small listening rooms.

Listening in automobiles, through earbuds or headphones, or through the speakers of a PC or laptop all have potential to transform the original recording into something very different. Radio formats, Internet delivery codecs, and compressed files for iPod and MP3 players provide more degradation to the sound qualities of the recording.

The music recording will not have the same characteristics when delivered to the consumer as it had in the recording studio. The recordist may hope the music recording will not be radically altered as it is delivered to each individual listener, but there is no reason to believe this will actually happen. The recordist must acknowledge the reality that such alterations will take place, to varying degrees, much more often than not. A well-crafted and mastered recording might transfer reasonably well to a wide variety of formats and systems, by anticipating some of the most prominent changes in sound quality that will result from delivery formats and playback systems. Still, the recording will change. Ultimately the recordist is not in control of how their art is heard.

## Concluding Remarks

This book will not end with a fatalistic and negative tone, that recordists ultimately do not control how their artwork is heard. It is not fitting, and the tone is not warranted. The work of recordists enriches the lives of many, brings great enjoyment to masses of people, and so much more.

Indeed, when listeners care enough about a recording to alter it to make it more their own, it indicates they care deeply about the recording, its music, and its message. In a not-so-odd way, it is a compliment to the artist and the recordist that the listener wishes to make the music even more aligned with their personal taste. Although many recordists might wish this alteration of sound qualities (which they poured much into) would not happen, we learn to anticipate it will. We learn the sounds of different formats or common circumstances, and we mix to create a sound that will transfer well to different situations. It is a great privilege to have one's mixes heard in so many different ways, different circumstances, formats, conditions, and to be embraced by an audience of listeners—whether they are in a car, living room, walking with earbuds, or in an elevator or dentist chair. The recordist makes recordings that have the potential to bring great meaning to people's lives, to make a difference to the quality of life of many.

I wish for you the immensely gratifying and surreal experience of chancing upon one of your recordings being played in public, on a poor, inaccurate sound system, where you might observe the positive responses of those around you, though they may not be really listening. May you be so fortunate.

Recording music can be a profound experience. Being part of the creation of a piece of music is often a very immense, intense, intimate, and unique privilege. One can witness magic, and be part of something that greatly surpasses the sum of the talents of all individuals of the project. One can certainly feel as though they have contributed to the making of great music, whatever the type of music or the level of accomplishment of the musicians. One can also be blessed with opportunities to use their skills to help others realize their dreams, or to use their skills to create music recordings of their own and pursue their own visions.

As we have learned, recordists do shape and create art. We create, compose, perform, polish, invent, inspire, facilitate, execute, arrange, shape, capture, and craft performances that are compositions.

This is *the Art of Recording.*

## Exercises

### 17.1

*Mixing Exercises*

Mixing requires sources to be compared to one another and relies on analytical-listening observations. Bringing one's attention to the level of perspective just above the individual sound source, where the sources are perceived as being of equal importance and can be accurately compared to one another, will allow this skill to develop correctly.

For these exercises, the reader will need to compile or obtain source tracks that can be mixed. Many DAWs come with tutorials that supply such tracks, and some can generate synth tracks easily. It will be most beneficial to record a few performers playing a simplified cover of a well-known song; this is a very rewarding (and educational) way of securing these tracks.

Start your exercises with 2, 3, or 4 tracks and gain confidence with results before moving on. As before, start with extreme settings, to become accustomed to where (the perspective and artistic element) the center of your attention must be. Listen long enough at the level of perspective to acclimate yourself, getting used to listening to those relationships. Once confident, deliberately shift your attention to other dimensions and different perspectives, then attempt to return to this perspective and focus. Once you recognize the extreme settings and the correct perspective, you will be able to effectively move to more subtle changes. Remember, do not burden yourself thinking about the musical result of your mixes at this point; allow yourself the opportunity to simply learn sound qualities and relationships. These mixes need not be shared with others. Record your mixes, and make new mixes daily for several days. After four or five days and new mixes, listen to compare your mixes; review your progress, successes; and areas for improvement.

The reader might want to use some of the graphs and figures from Part Three, used for evaluating the recordings of others, to keep track of their own mixes. They can be especially helpful in beginning musical balance, stereo location, and distance location exercises.

### Exercise 17.1.a

*Musical Balance*

1. Try to align two sources at the same loudness level. Once you believe they are at the same loudness level, check this impression by changing focus on the level of perspective.
2. Now notice how shifts in loudness level change loudness relationships of sources. Alter the loudness levels to make one sound much louder than the other; then notice that sounds can be soft and still be prominent. Move from substantial differences in loudness to almost imperceptible differences.
3. Next, try to create a musical balance that is contrary to performance-intensity cues (for instance, loud sounds appearing soft in the mix).
4. Now work through these issues again with 3, 4, then more sources.

### Exercise 17.1.b

*Timbral Balance*

1. Using the first mix you established for musical balance above, change the timbral balance by adding EQ to make a substantial increase in a portion of the spectrum of one of the sounds; now decrease the amount of EQ until you barely notice a shift in the timbral balance of the mix.
2. Next, repeat this process to subtract the same portion (frequency area) of the spectrum of one of the sounds; now increase the amount of EQ until you barely notice a shift in the timbral balance of the mix. Toggle the EQ in and out to compare the sound with and without EQ.
3. Work through these two concepts again with 3, 4, or more sources.
4. Bring your attention to how this shift of timbral balance alters the prominence of one instrument compared to the other(s).

### Exercise 17.1.c

*Stereo Location*

1. Returning to the first musical-balance mix, pan one source of the far right side of the sound stage and the other far left; listen to identify the location and widths of the sources. Next, bring them both to the center. As a third sound impression, move both sounds halfway between center and left or right, respectively. Listen carefully to the character of the sounds at these various locations.
2. Now, perform this same process for your second musical-balance mix.
3. Afterwards, add two other sources to the second mix and place all sounds in distinctly different locations to make a wide sound stage. Once completed, move the sources to separate locations near the center to create a narrow sound stage. Listen carefully to the character of the sounds at these various locations; evaluate widths and locations of sources as well as the extent of any blending or overlapping of images.
4. As a last step, listen carefully to the spectrum of the sounds. Place the sounds in locations where (1) spectral components are not covered by other instruments, and then (2) where the spectra on two or more instruments blend together. Evaluate the extent of the blending of images and sources, as well as image widths and locations.
5. Bring your attention to how image size and the placement of images can alter the prominence of an instrument compared to others.

### Exercise 17.1.d

*Distance Location*

1. Again returning to the first musical-balance mix, pan one source of the far right side of the sound stage and the other far left. After listening to this to become acquainted with the subtleties of the sounds, bring them both to the center; consider any changes to distance location that may have been created by masking of spectral information in overlapping stereo locations. Now vary the loudness of one source by 10 percent and listen for masking and a shift in distance; repeat this with the other sound. Change the percentage of loudness to notice how this might impact perceived distance.
2. Separate the sources far left and right again and begin changing the EQ of one of the sources to decrease timbral detail and therefore increase the perceived distance of the source from the listener. Return the EQ to a flat setting and add a reverb program to the source until timbral detail blurs and distance is increased. Remain

clearly focused on the level of timbral detail as you calculate the potentials and degrees of any change of distance location.

3. Consider how distance location impacts the quality of the musical materials, and the prominence of one sound over others as distance shifts.

## 17.2

### *Modifying Completed Mixes—Mastering-Related Exercises*

The reader will benefit greatly from methodically exploring the activities below. Each activity should become the focus of an exercise that would have the reader explore:

- How does the device or process transform the waveform?
- What changes to the waveform can I recognize?
- In what ways do these transformations impact the audio signal and the recording?

The accompanying companion website provides a variety of completed mixes that could be used for these exercises. Other source material can certainly be substituted, whether commercial recordings or your own mixes.

In all exercises, the reader should begin work with exaggerated settings. These settings are certain to be far from artistically pleasing, but will cause the changes to be readily apparent and easier to perceive in first encounters where subtle alterations or changes in level might well go undetected. Upon repetitions, make changes smaller and smaller to refine your ability to identify the changes and observe the more subtle alterations created to the qualities of the overall sound and the sound sources within the mixes. Repeating each exercise with different program material will benefit the reader.

When performing these exercises, do not think about the musical result of your processing at this point. Allow yourself the opportunity to simply learn the sound qualities and impacts of your actions. When you have control of these concepts and processes, you can start bringing your attention to making artistic decisions.

## Exercise 17.2.a

### *Compression*

Listen to the changes in dynamics and dynamic range created by applying compression to a completed mix, such as track 53; bring your attention to changes in the overall character of the mix, as well as how the compression alters individual sound sources and the musical balance.

### Exercise 17.2.b

*Equalization*

We will now add one band of EQ to track 52 or 38 (tracks with all instruments in the same or a similar environment); bring your attention to changes in the timbral balance of the mix; next, carefully follow each instrument in the mix to perceive how the EQ impacts their individual sound qualities and how each instrument might have changed in musical balance, stereo location, or distance location. Now add the same band of EQ to track 53 (where sources have different environments and a wider and more active sound stage); bring your attention to the same changes in the timbral balance of the mix; next, carefully follow each instrument in the mix to perceive how the EQ impacts their individual sound qualities and how each instrument might have changed in musical balance, stereo location, or distance location. Try these same exercises with different frequency bands and observe the results; then work toward perceiving the results of performing these exercises using four or five narrow and subtle EQ settings.

### Exercise 17.2.c

*Reverberation*

Apply a reverb preset or plug-in to track 53, and bring your attention to how a new overall, perceived performance environment was created and how it impacts the mix; apply a distinctly different reverb preset next, and bring your attention to the perceived performance environment and its impacts on the mix.

### Exercise 17.2.d

*Limiting*

Apply a limiter to track 53 (or some other mix), and prepare to raise the level of the program markedly; listen for any changes in dynamics and dynamic range created by limiting; bring your attention to changes in the overall character of the mix. Zero the settings and apply the limiter again; this time cause limiting to just slightly suppress the loudest peak of the mix.

# Glossary

*Absolute pitch (perfect pitch)* is the ability to recognize specific pitch levels (in relation to specific tuning systems).

*Accent microphones* are microphones that are dedicated to capturing a single sound source, or a small group of sound sources, within the total ensemble being recorded by a stereo microphone technique, and are used to supplement the stereo array.

*Active listening* is the listening process with the listener focused and intent on extracting certain specific types of information from music or other sounds.

*Amplitude* is the amount of displacement of the medium at any moment, within each cycle of the waveform (measured as the magnitude of displacement in relation to a reference level, or decibels).

*Analytical-listening* techniques are used to evaluate the artistic elements of sound; sound is evaluated within musical contents and these techniques seek to understand the function of the sound in relation to the musical or communication context in which it exists; it evaluates sound over time and uses the concept of *sound event*.

*Artistic elements* are characteristics of sound used to communicate artistic ideas and provide a resource for artistic expression.

*Blend* is the bringing together of all of the sonic components of an acoustic sound source during recording.

*Body* of a sound is the primary portion of the sound that is markedly different in dynamic contour, spectrum, and/or spectral envelope from the initial portion of the sound (up to the first 20–30 ms).

*Bridge* functions to connect two sections, contributing motion and direction; in instrumental bridges its materials add little substance to the song.

*Chord* is two or more simultaneously sounding pitches.

*Chorus* is a song section that is repeated regularly throughout the song and retains the same lyrics each time it appears; its message is typically more objective and wide reaching than the verse.

*Coda* is an ending section of a song, typically containing material from either the verse or the chorus, sometimes both simultaneously; in rare instances it is contrasting.

*Comped vocal* is a final, master vocal track that is a *comp*ilation of any number of takes, created by editing together portions of the best takes; these are selected for correct performance, vocal tone, tuning, expression, attitude, character, etc.

*Complete evaluation* is an examination of all of the artistic elements in a particular recording/piece of music.

*Complete evaluation graph* contains elements at the perspective of the overall texture that appear in the top tier and the elements at the perspective of the individual sound source on the lower tier against the timeline of the work; notes about important characteristics and qualities of elements appear on each tier to assist the complete evaluation, following a stream of shifting importance of materials and elements, and to improve skill in shifting focus and perspective.

*Critical-listening* techniques are used to evaluate the perceived parameters of sound; it is evaluating sound for its own content, out of the context of a piece of music, and out of time; it makes use of the concept of sound as an abstract idea, or a *sound object*.

*Depth of sound stage* is the area created by the distance of all sound sources from the perceived location of the listener; its boundaries create the depth of the sound stage and extend from the nearest source to the furthest sound sources (and their environments).

*Digital audio workstation (DAW)* is a computer-based recording production system comprised of an I/O interface, containing A/D and D/A conversion, monitor levels and routing, and software for potentially all production functions.

*Direct sound* is sound that travels on a direct path from a sound source to the listener or microphone.

*Directional sensitivity* of a microphone is its sensitivity to sounds arriving at various angles to the diaphragm.

*Distance* dimensions in recorded music are: (1) the distance of the listener to the sound stage, and (2) the distance of each sound source from the listener.

*Distance location continuum* is a scale for evaluating distance location based on the listener's sense of proximity, extending from immediately adjacent to the listener, to infinity, and comprised of the areas "Proximity," "Near," and "Far."

*Distance location graph is* used for plotting the distance locations of all sources in a mix; it utilizes the distance location continuum areas against a timeline.

*Distance perception* is the perception of the distance of a sound source from the listener, and is defined by: (1) the ratio of the amount of direct sound to reverberant sound, and (2) the primary determinant, the loss of low-amplitude (usually high-frequency) partials from the sound's spectrum with increasing distance (see definition of *timbre* or *timbral detail*).

*Distance sensitivity (reach)* is the ability of a microphone to accurately capture the detail of a source's timbre in relation to its distance from the sound source.

*Duration* is the perception of time.

*Dynamic contours* are changes in dynamic levels over time.

*Dynamic envelope* of a sound is the contour of the changes in the overall dynamic level of the sound throughout its existence.

*Early reflections* are those reflections that arrive at the ear or microphone within around 50 ms of the direct sound.

*Early sound field* is comprised of the reflections that arrive at the listener or microphone within the first 50 ms after the arrival of the direct sound.

*Effective listening zone* is an area in the control room where the spatial characteristics of reproduced sound can be accurately perceived.

*Environmental characteristics* are the sound characteristics of an environment; an overall sound quality that is comprised of a number of component parts; individual sound sources and the complete sound stage (perceived performance environment) have individual environmental characteristics.

*Environmental characteristics evaluation* defines the characteristics of the environment itself.

*Environmental characteristics graph* is composed of (1) the reflection envelope, (2) the spectrum, and (3) the spectral envelope plotted against a timeline; it seeks to define the characteristics of the environment itself.

*Equivalence* is the concept that all of the artistic elements of sound have an equal potential to carry the most significant musical information; any artistic element has the potential to be the central carrier of the musical idea at any moment in time; any element may also be a secondary element at any moment in time or at any level of perspective, and the importance of any element might shift at any moment.

*Focus* is the act of bringing some aspect of sound to the center of one's attention and at a specific level of detail *(perspective).*

*Form* is the piece of music as if perceived, in its entirety, in an instant; it is the substance and shape of the piece of music, perceived from conceptualizing the whole; it is a global quality, as an overall concept and essence.

*Formants (formant regions)* are an individual frequency or certain ranges of frequencies within the spectrum that are emphasized consistently, no matter the fundamental frequency.

*Frequency* is the number of similar, cyclical displacements in the medium, air, per time unit (measured in cycles of the waveform per second, or Hz).

*Frequency bands* are areas of frequency activities defined as a bandwidth between an upper and lower boundary.

*Frequency response* is a measure of how a device responds to the same sound level at different frequencies.

*Fundamental frequency* is the periodic vibration of a waveform, producing the sensation of a dominant frequency; it is measured by the number of periodic vibrations, or cycles of the waveform, that repeat its characteristic shape during a second.

*Harmonic progression* is the movement from one chord to another, in a stylized sequence.

*Harmonics* are frequencies in the spectrum that are whole-number multiples of the fundamental frequency.

*Harmony* is created by patterns of the harmonic progression.

*Hierarchy* is the organization of materials by levels of importance.

*Host environment* is the environment in which an individual or groups of sound sources are sounding.

*Imaging* is the lateral location and distance placement of the individual sound sources within the sound stage, and provides depth and width to the sound stage.

*Inherent sound quality* is the unique sonic imprint of any device or process in the signal chain that is the result of its normal performance characteristics and sound qualities.

*Interval* is the distance between the perceived levels of two (or more) pitches, played either in succession (as melody) or simultaneously (as harmony).

*Interaural amplitude differences (IAD)* are arrivals of the same sound at each ear at a different sound pressure level (amplitude) and are critical to determining localization for sounds at high frequencies.

*Interaural spectral differences (ISD)* result when the head of the listener blocks certain frequencies from the furthest ear (when the sound is not centered).

*Interaural time differences (ITD)* are arrivals of the same sound at each ear at a different time; a sound that is not precisely in front or in back of the listener will arrive at the ear closest to the source before it reaches the furthest ear.

*Introduction* is any instrumental music that occurs before the entrance of the voice; most often it foreshadows material from the verse or chorus, but on rare occasions it borrows from another section.

*Key* is a summary of all of the sound sources plotted on a graph, showing their individual colors, labeling, or formats.

*Loudness* is the perception of the overall excursion (acoustic energy) of the waveform (amplitude).

*Masking* is the covering of the qualities or activities of a sound source by another sound source.

*Mastering* is the process that fills the void between mixing and replication, where sound can be enhanced one last time and any problems repaired before the recording is finalized; further functions of sequencing, levels, spacing, compression, equalization, etc., might also be performed.

*Melodic contour graph* places a melodic line on an *X-Y* graph, mapping its activity in pitch/frequency register against a timeline; this is an excellent way to acquire skill in hearing pitch levels against pitch register designations, and in placing sources against a timeline; it is an important bridge between traditional music dictation (and notation) and the graphs and processes of this system.

*Metric grid* is the underlying pulse of a piece of music and is a reference pulse against which all durations can be defined, thereby allowing the listener to make rhythmic judgments in a precise and consistent manner.

*Middle-eight* is a bridge with vocals; it is musically contrasting to the chorus and verses, providing a different impression of the subject.

*Mixing (mixdown, the mix)* is where the individual sound sources that are being recorded or were previously recorded onto a multitrack are combined into a two-channel or a surround-sound recording that will become the final version of the piece after the mastering process.

*Monitor level* is the sound pressure level (SPL) at which the recordist listens to sound reproduced through the monitor system.

*Monitor system* is comprised of all factors that impact reproduced sound: power amplifier(s), loudspeaker drivers, crossover networks, loudspeaker enclosures, connector cables, and the listening room itself.

*Musical balance* is the interrelationship of the dynamic levels of each sound source, to one another and to the entire musical texture.

*Musical balance graph* plots the individual sound sources of a musical texture by their dynamic contours, against a single timeline, with the work's RDL as a reference.

*Multitier graphs* contain a *Y*-axis divided to allow several components to be represented against the same timeline; it allows different characteristics to be graphed against the same timeline.

*Nominal level* is used to establish a reference for plotting the dynamic contours of the spectral components of environmental characteristics; it represents the changing dynamic envelope of the environment, where the sound source's frequency components are unaltered.

*Objective descriptions* describe the states and activities of the physical characteristics of sound.

*Off-axis* is any deviation from a microphone's 0° center point for sound arrival.

*Off-axis coloration* is a change in the timbre of a sound source, caused by change in the frequency response of a microphone to sounds arriving off-axis.

*Onset (prefix)* is the initial portion (up to 20–30 ms) of the sound that is markedly different in dynamic contour, spectrum, and/or spectral envelope from the remainder of the sound (the body).

*Overall texture* is the highest level of perspective, bringing the listener to focus on the composite sound of a recording or piece of music; dimensions are the piece of music/recording's form, perceived performance environment, sound stage, reference dynamic level, program dynamic contour, and its timbral balance.

*Overtones* are those frequencies in the spectrum that are not proportionally related to the fundamental frequency.

*Partials* are all of the frequencies of the spectrum: overtones and harmonics, subharmonics and subtones, and formats and formant regions.

*Passive listening* is the process of listening while the listener is consciously focused on some activity other than the music (perhaps eating, a conversation, a dentist's drill, etc.); in fact, the listener might not actually be conscious of the music or sound.

*Perceived parameters (five) of sound* are pitch, loudness, duration, timbre (perceived overall quality), and space (perceived characteristics).

*Perceived performance environment* is the environment of the sound stage, and is the overall environment where the performance (recording) is heard as taking place; its environmental characteristics shape the entire recording and bind all the individual sound sources and their spaces together into a single performance area; a characteristic of the overall texture.

*Perfect pitch (absolute pitch)* is the ability to recognize specific pitch levels (in relation to specific tuning systems).

*Performance intensity (perceived)* is the timbre/sound quality of the sound source created during a performance; the dynamic level at which the sound source was performing when it was recorded; it is comprised of the loudness, energy exerted, performance technique, and the expressive qualities of the performance.

*Performance intensity versus musical balance graph* contrasts performance intensity plotted as the dynamic levels of the original performance against the loudness levels of the sources as they exist in the final recording (musical balance).

*Perspective* is the level of detail at which the sound material is heard, and is related to a specific level of the structural hierarchy; it brings the listener's *focus* to a specific level of detail.

*Phantom images* are perceived sound-source locations that are sounding at locations where a physical sound source (loudspeaker) does not exist.

*Phon* is the unit of measure for perceived loudness established at 1 kHz, based on subjective listening tests.

*Physical dimensions (five) of sound* are frequency, amplitude, time, timbre, and space.

*Pitch* is the perception of the frequency of the waveform, defined as the perceived position of a sound on a scale from low to high, and as an attribute of hearing sensation by which sounds may be ordered on a musical scale.

*Pitch-area* is a defined area between an upper and a lower pitch level, in which a significant and prominent portion of the source's spectrum exists.

*Pitch-area analysis* graph is used to plot the pitch areas of nonpitched sounds, showing the location, widths, and densities of the sound's pitch areas and the relative loudness levels of the pitch areas; this is a rudimentary sound-quality evaluation graph.

*Pitch definition (definition of fundamental frequency)* is the amount of presence of a sense of pitch, or the dominance of the fundamental frequency of the source; pitch definition is placed on a continuum between the two boundaries of well-defined in pitch, or precisely pitched (as a sine wave), through completely void of pitch or nonpitched (as white noise).

*Pitch density* is the range of pitches spanned by a musical idea plus the spectrum of the sound source playing it; it is the amount and placement of pitch-related information of a single sound source within the overall pitch range of the musical texture; it is at the perspective of the individual sound source.

*Pitch/frequency registers* are divisions of the hearing range that were established to estimate the relative level of the pitch material.

*Pitch-level estimation* is the ability to consistently identify the value (pitch level or name) and register placement level of a pitch or frequency.

*Pitch reference (internal)* is a sense of pitch level that is present (usually unconsciously) within each individual.

*Point source* is a phantom image that occupies a focused, precise point in the sound stage.

*Polar pattern* of a microphone illustrates its sensitivity to sounds at various frequencies in front, in back, and to the sides, and the actual pattern is spherical around the microphone.

*Prechorus* is a song section that provides a link between verses and choruses when they have materials that are significantly contrasting; transitional in character, this section is often not present.

*Prefix (onset)* is the initial portion (up to 20–30 ms) of the sound that is markedly different in dynamic contour, spectrum, and/or spectral envelope from the remainder of the sound (the body).

*Present* is our consciousness of "now," where we are at once experiencing the moment of our existence, evaluating the immediate past of what has just happened, and anticipating the future of our window of consciousness; a window of time through which we perceive the world, and listen to sound.

*Primary elements* are the aesthetic and artistic elements of sound that directly contribute to the basic shape or characteristics of a musical idea.

*Primary musical materials* are musical materials that are perceived as being more important than others; they will carry the weight of communicating the musical message and expression of the music.

*Primary phantom images* are any of the five source locations in surround sound that exist between adjacent pairs of speakers.

*Program dynamic contour* is the dynamic contour of the overall program, a single dynamic level/contour of the composite sound, the result of combining all sounds in the program; it is a dimension of the overall musical texture.

*Prominence* is what is most noticeable or conspicuous at a particular moment in time, and what draws the listener's attention in a recording; it is not necessarily the most important or most significant.

*Proximity* is the space that immediately surrounds the listener, the listener's own personal space; it is used as a reference for judging distance location.

*Range* is the complete span of an artistic element or perceived parameter, such as the range of hearing covering all audible frequencies from the lowest to the highest frequencies humans can hear.

*Recording aesthetic* is the relationship of the qualities of a recording to the qualities of the original live performance in a performance space.

*Recording sessions* is a term used here to describe recordings where all the parts (sound sources and musical materials) are played at once; the entire musical texture is recorded simultaneously as a single sound, to stereo or surround.

*Recordist* is a person involved in the production of audio recordings: recording engineer, record producer, sound designer, sound synthesist, mastering engineer, and others.

*Reference dynamic level (RDL)* is the overall or global intensity level of a piece of music and is the reference level for evaluating dynamics for musical balance and program dynamic contour; it is the "perceived performance intensity" of the work as a whole, conceptualized as a single entity out of time; a dimension of the overall texture.

*Reflected sound* is sound that bounces (reflects) off surfaces or objects before arriving at the listener.

*Refrain* is similar to the chorus but is one line long and returns only after other material is presented; its line of text is nearly always the title of the song.

*Register* is a specific portion of the range, usually with a unique character (such as a unique timbre, or some other determining factor) that will differentiate it from all other areas of the source's range.

*Relative pitch* is the ability to consistently and reliably judge pitch level within about 10 percent of actual level.

*Reverberant sound* is a composite of many reflections of the sound arriving at the listener (or microphone) in close succession.

*Reverberation time (RT60)* is the length of time required for the reflections to reach an amplitude level of 60 dB lower than that of the original sound source.

*Rhythmic pattern* is a group of durations, and can be applied to any artistic element of sound.

*Scratch performances* or *scratch tracks* are performances recorded at the beginning of tracking; they are intended to be replaced later by overdubs, and are meant to be a guide for the band.

*Secondary elements* are those aspects of the sound that assist, enhance, or support the primary elements.

*Secondary musical materials* are musical materials perceived as being subordinate to others; they will in some way enhance the presentation of the primary materials by their presence and activity in the music, and usually function to support the primary musical ideas.

*Secondary phantom images* are source locations in surround sound that exist between nonadjacent pairs of speakers.

*Signal chain* is the flow of signal through a chain of recording/reproduction stages or devices.

*Sound event* is the shape or design of the musical idea (or abstract sound) as it is experienced over time; it is understood as activity unfolding and evolving over time, and is used in *analytical-listening* observations.

*Sound object* is the perception of the whole musical idea (or abstract sound) at an instant, out of time; it is understood as the qualities of a sound itself in its many variables and as it exists as a global quality or "object"; it is used for *critical-listening* applications and is always considered without relationship to another sound.

*Sound quality* is the artistic element that uses perceived timbre qualities for artistic expression.

*Sound-quality evaluation* seeks to define and describe the states and activities of the sound source's (1) dynamic envelope, (2) spectral content, and (3) spectral envelope, and will also make use of the listener's carefully evaluated perception of (4) pitch definition.

*Sound stage* is a single area within which all sound sources are perceived as being located in providing the "performance" that is the recording; it has an apparent physical size of width and depth.

*Space* means dimensions in audio recording: distance of the sound source to the listener, angle of the sound source to the listener, geometry of the environment in which the sound source is sounding, and location of the sound source within the host environment.

*Space within space* is the concept where sound sources and their individual environments can exist within the overall environment of the recording (perceived performance environment), or spaces can exist within another space; a hierarchy of environments existing within other environments; it also allows for different environments to be contained within others and to also possibly coexist within the same recording.

*Spatial relationships* perceived in current recording reproduction are the location of the sound source being at an angle to the listener (in front, behind, to the left, to the right, etc.), the location of the sound source being at distance from the listener, and an impression of the type, size, and acoustic properties of the host environment.

*Spectral envelope* is the composite of each individual dynamic level and dynamic envelope of all of the individual partials of the spectrum.

*Spectrum (spectral content)* is the composite of all of the frequency components of a sound, and is comprised of the fundamental frequency, harmonics, and overtones, sometimes including sub-harmonics and subtones.

*Spread image* is a phantom image that appears to occupy an area; it has a width or size that extends between two audible boundaries.

*Stage-to-listener distance* establishes the front edge of the sound stage with respect to the listener and determines the level of intimacy of the music/recording.

*Stage width* (sometimes called *stereo spread*) is the width of the entire sound stage.

*States of sound* (three) are physical dimensions, human perception, and sound as idea.

*Stereo location* is the perceived placement/location of a sound source within the stereo playback array.

*Stereo microphone techniques (arrays)* are composed of two or more microphones (or diaphragm assemblies) in a systematic arrangement; they are designed to record sound in such a way that upon playback (through two channels) a certain sense of the spatial relationships of the sound sources present during the recorded performance is reproduced.

*Stereo sound* is a two-channel playback format that attempts to reproduce all spatial cues through two separate loudspeakers.

*Stereo-sound location graph* plots the locations of all sound sources against the timeline of the work.

*Structure* is the architecture of the musical materials and the interrelationships of a composition.

*Surround sound* is a multichannel playback format that reproduces spatial cues, etc., through five separate loudspeakers and a subwoofer.

*Surround-sound location graph* allows the reader to plot the locations of all sound sources against the timeline of the work, utilizing "left," "right," "center," "left surround," "right surround," and "rear center" locations in various formats as the *Y*-axis.

*Tempo* is the rate or speed of the pulses of the metric grid, measured in metronome markings (pulses per minute, abbreviated "MM"); in a larger sense, it can be the rate of activity of any large or small aspect of the piece of music (or of some other aspect of audio—for example, the tempo of a dialogue).

*Temporal fusion* is the perception of reverberant sound used with the direct sound to create a single impression of the sound in its environment.

*Timbral balance* is the combination of all of the pitch densities of all of the recording's sounds; it is the distribution and density of pitch/frequency information in the recording/music; it is a characteristic of the overall texture that represents the recording's "spectrum."

*Timbral detail (definition of timbre)* is the subtle components and/or changes in the content of a sound's timbre (dynamic envelope, spectral content, and/or spectral envelope).

*Timbre* is the overall quality of a sound comprising a multitude of functions of frequency and amplitude displacements; its primary component parts are the dynamic envelope, spectrum, and spectral envelope.

*Timbre perception* is the perception of the mixture of all of the physical aspects that comprise a sound, as a global form, or the overall character of a sound, which we recognize as being unique.

*Time perception* is the estimation of elapsed clock time, significant to the perception of the global qualities of a piece of music and to the estimation of durations when a metric grid is not present in the music.

*Timeline* is the time axis (or *X*-axis) of an *X-Y* graph, and represents the length of an example, divided into some appropriate time unit.

*Tracking* is the recording of individual instruments or voices (sound sources), or small groups of instruments or voices, on separate tracks in a multitrack recorder.

*Transient response* is the time required for a microphone to accurately track the waveform of a sound source.

*Triad* is a chord composed of three pitches, combining two intervals of a third.

*Verse* is a reappearing section of a song that tells the plot or story; verses present a different text at each appearance, except the occasional repeated verse.

*X-Y graph* is a two-axis, two-dimensional graph used herein to notate or represent the qualities and activities of various artistic elements.

# Bibliography


Alten, Stanley R., *Audio in Media*, Tenth Edition, Belmont, CA: Wadsworth Publishing Company, 2013.

Backus, John, *The Acoustical Foundations of Music*, Second Edition, New York: W.W. Norton & Co., Inc., 1977.

Ballou, Glen, *Handbook for Sound Engineers*, Fourth Edition, Oxford: Focal Press, 2008.

Bartlett, Bruce and Jenny Bartlett, *Practical Recording Techniques*, Sixth Edition, Oxford: Focal Press, 2012.

Beatles, The, *The Beatles Anthology*, San Francisco: Chronicle Books, 2000.

Beatles, The, *The Beatles Complete Scores*, Milwaukee: Hal Leonard Corporation, 1993.

Bech, Søren and O. Juhl Pedersen, editors, Proceedings of a Symposium on *Perception of Reproduced Sound*; Gammel Avernæs, Denmark, 1987, Peterborough, NH: Old Colony Sound Lab Books, 1987.

Beranek, Leo L., *Acoustics*, New York: American Institute of Physics, Inc., 1986.

Bergson, Henri, *Matter and Memory*, New York: Humanities Press, 1962.

Berry, Wallace, *Form in Music*, Englewood Cliffs, NJ: Prentice-Hall, 1966.

Blauert, Jens, "Sound Localization of the Median Plane," *Acustica* 22, 1969/70: pp. 205–213.

Blauert, Jens, *Spatial Hearing*, Cambridge, MA: The MIT Press, 1997.

Blaukopf, Kurt, "Space in Electronic Music," in *Music and Technology, Stockholm Meeting June 8–12, 1970*, pp. 157–172, New York: Unipub, 1971.

Borwick, John, *Loudspeaker and Headphone Handbook*, Third Edition, Oxford: Focal Press, 2001.

Borwick, John, *Sound Recording Practice*, Fourth Edition, Oxford: Oxford University Press, 1996.

Burgess, Richard James, *The Art of Music Production*, Third Edition, London: Omnibus Press, 2001.

Burgess, Richard James, *The Art of Music Production: The Theory and Practice*, Fourth Edition, Oxford: Oxford University Press, 2013.

Butler, David, *The Musician's Guide to Perception and Cognition*, New York: Schirmer Books, 1992.

Camras, Marvin, *Magnetic Recording Handbook*, New York: Van Nostrand Reinhold Company, 1988.

Case, Alexander, *Mix Smart: Pro Audio Tips for Your Multitrack Mix*, Boston: Focal Press, 2011.

Case, Alexander, *Sound FX: Unlocking the Creative Potential of Recording Studio Effects*, Boston: Focal Press, 2007.

Chowning, John, "The Simulation of Moving Sound Sources," *Computer Music Journal* 1 (3), 1977: pp. 48–52.

Clifton, Thomas, *Music As Heard: A Study in Applied Phenomenology*, New Haven, CT: Yale University Press, 1983.

Cooper, Grosvenor W. and Leonard B. Meyer, *The Rhythmic Structure of Music*, Chicago: The University of Chicago Press, 1960.

Cooper, Paul, *Perspectives in Music Theory*, New York: Dodd, Mead & Company, 1973.

Crich, Tim, *Recording Tips for Engineers*, Third Edition, Boston: Focal Press, 2010.

Davis, Don and Eugene Patronis, *Sound System Engineering*, Third Edition, Oxford: Focal Press, 2006.

Davis, Don, and Chips Davis, "The LEDE(tm) Concept for the Control of Acoustic and Psychoacoustic Parameters in Recording Control Rooms," *Journal of the Audio Engineering Society*, 28 (9), 1980: pp. 585–595.

Davis, Gary and Ralph Jones, *Sound Reinforcement Handbook*, Second Edition, Milwaukee: Hal Leonard Publishing Corporation, 1989.

Deutsch, Diana, *The Psychology of Music*, Third Edition, Orlando, FL: Academic Press, Inc., 2012.

Deutsch, Diana and J. Anthony Deutsch, *Short-Term Memory*, New York: Academic Press, 1975.

Dockwray, Ruth and Allan F. Moore, "Configuring the Sound-Box 1965–1972," *Popular Music*, 29 (2), 2010: pp. 181–197.

Dowling, William J., *Beatlesongs*, New York: Fireside, 1989.

Doyle, Peter, *Echo & Reverb: Fabricating Space in Popular Music Recording 1900–1960*, Middletown, CT: Wesleyan University Press, 2005.

Droney, Maureen, *Mix Masters: Platinum Engineers Reveal Their Secrets for Success*, Boston: Berklee Press, 2003.

Eargle, John, *Handbook of Recording Engineering*, Third Edition, New York: Chapman & Hall, 1996.

Eargle, John, *The Microphone Book*, Second Edition, Boston: Focal Press, 2004.

Eargle, John, *Music, Sound and Technology*, New York: Van Nostrand Reinhold, 1995.

Eargle, John, editor, *An Anthology of Reprinted Articles on Stereophonic Techniques*, New York: Audio Engineering Society, Inc., 1986.

Eisenberg, Evan, *The Recording Angel: The Experience of Music from Aristotle to Zappa*, New York: Penguin Books, 1987.

Erickson, Robert, *Sound Structure in Music*, Berkeley, CA: University of California Press, 1975.

Emerick, Geoff and Howard Massey, *Here, There and Everywhere: My Life Recording the Music of The Beatles*, New York: Gotham Books, 2006.

Everett, Walter, *The Beatles as Musicians: The Quarry Men through Rubber Soul*, Oxford: Oxford University Press, 2001.

Everett, Walter, *The Beatles as Musicians: Revolver through the Anthology*, Oxford: Oxford University Press, 1999.

Everett, Walter, editor, *Expression in Pop-Rock Music: Critical and Analytical Essays*, Second Edition, New York: Routledge, 2008.

Everett, Walter, *The Foundations of Rock: From Blue Suede Shoes to Suite Judy Blue Eyes*, Oxford: Oxford University Press, 2009.

Fay, Thomas, "Perceived Hierarchic Structure in Language and Music," *Journal of Music Theory* 15 (1–2), 1971: pp. 112–137.

Federkow, G., W. Buxton, and K. Smith, "A Computer-Controlled Sound Distribution System for the Performance of Electronic Music," *Computer Music Journal* 2 (3), 1978: pp. 33–42.

Flitner, David, editor, *Less Noise, More Soul: The Search for Balance in the Art, Technology, and Commerce of Music*, Milwaukee: Hal Leonard Corp., 2013.

Frith, Simon, *Performing Rites: On the Value of Popular Music*, Cambridge: Harvard University Press, 1996.

Frith, Simon and Simon Zagorski Thomas, editors, *The Art of Record Production: An Introductory Reader for a New Academic Field*, Surrey: Ashgate Publishing Limited, 2012.

Gibson, David, *The Art of Mixing*, Second Edition, Boston: Course Technology, 2005.

Hall, Donald E., *Musical Acoustics*, Third Edition, Belmont, CA: Brooks/Cole, Cengage Learning, 2001.

Handel, Stephen, *Listening: An Introduction to the Perception of Auditory Events*, Cambridge, MA: MIT Press, 1993.

Harley, Robert, *The Complete Guide to High-End Audio*, Albuquerque, NM: Acapella Publishing, 1994.

Hatschek, Keith, *The Golden Moment: Recording Secrets from the Pros*, San Francisco: Backbeat Books, 2005.

Hawking, Stephen W., *A Brief History of Time: From the Big Bang to Black Holes*, New York: Bantam Books, 1988.

Helmholtz, Hermann, *On The Sensations of Tone*, New York: Dover Publications, Inc., 1967.

Hertsgaard, Mark, *A Day in the Life: The Music and Artistry of the Beatles*, New York: Delacorte Press, 1995.

Hodgson, Jay, *Understanding Records: A Field Guide to Recording Practice*, New York: Continuum, 2010.

Holman, Tomlinson, *Surround Sound Up and Running*, Second Edition, Boston: Focal Press, 2008.

Holman, Tomlinson, *Sound for Film and Television*, Boston: Focal Press, 1997.

Howard, David M. and James Angus, *Acoustics and Psychoacoustics*, Fourth Edition, Boston: Focal Press, 2009.

Huber, David Miles and Robert E. Runstein, *Modern Recording Techniques*, Eighth Edition, Boston: Focal Press, 2014.

Karkoschka, Erhard, "Eine Hörpartitur elektronischer Musik," *Melos* 38 (11), 1971: pp. 468–475.

Karkoschka, Erhard, *Neue Musik / Analyses*, Herrenberg: Doring, 1976.

Katz, Bob, *Mastering Audio: The Art and the Science*, Oxford: Focal Press, 2002.

Katz, Mark, *Capturing Sound: How Technology Has Changed Music*, Berkeley, CA: University of California Press, 2004.

Kefauver, Alan P., *The Audio Recording Handbook*, Middleton, WI: A-R Editions, Inc., 2001.

Kefauver, Alan P., *Fundamentals of Digital Audio*, Middleton, WI: A-R Editions, Inc., 1999.

Koffka, Kurt, *Principles of Gestalt Psychology*, New York: Harcourt, Brace, and World, 1963.

Kuttruff, Heinrich, *Room Acoustics*, Fifth Edition, London: Applied Science Publishers Ltd., 2009.

Lacasse, Serge, "Persona, Emotions and Technology: the Phonographic Staging of the Popular Music Voice," *The Proceedings of the 2005 Art of Record Production Conference* (www.artofrecordproduction.com), 2005.

LaRue, Jan, *Guidelines for Style Analysis*, Second Edition, Detroit: Harmonie Park Press, 1996.

Letowski, Tomasz, "Development of Technical Listening Skills: Timbre Solfeggio," *Journal of the Audio Engineering Society* 33 (4), 1985: pp. 240–244.

Lewisohn, Mark, *The Beatles Recording Sessions: The Official Abbey Road Studio Session Notes 1962–1970*, New York: Harmony Books, 1988.

Lewisohn, Mark, *The Complete Beatles Chronicle*, London: Hamlyn, 2003.

Martin, George, *All You Need Is Ears*, New York: St. Martin's Press, 1979

Martin, George, with William Pearson, *With a Little Help From My Friends: The Making of Sgt. Pepper*, Boston: Little, Brown and Company, 1994.

Massey, Howard, *Behind the Glass: Top Record Producers Tell How They Craft the Hits*, San Francisco: Miller Freeman Books, 2000.

McAdams, Stephen and Albert Bregman, "Hearing Musical Streams," *Computer Music Journal* 3 (4), 1979: pp. 26–43.

Meyer, Leonard B., *Emotion and Meaning in Music*, Chicago: The University of Chicago Press, 1956.

Meyer, Leonard B., *Explaining Music: Essays and Explorations*, Berkeley, CA: University of California Press, 1973.

Meyer, Leonard B., *Music, the Arts and Ideas*, Chicago: The University of Chicago Press, 1967.

Miller, George, "The Magical Number Seven, Plus or Minus Two," *Language and Thought*, ed. Donald C. Hildum, pp. 3–31, Princeton, NJ: Van Nostrand Company, Inc., 1967.

Mills, A. W., "On the Minimum Audible Angle," *Journal of the Acoustical Society of America* 30, 1958: pp. 237–246.

Moore, Allan F., *Rock: The Primary Text: Developing a Musicology of Rock*, Second Edition, Surrey: Ashgate Publishing Limited, 2001.

Moore, Allan F., *Song Means: Analysing and Interpreting Recorded Popular Song*, Surrey: Ashgate Publishing Limited, 2012.

Moore, Brian C. J., *An Introduction to the Psychology of Hearing*, Sixth Edition, Bingley: Emerald Group Publishing Limited, 2012.

Moorefield, Virgil, *The Producer as Composer: Shaping the Sounds of Popular Music*, Cambridge, MA: The MIT Press, 2005.

Moulton, David, *Golden Ears: Know What You Hear*, Sherman Oaks, CA: KIQ Production, Inc., 1995.

Moulton, David, *Total Recording: The Complete Guide to Audio Production and Engineering*, Sherman Oaks, CA: KIQ Production, Inc., 2000.

Moylan, William, *An Analytical System for Electronic Music*, Ann Arbor, MI: University Microfilms, 1983.

Moylan, William, "Aural Analysis of the Characteristics of Timbre," Paper presented at 79th Convention of the Audio Engineering Society, New York, NY, 1985.

Moylan, William, "Aural Analysis of the Spatial Relationships of Sound Sources as Found in Two-Channel Common Practice," Paper presented at 81st Convention of the Audio Engineering Society, Los Angeles, CA, 1986.

Moylan, William, "A Systematic Method for the Aural Analysis of Sound Sources in Audio Reproduction/Reinforcement, Communications, and Musical Contexts," Paper presented at 83rd Convention of the Audio Engineering Society, New York, NY, 1987.

Moylan, William, *The Art of Recording: the Creative Resources of Music Production and Audio*, New York: Van Nostrand Reinhold, 1992.

Neuhoff, John, *Ecological Psychoacoustics*, London: Elsevier Academic Press, 2004.

Neve, Rupert, "Design and the Designer: A Point of Reference," Paper presented at the 99th Convention of the Audio Engineering Society, New York, NY, 1995.

Newell, Philip, *Recording Spaces*, Oxford: Focal Press, 2000.

Newell, Philip, *Recording Studio Design*, Oxford: Focal Press, 2003.

Nisbett, Alec, *The Sound Studio*, Seventh Edition, Boston: Focal Press, 2003.

Olson, Harry F., *Music, Physics and Engineering*, Second Edition, New York: Dover Publications, Inc., 1967.

Plomp, Reinier, *Aspects of Tone Sensation: A Psychophysical Study*, New York: Academic Press Inc., 1976.

Pohlmann, Ken, *Principles of Digital Audio*, Third Edition, New York: McGraw-Hill, Inc., 1995.

Polanyi, Michael, *Personal Knowledge: Towards a Post-Critical Philosophy*, Chicago: University of Chicago Press, 1962.

Pousseur, Henri, "Outline of a Method," in *die Reihe, Nr. 3,* ed. Herbert Eimert and Karlheinz Stockhausen, pp. 44–88, Bryn Mawr, PA: Theodore Presser, Co., 1959.

Ramone, Phil and Charles L. Granata, *Making Records: the Scenes Behind the Music*, New York: Hyperion, 2007.

Randall, J. K., "Three Lectures to Scientists," *Perspectives of New Music* 3 (2), 1967: pp. 124–140.

Rayburn, Ray, *Eargle's The Microphone Book: From Mono to Stereo to Surround – A Guide to Microphone Design and Application*, Boston: Focal Press, 2011.

Reynolds, Roger, "It(')s Time," *Electronic Music Review* 7, 1968: pp. 12–17.

Reynolds, Roger, *Mind Models: New Forms of Musical Experience*, New York: Praeger Publishers, 1975.

Reynolds, Roger, "Thoughts of Sound Movement and Meaning," *Perspectives of New Music* 16 (2), 1978: pp. 181–190.

Risset, Jean-Claude, *Musical Acoustics*, Paris: Centre George Pompidou Rapports IRCAM No. 8, 1978.

Roads, Curtis, editor, *The Music Machine*, Cambridge, MA: The MIT Press, 1989.

Roederer, Juan G., *Introduction to the Physics and Psychophysics of Music*, Second Edition, New York: Springer-Verlag, 1979.

Rossing, Thomas D., *The Science of Sound*, Second Edition, Reading, MA: Addison Wesley Publishing Company, 1990.

Rumsey, Francis, *Spatial Audio*, Oxford: Focal Press, 2001.

Rumsey, Francis and Tim McCormick, *Sound and Recording: An Introduction*, Fifth Edition, Oxford: Focal Press, 2006.

Russ, Martin, *Sound Synthesis and Sampling*, Oxford: Focal Press, 1996.

Savage, Steve, *Bytes & Backbeats: Repurposing Music in the Digital Age*, Ann Arbor, MI: The University of Michigan Press, 2011.

Savona, Anthony, Editor, *Console Confessions: The Great Music Producers in their Own Words*, San Francisco: Backbeat Books, 2005.

Schaeffer, Pierre, *A la recherche d'une musique concrète*, Paris: Editions du Seuil, 1952.

Schaeffer, Pierre, and Guy Reibel, *Solfège de l'objet sonore*, Paris: Editions du Seuil, 1966.

Schaeffer, Pierre, *Traité des objets musicaux*, Paris: Editions du Seuil, 1966.

Schouten, J.F., "The Perception of Timbre," *Report of the 6th International Congress on Acoustics*, 90, 1968: pp. 35–44.

Senior, Mike, *Mixing Secrets for the Small Studio*, Boston: Focal Press, 2011.

Smith, F. Joseph, *The Experiencing of Musical Sound: Prelude to a Phenomenology of Music*, New York: Gordon and Breach Science Publishers, Inc., 1979.

Stephenson, Ken. *What to Listen for in Rock: A Stylistic Analysis*, New Haven: Yale University Press, 2002.

Stevens, Stanley Smith and E. B. Newman, "The Localization of Actual Sources of Sound," *American Journal of Psychology* 48, 1936: pp. 297–306.

Stockhausen, Karlheinz, "The Concept of Unity in Electronic Music," *Perspectives of New Music* 1 (1), 1962: pp. 39–48.

Stravinsky, Igor, *Poetics of Music: In the Form of Six Lessons*, Cambridge, MA: Harvard University Press, 1970.

Streicher, Ron and F. Alton Everest, *The New Stereo Soundbook*, Pasadena, CA: Audio Engineering Associates, 1998.

Talbot-Smith, Michael, editor, *Audio Engineer's Reference Book*, Second Edition, Oxford: Focal Press, 1999.

Tenney, James, *META≠HODOS and META Meta≠HODOS*, Oakland, CA: Frog Peak Music, 1986.

Varèse, Edgard, "The Liberation of Sound," *Perspectives of New Music* 5 (1), 1966: pp. 11–19.7

Warren, Richard M., *Auditory Perception: A New Synthesis*, New York: Pergamon Press, Inc., 1982.

Watkinson, John, *The Art of Digital Audio*, Third Edition, Oxford: Focal Press, 2001.

Watkinson, John, *The Art of Sound Reproduction*, Oxford: Focal Press, 1998.

Wertheimer, Max, "Laws of Organization in Perceptual Forms," in *A Source Book of Gestalt Psychology*, ed. Willis Ellis, pp. 71–88, London: Routledge & Kegan Paul, 1938.

Winckel, Fritz, *Music, Sound and Sensation: A Modern Exposition*. New York: Dover Publications, Inc., 1967.

Winckel, Fritz, "The Psycho-Acoustical Analysis of Structure as Applied to Electronic Music," *Journal of Music Theory* 7 (2), 1963: pp. 194–246.

Winer, Ethan, *The Audio Expert: Everything You Need to Know About Audio*, Boston: Focal Press, 2012.

Woram, John M., *Sound Recording Handbook*, Indianapolis: Howard W. Sams & Company, 1989.

Zak III, Albin J., *I Don't Sound Like Nobody: Remaking Music in 1950s America*, Ann Arbor, MI: The University of Michigan Press, 2010.

Zak III, Albin J., *The Poetics of Rock: Cutting Tracks, Making Records*, Berkeley, CA: University of California Press, 2001.

# Discography

Beatles, The.

"Across the Universe," *Let It Be*, EMI Records Ltd., 1970, 1987. CDP 7 46447 2.

"Being for the Benefit of Mr. Kite!" *LOVE*, EMI Records Ltd., 2006. 0946 3 79810 2 3/0946 3 79810 9 2.

"Carry That Weight," *Abbey Road*, EMI Records Ltd., 1969, 1987. CDP 7 46446 2.

"Come Together," *Abbey Road*, EMI Records Ltd., 1969, 1987. CDP 7 46446 2.

"Come Together," *1*, EMI Records Ltd., 1969, 2000. CDP 7243 5 29325 2 8.

"The Continuing Story of Bungalow Bill," *The Beatles* (White Album), EMI Records Ltd., 1968, 1987. CDP 7 46443 2.

"A Day in the Life," *Sgt. Pepper's Lonely Hearts Club Band*, EMI Records Ltd., 1967, 1987. CDP 7 46442 2.

"Day Tripper," *1*, EMI Records Ltd., 1965, 2000. CDP 7243 5 29325 2 8.

"The End," *Abbey Road*, EMI Records Ltd., 1969, 1987. CDP 7 46446 2.

"Every Little Thing," *Beatles for Sale*, EMI Records Ltd., 1964, 1987. CDP 7 46438 2.

"Golden Slumbers," *Abbey Road*, EMI Records Ltd., 1969, 1987. CDP 7 46446 2.

"A Hard Day's Night," *A Hard Day's Night*, EMI Records Ltd., 1964, 1987. CDP 7 46437 2.

"Help!" *Help!*, EMI Records Ltd., 1965, 1987. CDP 7 46439 2.

"Help!" *LOVE*, EMI Records Ltd., 2006. 0946 3 79810 2 3/0946 3 79810 9 2.

"Here Comes the Sun," *Abbey Road*, EMI Records Ltd., 1969, 1987. CDP 7 46446 2.

"Here Comes the Sun," *Abbey Road*, EMI Records Ltd., 1969, digital re-master 2009. 0946 3 82468 2 4.

"Here Comes the Sun," *LOVE*, EMI Records Ltd., 2006. 0946 3 79810 2 3/0946 3 79810 9 2.

"Hey Jude," *1*, EMI Records Ltd., 1968, 2000. CDP 7243 5 29325 2 8.

"I am the Walrus," *Magical Mystery Tour*, EMI Records Ltd., 1967, 1987. CDP 7 48062 2.

"I am the Walrus," *LOVE*, EMI Records Ltd., 2006. 0946 3 79810 2 3/0946 3 79810 9 2.

"I Saw Her Standing There," *Please Please Me*, EMI Records Ltd., 1963, digital re-master 2009. 0946 3 82416 2 1.

"I Want to Hold Your Hand," *LOVE*, EMI Records Ltd., 2006. 0946 3 79810 2 3/0946 3 79810 9 2.

"It's All Too Much," *Yellow Submarine* Songtrack, EMI Records Ltd., 1999. CDP 7243 5 21481 2 7.

"I've Just Seen A Face," *Help!*, EMI Records, Ltd., 1965, 1987. CDP 7 46439 2.

"Let It Be," *1*, EMI Records Ltd., 1970, 2000. CDP 7243 5 29325 2 8.

"Let It Be," *Let It Be*, EMI Records Ltd., 1970, 1987. CDP 7 46447 2.

"Let It Be," *Let It Be*, EMI Records Ltd., 1970, digital re-master 2009. 0946 3 82472 2 7.

"Let It Be," *Let It Be . . . Naked*, Apple Corps Ltd./EMI Records Ltd., 2003. CDP 7243 5 95713 2 4.

"Let It Be," *Past Masters, Volume Two*, EMI Records Ltd., 1970, 1988. CDP 7 90044 2.

"Love You To," *Revolver*, EMI Records Ltd., 1966, 1987. CDP 7 46441 2.

"Lucy in the Sky With Diamonds," *Sgt. Pepper's Lonely Hearts Club Band*, EMI Records Ltd., 1967, 1987. CDP 7 46442 2.

"Lucy in the Sky With Diamonds," *Yellow Submarine* Songtrack, EMI Records Ltd., 1999. CDP 7243 5 21481 2 7.

"Lucy in the Sky With Diamonds," *Yellow Submarine* (DVD 5.1 Surround), Subafilms Ltd., 1968, 1999. 0 27616 7508 22.

"Maxwell's Silver Hammer," *Abbey Road*, EMI Records Ltd., 1969, 1987. CDP 7 46446 2.

"Paperback Writer," *Past Masters, Volume Two*, EMI Records Ltd., 1966, 1988. CDP 7 90044 2.

"Penny Lane," *Magical Mystery Tour*, EMI Records Ltd., 1967, 1987. CDP 7 48062 2.

"Rain," *Past Masters, Volume Two*, EMI Records Ltd., 1966, 1988. CDP 7 90044 2.

"She Came in Through the Bathroom Window," *Abbey Road*, EMI Records Ltd., 1969, 1987. CDP 7 46446 2.

"She Loves You," *1*, EMI Records Ltd., 1963, 2000. CDP 7243 5 29325 2 8.

"She Said She Said," *Revolver*, EMI Records Ltd., 1966, 1987. CDP 7 46441 2.

"Something," *1*, EMI Records Ltd., 1969, 2000. CDP 7243 5 29325 2 8.

"Something," *Abbey Road*, EMI Records Ltd., 1969, 1987. CDP 7 46446 2.

"Something," *LOVE*, EMI Records Ltd., 2006. 0946 3 79810 2 3/0946 3 79810 9 2.

"Strawberry Fields Forever," *Magical Mystery Tour*, EMI Records Ltd., 1967, 1987. CDP 7 48062 2.

"Strawberry Fields Forever," *Magical Mystery Tour*, EMI Records Ltd., 1967, digital re-master 2009. 0946 3 82465 2 7.

"Ticket to Ride," *Help!*, EMI Records Ltd., 1965, 1987. CDP 7 46439 2.

"Tomorrow Never Knows," *Revolver*, EMI Records Ltd., 1966, 1987. CDP 7 46441 2.

"While My Guitar Gently Weeps," *The Beatles* (White Album), EMI Records Ltd., 1968, 1987. CDP 7 46443 2.

"While My Guitar Gently Weeps," *LOVE*, EMI Records Ltd., 2006. 0946 3 79810 2 3/0946 3 79810 9 2.

"Wild Honey Pie," *The Beatles* (White Album), EMI Records Ltd., 1968, 1987. CDP 7 46443 2.

"Yesterday," *Help!*, EMI Records Ltd., 1965, 1987. CDP 7 46439 2.

"You Never Give Me Your Money," *Abbey Road*, EMI Records Ltd., 1969, 1987. CDP 7 46446 2.

Yes, "Every Little Thing," *Yes*, Atlantic Recording Corporation, 1969. 8243-2.

# Index

Note: Page numbers in **bold** are for figures, those in *italics* are for tables.